The Dynamics of Dredging

Publications by Placer Management Corp.

OECON-Offshore Exploration Conference-- Proceedings	
I- Long Beach, CA	1966
II- Long Beach, CA	1967
III- New Orleans	1968
IV- Athens, Greece: Mideast	1968
V- San Diego, CA	1969
OTRAN-Ocean Test Ranges & Instrumentation Conference- Proceedings	
Honolulu, Hawaii	1968
SOCCO-Super Ocean Carrier Conference- Proceedings	
New York	1974
WODCON- World Dredging Conference- Proceedings	
I- New York	1967
II- Rotterdam	1968
III- Singapore	1970
IV- New Orleans	1971
V- Hamburg	1973
VI- Taipei	1974
VII- San Francisco	1976
VIII- Vancouver, BC, Canada	1980
XII- Orlando, FL	1989
OCEAN MINING SYMPOSIUM- Proceedings	
I- Los Angeles	1967
II- Houston	1973
"Bang! Bang! In AMPANG; Dredging Tin During Malaya's Communist Emergency"	
Norman Cleaveland	1973
"DREDGING MARKET in the US, Results of the National Dredging Study"	
A. D.Little, Inc.	1976
"MODERN PLACER MINING"	
M. J. Richardson	1992

The Dynamics of Dredging

Mort J. Richardson
Placer Management Corp.
WORLD DREDGING/Mining & Construction
Irvine, California

Steve L. Richardson, Editor
WORLD DREDGING/Mining & Construction

Placer Management Corp.
Irvine, California, USA

Library of Congress Cataloging-in-Publication Data

The Dynamics of Dredging : overview of international dredging industry, history and technological development in equipment and operations / Mort J. Richardson Editor: Steve L. Richardson.
Includes bibliographical references and index.
ISBN 0-9665413-1-6

PLACER MANAGEMENT CORP.

1 2 3 4 5 6 7 8 9 0 DOC/DOC 1 3 0 7 7 9

ISBN 0-9665413-1-6

Placer Management Corp. is the publisher of
WORLD DREDGING/Mining & Construction, and
other technical books.
17951-C Skypark Circle, Irvine, CA 92614, USA

CONTENTS

DEDICATION

To the Memory of the following individuals whose personal friendship provided the foundation for my involvement in the Dredging Industry and the preparation of this book.

W.H. "Bill" Bauer
First Chairman, World Dredging Association, (WODA)-1968. Founder,Bauer Dredging Co., Port Lavaca, Texas, USA.

Norman Cleaveland
Chairman, First World Dredging Conference (WODCON), New York-1967. Former president, Pacific Tin Consolidated Corp.; Malaysia & New York; dredge mining engineer.

W.R. "Bill" Murden
Chairman, WODA & WEDA: 1978. Former Chief, Dredging Division, US Army Corps of Engineers.

MEMORIAL
To those of the New York/New Jersey Port Authority who perished in the terrorist attack on the World Trade Center, New York City, September 11, 2001.

Mort J. Richardson

ACKNOWLEDGEMENTS

Many companies and individuals have contributed to the documentation and guidance that have made this book possible. The history of dredging, its equipment and management, illustrates the innovations that produced modern dredges and their operations. Also, a tribute is given to the historical research and preparation of European and Belgian dredging contributed by a Belgian specialized writer, Mr. Mon Vanderostyne.

A special thanks for the excerpt used from the 50th anniversary publication by the Dutch Association of Dredging Contractors, Vereniging "Centrale Baggerbedrijf; "Fashioning Nature's Frontiers, 1000 Years of Dutch Dredging," included in **Appendix C.**

Individuals who have shown leadership in the industry are listed in **Appendix D** under **Personalities in Dredging**. Authors of papers whose material is included in this book are cited in the **Bibliography**. A valuable source of documentation was from the Proceedings of **WODCON** beginning in 1967, including **WODA** (**WEDA, CEDA, & EADA**) conferences through 2001; in addition to **World Dredging/Mining & Construction** magazine. I gratefully recognize the indispensible contribution as Editor of this book made by my son, Steven L. Richardson.

Front cover: The 17,000 m^3 Jumbo Trailing Suction Hopper (TH) dredge, *Nile River*, conducting deepening and reclamation below the famous Casino, Monte Carlo, Principality of Monaco. Performing extension of the Port of Condamine.

PREFACE

In preparation for the January 2000 issue of **World Dredging/Mining & Construction** magazine, we researched information from the industry to select the 10 major dredging programs of the 20th Century and the most significant advancements in Dredge Technology. This book is a result of that effort out of which I developed the theme: **"The Dynamics of Dredging."**

Necessity is the mother of invention, an old adage that is appropriate to the development of the Dredging Industry. Human activity in preserving life and property from flooding has been a stimulus to dredging, coupled with the need for agriculture and defenses. The historical aspect is covered in areas of the world where early dredging was dominant. The dredging industry has enjoyed a common heritage that transcended national boundaries. The activities of man to innovate and develop techniques to excavate and move water-bound soil, accompanied the early civilizations of Europe, China, and the Middle East, followed by the New World of North and South America.

Mechanical dredges began to evolve in the 15th century from crude forms of dipper, bag and bucket designs. The 18th century produced more refined equipment though still dependent on man and horse to provide the power to conduct excavation. With the invention of steam power in the early 19th century, followed by diesel power, more sophisticated and powerful dredges began to appear.

My criterion has been to illustrate the role of dredging equipment, focusing upon the industrial and commercial impact of the industry. Recognition is given to the engineers, management and personnel that have made it all possible. The essence of the theme is that dredging has emerged into a vital role in the development of worldwide economies.

The ability to create new land from the sea, mining and transporting sand from distant locations by *Jumbo* Trailing Suction Hopper (TH) dredge, has been one of the more significant advances in dredging that is contributing to major infrastructure development. They have also opened up new capabilities to support the Offshore Petroleum Industry through deep-water trenching for the laying of offshore pipelines.

Mort J. Richardson
Irvine, California

Irvine, California
July 2002

CHAPTER 1

INTRODUCTION

Multi-discipline Industry:

Dredge design and construction and the function of dredging are involved in the following engineering and scientific fields:

Civil Engineering: Foundations, soils, reclamation, coastal hydraulics, harbor and waterway maintenance, and hydrology;

Mechanical Engineering: Equipment design, analysis, experimentation and manufacturing;

Marine Engineering: Machinery, deck equipment, power and propulsion systems;

Ocean Engineering: Dredging, offshore and coastal structures, sub-sea systems, and beach nourishment;

Naval Architecture: Hull design, construction, and systems integration;

Electrical & Electronics Engineering: Motors, controllers, computers, programming, instrumentation, GPS positioning and navigation;

Mining Engineering: Exploration, design and operations relating to mining of ores in unconsolidated materials;

Geography: Geology and geomorphology;

Environment: Ecology, aquatic biology, wetlands, animal habitat, evaluation and design;

Oceanography: Littoral drift, shoreline erosion, wave forces and tides.

Terminology

The United States of America settled upon the following term for a dredge, different from the rest of the world. The US named a dredge a **Dredge.** Other countries call a dredge a **Dredger**. That name was perhaps derived from the Dutch term**, Bagger-machine,** or, **Baggerbedrijf**; **Bagger** being translated ***mud.*** The U.S. defines a **dredger as a person** who operates a dredge and that the dredge operates by **dredging**.

Each type of dredge is described in the following chapters where specific applications have exemplified their various attributes. Applications of dredges include **New Construction, Maintenance, Reclamation and Mining** with descriptions of their capabilities and performance on projects. The choice of which dredge for each application is a perennial problem of the dredge owner.

The dredger is the tool of the civil engineer. He selects his preferred tool which is suitable to execute the assigned job. IHC HOLLAND

"For many centuries, dredging was a trade and its traditional skills were handed down from father to son. The emergence of family businesses in the second half of the nineteenth century coincided with the need of more capital and the development of new technology and powerful plant. For the first time in history, this accumulation of capital, technology and human resources made hydraulic engineering projects feasible that until then could only be dreamed about. At the same time, this evolution caused a strong drive for tendering abroad - the only way to get a reasonable return on capital, once major infrastructure projects in their own country had come to an end.

In the same way as strong family businesses had emerged in the early days of capitalism, dredging companies merged into major groups as from the seventies of the twentieth century. The same trigger would unleash the same process. The increase in scale of projects and plant, the technological innovations, the required accumulation of capital and the growing risks led to teaming up and ultimately to the establishment of world-spanning corporate groups."

Improvements in dredging have broadened its contributions to an elevated role in **Project Management**, moving the function of dredging into a more prestigious role within "wet" Civil Engineering and Construction. Dredging companies have made new strides in meeting the demands of major infrastructure development. This role has also been expanded through new forms of contracts for dredging such as **Design & Construct (D&C),** discussed further in Construction Dredging.

Examples of new leadership from the dredging industry are discussed under different types of dredging, such as major port and reclamation projects in SE Asia, the Middle East, Europe, the US and South America. Wherever development is dependent upon maritime trade, leaders in the dredging industry have participated in the planning and execution of infrastructure development.

Definitions

Abbreviations that are used in this book for different categories of dredges originated in **WORLD DREDGING/Mining & Construction** magazine. They are cited in each **Annual Directory of Worldwide Dredge Fleets, Suppliers & Engineers.**

Dredge Types

In order to bring clarity to the definition of Dredges, I have simplified the process with a reduction to fundamentals. On that premise there are two basic dredge subdivisions.

Mechanical:

Bucket Ladder-BL
Bucket Grab/Clam-BG/C
Bucket Backhoe/Excavator & Dipper-BB
Bucket Dragline-BD

Suction/Hydraulic:

Cutter Suction-CS
Trailing Suction Hopper-TH

Plain Suction-S
Dustpan Suction-SD
Bucket-wheel Suction-BWS
Bucket Suction-BB/S
(See Chapter 4 for dredge type descriptions)

Categories of Dredging

There are two basic categories of dredging: **Civil Construction** and **Mining**. Each category can be subdivided into defined activities and is discussed in detail.

Civil Construction Dredging

Maintenance dredging: Maintains the navigable depth of ports and waterways;

New Construction dredging: Deepening and expanding of existing ports and waterways;

Reclamation dredging: Produces and places material for building of new land and restoration of wetlands, beaches, and shorelines;

Environmental dredging: Removal of polluted material for remediation of harbors, waterways, lakes and inland deposits;

Pipeline trenching: Offshore pipeline laying and burial; depths to +600 ft.

Dredge Mining

The Mining function of dredging may be divided into two basic divisions:

Precious minerals: Gold, platinum and diamond that are dredged from alluvial and eluvial deposits (i.e., unconsolidated material), found in rivers, lakes, inland weathered deposits and ocean littoral zones.

Industrial minerals: The largest segment of Dredge Mining is in the excavation or production of sand and gravel. Other minerals include tin, mineral sands, phosphates, and rutile/titanium. Stripping overburden and excavating tailings from hard rock mining, such as copper, iron and coal is a common function of dredge mining.

CHAPTER 2

HISTORY OF DREDGING

Early Development

Dredging by mechanical means can be traced to medieval times (i.e., 500-1500 AD) when it grew from manual and animal labor into vessels and equipment. The *Spoon and Bag* dredge was the original form of dredge type dating back many centuries; to the **Phoenicians, Romans** and the **Chinese.** Early tribes along the **Tigris** and **Euphrates** rivers are recorded to have used this type of dredge. Sketches from the 15th century by the famous Italian artist, **Leonardo Da Vinci** show two types of dredge in that period. They are titled: ***Lagoon dredge*** dated 1486 and ***Sea Dredger*** in 1490.

Europe

Belgium: Dredging & Hydraulic Engineering

By the fate of history and the bare facts of geography, **Belgium** can be considered as the **cradle of dredging and hydraulic engineering**. At the very heart of the so-called **Golden Triangle of the Low Countries**, two facts may explain the peculiar concentration of dredging and hydraulic engineering know-how in this tiny kingdom: the need for keeping open the access routes to the sea and the eternal fight against flooding which has marked its history.

Belgium has a tradition of one thousand years of dealing with water. Located along the North Sea coast, its 1100 km of inland waterways have always made for a unique geographic situation - at the crossroads between England and the Rhineland, or between the Mediterranean and the Scandinavian countries. From early medieval days, the sea brought prosperity to **Flanders**, the coastal region of Belgium. Yet, the shallow waters of the Flemish Banks have always required innovation and creativity for finding solutions against siltation. The closure for more than 200 years of the river **Scheldt**, due to the Wars of Religion and diplomatic plays on the old continent, increased the awareness of how vital it is to keep this "artery of prosperity" open. Keeping open its sea lanes has been a prerequisite of Belgian prosperity for centuries. Developing dredging and hydraulic engineering know-how was the obvious solution.

Early dredging know-how

Dredging works in the canals, linking the medieval city of **Brugge** to the North Sea, are chronicled as early as **c.1292**. A full century earlier, **c.1180**, a primitive form of lock was already in operation at **Damme** on the canal from Brugge to the sea - a remarkable historic *first* in a country where dredging companies recently built the biggest lock in the world at **Berendrecht** near **Antwerp**.

Another contribution to early dredging techniques was the use of **stanches** or **flash locks**, which could be opened suddenly, releasing a torrent that carried a vessel over a shallow place. It is a technique that was developed in Belgium **c.1251**, on the **Lieve-canal** between the land-locked city of **Gent** and the sea.

Dredging works in Belgium are also described in **c.1434**, when Philip the Good, count of Burgundy, ordered the broadening of the river **Zenne** in Brussels. An ancient type of mud mill, which would later become known as the **Dutch- or Amsterdam mill**, can already be noticed on canvas of the Flemish painter **Pieter Bruegel the Elder (1525/1530-1569).** This early dredging know-how in the Belgian provinces was passed down through the ages, and statesmen of the time were very well aware of that tradition. When the **Russian czar, Peter the Great**, travelled the country in 1717, he used the opportunity to study locks and harbor engineering in three or four Belgian sea ports of the time.

At a time when modern materials were not yet invented, when heavy plant was not yet developed and when brute, cheap labour was abundantly available, early dredging works were mainly based on draining sluices, creating a counter current, or used the ancient, traditional hand-drag. In the long process of slow technological development during the years before the industrial revolution marked a decisive breakthrough, it was the **Flemish mathematician Simon Steven (1548-1620)** who brought about a major improvement of the hand-drag, when he applied his ***theorem of the triangle of forces*** for hoisting and levering this extremely heavy piece of equipment. As one of the famous examples of the massive brain drain from Belgium to Holland following the 16th century Wars of Religion, Simon Steven would later become quartermaster general of the army of Prince Maurice of Nassau, and make significant contributions for dredging technology during the **Golden Age of Holland.**

In the collective memory of the Belgian people, the rise and fall of Antwerp during its age of splendour in the 16th century has remained a textbook case for the importance of keeping open the sea lanes. Just like Brugge in the 13th century, **Antwerp was one of the busiest and richest ports of the old continent until the unfortunate closure of the river Scheldt in c.1585.** The important effort in our days for maintaining to depth the 75 km access route to open water, may be understood by the economic disaster that happened four centuries ago.

In mid 16th century, Antwerp was the second biggest city of Europe - enjoying commerce, wealth and prosperity thanks to the river Scheldt outlet to the sea. When the war against Spain came to an end, the river was closed off and would not reopen before the French revolution. Even after navigation was declared free for all riparian states in the 19th century, a tax was levied on all shipping traffic until c.1863. **When the Scheldt was closed, the result was felt almost immediately. Within four years, Antwerp's population had fallen to 42 percent of what it had been - causing a brain drain that would set the scene for the Golden Age of its rival Holland.**

The building of the port of **Zeebrugge** was directly inspired by the calamitous experience that had once hit Antwerp and the Belgian economy as a whole. In the years before the First World War, the sheltered harbour of Zeebrugge was purposefully built on open North Sea water. Never again would siltation or closure of a river threaten the Belgian economy. Zeebrugge would develop into large scale extensions at the end of the 20th century, a dredging and hydraulic engineering project that is rated among the top achievements in the industry. Even then, **maintenance dredging** in the Scheldt river and in the maritime access routes currently amounted to about **20 to 40M m^3** of spoil per year. Total volume of dredging on the Scheldt river alone, has been estimated at **500M m^3** since the end of the Second World War.

History of flooding and land reclamation

Securing the access to open water and the traumatic memory of being closed off from the sea explain the millenary tradition of dredging know-how in Belgium. There is a second explanation: the history of flooding since time immemorial. Taken together, it may seem as no coincidence that the current strong position of Belgian dredging companies on world markets found its roots in this very country. The facts of geography and the fate of its history just had to cause this kind of specialization in hydraulic engineering.

The Flemish coastal region of Belgium was born in the salty lowlands behind the sea. Protection of the population against the ever present danger of flooding not only created a collective awareness, but it also brought about a tradition of land reclamation. The **first sea dikes** were already thrown up in the **10th century**, other polder land was created at the end of the 11th and the end of the 13th century. In the very beginning, these embankments were purely defensive, but by the end of the 12th century local communities initiated a more daring, offensive approach that was carried out "in the wet." Sounds familiar, considering the massive land reclamations that Belgian dredging companies execute these days all over the globe?

The history of the Belgian lowland is marked by the memory of apocalyptic flooding, in which sometimes tens of thousands of people drowned and land and belongings disappeared under water for many years. The worst flood disasters were called after the saint of the calendar, such as the St.-Elisabeth-flood of 1421, the St.-Felix-flood of 1530, the All Saints flood of November 1st, 1570. In the latter one, the ancient community of **Saeftinghe** would forever vanish in the river Scheldt. Even in our days, more than one quarter of the Flanders region in Belgium is likely to be flooded by the fact of its geographic location.

Modern time flooding happened in 1953, 1976, 1990 and 1993, leading to the ambitious **Sigma-programme** for protection of the **Scheldt-region** in which all Belgian dredging companies participate. In other times, man himself created artificial inundation, by cutting dikes for strategic reasons. In a country that has been the battlefield of Europe for many centuries, people learnt to keep the enemy off by flooding the land. It happened during the siege of Antwerp in 1585, and the weapon was reused in 1914 when the IJzer river lowlands were flooded, leading to the freezing of the First World War frontlines and the subsequent four years war of attrition.

Throwing up embankments, the colonization of uncultivated land and early land reclamation was a family activity and a feudal duty alike. In the eternal fight against the water, abbeys and monasteries played a crucial role, as they were virtually the only institutions that were capable of mobilizing people and labor force. Land that was painstakingly reclaimed behind dikes, was lost to nature after flooding, and yet re-conquered by people fighting for their survival. It was this process that molded the character of the Belgian people, and that ultimately lead to the establishment of specialized dredging companies. It is a know-how that has been recognized for centuries, also in literature. When the Italian writer **Dante Alleghieri** described his descent into hell on **Good Friday 1300**, he compared the hard embankment on which he was sitting to the dikes that Flemish people used to build between Brugge and the Dutch border: ***Quale i Fiamminghi tra Guzzante e Bruggia ... Fanno lo schermo perché 'l mar si fuggia.*** It is that tradition that has come to our times. It is that ancient tradition that would be further developed, especially in this very country, when the industrial revolution would change the dredging world.

Early industrialization and Belgian dredging technology

In the field of dredging technology, the transition from manpower to steam power and from wood to steel probably was the single most significant evolution of all times. It happened during the industrial revolution, a process of innovation from late 18th to mid 19th century. The importance of the Belgian contribution in this respect cannot be over estimated. Already in the mid 19th century, Belgian industry led the way in innovative dredging technology.

It is a fact of history that the industrial revolution, which originated in England, found its way on the European continent through Belgium. There were several reasons for it: the political context in Europe had driven France into isolation, as a defeated power. Holland which was still largely an agrarian country did not need the new steam power, as it disposed of cheap energy from its 25,000 windmills. Until 1870, Germany remained fragmented in toll-protected principalities. Russia was heavily dependent on West-European capital and skills. The presence of coal, the centuries-old tradition of iron forging, the abundance of a skilled workforce and the traditional commercial links with England, made Belgium a particularly suited entry gate for introducing the new technology on the European continent.

In the first half of the 19th century, an unsurpassed economic boom took place in the **Liège region** of Belgium - mainly through the very diversified emerging industrial empire of **John Cockerill**, the son of an English mechanic and entrepreneur who had settled in the country. His applications of the newly invented steam power seemed endless, and were exported all over Europe. **The *Liégeoise* dredging vessel would remain the plant of choice and the workhorse of the dredging industry for almost 150 years.**

The ***Liégeoise* was a bucket ladder dredge (BL),** which has been built in several models through the years - in a towed and a sailing version. Responding to increasing demand, Cockerill moved his Liège-based shipyard along the river **Meuse** to Antwerp, where more space for expansion and better depth was available on the river Scheldt. But other companies in the Liège region continued building the famous bucket dredge. They included entrepreneurs such as **Tilkin-Mention, Bonivert, Spiroux, Jacques Piedboeuf** and **Perin.** Their vessels would be found all over Europe, as well as on other continents.

According to **David McCullough** in his epic chronicle about the creation of the Panama Canal ***(The Path Between the Seas)***, twenty *extremely well built* ladder dredges, constructed in Belgium, were used by the contractor **Artigue et Sonderegger**. As an anecdote, it is worth recalling that colonel **George Washington Goethals,** the man who would ultimately succeed in cutting the Panama Isthmus, was the son of Belgian immigrants, who had their roots in the very region of what is currently the heart of Belgian dredging industry.

The first steam dredges in Holland were constructed in Belgium as well. They were bought by state authorities. It was not before 1864 that a private company in Holland would acquire a steam driven bucket dredge of its own - it was a *Liégeoise* vessel. In the **mid 19th century, 22 dredges of Tilkin-Mention were operating on Dutch and Belgian rivers alone.** Even in the early thirties of the 20th century, *Liégeoise* dredges were still being used in the Netherlands, at the time of the daring enclosure of the Zuiderzee.

As for the Cockerill company, their dredging vessels found their way in Russia. A complete dredging installation was sold to the **port of Loban** and in the sixties of the 19th century Cockerill operated a shipyard on **Gontouriefsky island off Saint-Petersburg**. In 1897, the famous American engineer**, Lindon Bates**, concluded an agreement with Cockerill for exporting his powerful ***Bêta*** **cutter suction dredge to Russia**. The contract had been prepared by a Belgian government mission, studying dredging techniques on the **Mississippi river**. Having seen the results of Bates' *Bêta* dredge, producing 5,946 yd^3/hour during 175 hours and a peak output of 7,789 yd^3 per hour, the Belgian government delegation concluded that this plant was "very superior" to anything else in the industry. The next year, Bates was invited to Brussels, where he became one of the most celebrated speakers on the **7th International Navigation Congress of PIANC**. Among the Lindon Bates dredges that Cockerill exported to Russia, one was christened ***Wolga***, and another was named ***Nicholaev*** after the reigning czar.

Emergence of Belgian dredging companies

Just like in other countries, the founding fathers of what are currently dredging conglomerates of world class were originally civil contractors in Belgium. Some of them owned a bucket dredge others installed a grab or a crane on a pontoon. Dredging and civil works melted into one another. Against the background of a centuries-old know-how in the country, the slow process of professionalization involved ever more capital, investments and industrial risks. In the course of time, dredging activity was split off and became a real separate business.

Baggerwerken Decloedt & Zoon

In mid 19th century, **Johannes De Cloedt** was a small contractor in the Belgian city of Brugge, building houses, a church, a wall, or tendering for maintenance works at locks in the inner city or on the many waterways behind the coast. It was his son, Emmanuel (1845-1919), who oriented the family business towards hydraulic engineering, which at that time was mainly restricted to maintenance dredging in the North Sea port of **Ostend.** In 1928, his son **Prosper De Cloedt** incorporated the company. At his death in 1955, he was succeeded by his son Raymond, who in turn would leave the helm to his son **Jean-Jacques De Cloedt**, the great-grandson of the founder. For more than 100 years, the Decloedt companies were managed as a family business, until their integration in April 1991 in the **Deme** holding structure.

The reputation of **Baggerwerken Decloedt** is traditionally linked with deepening works and maintenance dredging on the **Flemish Banks** off the Belgian coast - although the company was already tendering for maintenance dredging on the river Scheldt in Antwerp in 1894. Working together in the Deme-structure, the coastal experience of Decloedt resulted in excellent synergy with Antwerp based **Dredging International.** Already before the First World War, Emmanuel De Cloedt had established a business relation with **Jean Cousin**, a contractor in Brussels, and **Louis Cloiseau**, a colleague in Paris. In 1907, Cousin and Cloiseau were the builders of the sheltered **Zeebrugge outport**, at which Decloedt became involved.

A joint-venture between Baggerwerken Decloedt and Georges Cousin, the son of Jean Cousin, would be in charge of maintenance dredging works in Zeebrugge after the Second World War. At a very early stage, Decloedt took also part in wreck lifting - an early pre-figuration of the raising of the capsized ***Herald of Free Enterprise*** ferryboat in 1985 by Belgian contractors. As such, Decloedt was in charge of lifting wrecks in the English Channel and in Zeebrugge, where the German navy had operated a submarine base during the First World War.

The De Cloedt family extended its activities all over the world. At the end of the 20th century, a network of subsidiaries had been established in several countries. Until the independence of Congo, Decloedt kept a separate company in the former Belgian colony. Another subsidiary, **Overseas Decloedt & Fils**, was established in **Singapore** - many years before the rest of the dredging industry *discovered* the emerging **New Tiger countries**. Among the company's early assignments in Asia, the heroic **Bluff, New Zealand, contract of 1956,** the very first reclamation in **Malaysia's Port Klang** and the extension of the first runway on **Sydney's Kingsford-Smith airport** should be mentioned.

Ackermans & van Haaren

Currently an investment group and industrial industrial holding company, Ackermans & van Haaren was set up as a civil engineering company building roads, quay walls, port infrastructure, railroads and buildings in 1852. Its dredging activity started in the 1870s. Its name refers to **two Dutch contractors**, **Nicolaas van Haaren (1835-1904)** and **Hendrik-Willem Ackermans (1855-1945),** whose families became closely related by several marriages. For a short time around 1870, they managed a dredging company in the southern part of the Netherlands, which they moved to Belgium in 1880. For more than a century, Ackermans & van Haaren marked the profile of dredging business in Belgium.

On behalf of the **Corps of Engineers of the Belgian army**, both contractors carried out **gravel dredging** works in the river **Meuse** for building fortifications and military forts. Very soon, they switched their main activity to Antwerp, where **capital dredging was executed for the first time in history**. It was calculated that Ackermans & van Haaren**, between 1894 and 1911, dredged some 25M m³ when deepening the Western Scheldt from 7.5 to 8 m - an impressive volume at that time**. Current dredging programs, by the way, will allow safe navigation with a vessel draught of 13.1 m.

The company was formally incorporated on December, 30, 1924. From the very beginning, Ackermans & van Haaren held a leading position in technological innovation. **In 1895, the company designed a suction hopper dredge**, capable of unloading by its own. In the same year, a first vessel of that type was built and christened as ***Schelde II***. A second one, named ***Gelderland-1***, would follow in 1905.

It is interesting to note that, in the late 20th century, the Deme-group in which the dredging section of Ackermans & van Haaren would be integrated, paved the way for all other dredging companies in the world, by ordering a completely new generation of mega-dredges. In one single, daring move in 1994, the **17,000 m³ *Pearl River*** became the very first suction hopper dredge of a completely new generation - featuring twice as much capacity as its biggest predecessor.

For all its maintenance dredging work, it was the expansion of the port of Antwerp that made the Ackermans & van Haaren reputation in the first place. Between 1903 and 1908, the city of Antwerp and the Belgian state paid the company 13M gold franc for the construction of quay walls, the excavation of the first harbor dock and part of a connecting canal. **After the Second World War, the company was closely involved in the famous ten year infrastructure programme (1956-1967), digging docks and deepening fairways, which would make the port of Antwerp the second biggest in Europe**. In the same way, the creation of a completely new port on the left bank of the river Scheldt as from 1971 would not be imaginable without the unremitting commitment of Ackermans & van Haaren.

The expansion of the company was unheard of at the time. In 1894, Ackermans & van Haaren owned three dredging vessels and two pumps, with transport equipment. By the year 1910, the company used ten dredges, nine pumps and 22 trailers. As a result, early expansion abroad was sought, in order to achieve an optimal return on that expensive investment. In the years after the turn of the 19th century, and well before the First World War, Ackermans & van Haaren was an active player in dredging works on four continents. Dredging volume by Ackermans & van Haaren, in Belgium and abroad, between **1888 and 1967** has been calculated at no less than **593,254,000 m³**. That is without counting the **8,548,000 m³** of rock that was removed and some **1,941,000 m³** of civil constructions and bank protection that were realised.

Algemene Baggermaatschappij SGD

Apart from Ackermans & van Haaren, maintenance dredging works on the river Scheldt were traditionally awarded to the *Algemene Baggermaatschappij* (in French: ***Société Générale de Dragage** or **SGD)*** - the third major contractor, which would later become part of the current dredging configuration in Belgium. Algemene Baggermaatschappij SGD was established on April 5, 1930 but had older roots. The Brussels based contractor, Georges Cousin (who, as we saw, was the business partner of Prosper De Cloedt), used to carry out maintenance dredging works on the river Scheldt since 1918. Cousin hired a French director named **d'Haille** as his general manager for dredging works - a job that would later be taken over by his son Louis d'Haille. When Cousin fell ill, he sold his business to the Belgian major contracting group CFE and to the De Cloedt family, who together established SGD.

The ship with the biggest deadweight capacity that ever called at Antwerp, was the 335 m, 274,726 dwt bulk carrier ***Main Ore***. The draught record for Antwerp was set in 1985 at 14.95 m, with the ***Permeke***. Such achievements would clearly not be possible, without the continuous efforts that companies like "Algemene Baggermaatschappij" (SGD) have made.

Some ten bars are a permanent impediment to traffic on the Western Scheldt, a tidal river. Until the late 19th century, Antwerp-bound vessels sailed up the Scheldt as far as was possible. At that point they waited for the high tide, which brought them further in port. The appearance of bigger sea-going vessels, iron steamboats with a greater draught, made dredging inevitable. Maintaining to the required depth the 75 km from Flushing to Antwerp, has remained a Homeric undertaking.

Unlike Ackermans & van Haaren, which was specialised in cutter suction dredges used in digging docks, Algemene Baggermaatschappij (SGD) specialized in trailing suction hopper dredges, suited for maintenance dredging in a river environment. The SGD-engineering department had a reputation for strong technological innovation. Yet, the old but remarkably accurate bucket dredges were not forgotten, as dredging professionals from all over the world could witness between 1966 and 1969 when SGD realized, in joint venture, a much admired tunnel crossing under the river Scheldt: the Kennedy tunnel.

Ondernemingen Jan De Nul

The remarkable history of the De Nul dredging group has been described as a replay, in the timeframe of two generations, of the evolution of Belgian dredging business as a whole over more than a century. Ondernemingen Jan De Nul is still a family business and one of the very few that is not controlled by financial or industrial interests. As the owner of the biggest trailing suction hopper dredge and the strongest cutter suction dredge in the world, the Belgian company is one of the leaders in technology and scale. The current chief-executive, as well as his father, are said to have a particular talent for mechanical engineering. Some of the biggest dredging and hydraulic engineering contracts in the world have recently been awarded to De Nul: amongst others the building of Chek Lap Kok airport in Hong Kong and the long-term maintenance dredging works in Argentina on the Rio Parana and the Rio de la Plata.

The roots of the De Nul group, which has an estimated turnover for 2001 of more than **US$400M**, are modest and unpretentious. Between the two world wars, grandfather De Nul was running furniture works in the Belgian city of Aalst. In 1938 his son, Jan De Nul, set up business as a contractor. He became very successful, and very soon building a hospital, a bridge or road works would replace the original construction of private houses. Through the years, he found business opportunities in the port of Gent, such as the construction of a hangar - the very first in pre-stressed concrete.

With foresight and vision from the fifties, Jan de Nul was aware that hydraulic engineering would become one of the booming markets in the second half of the century. The construction of new waterways in Belgium and the broadening and deepening of existing ones would be the trigger. In the early fifties he built his first vessel, the ***Jan-Pieter-I.*** Others would follow, as more and bigger projects would successfully be completed. His son, **Jan-Piet De Nul**, took over the drive of his father.

Ondernemingen Jan De Nul began its activities in a rather limited geographic area, the West- and East-Flanders provinces of Belgium. Very soon however, he tried his luck in the then booming contracting business of the port of Antwerp, where he was considered by his competitors as an outsider. From his underdog position, he succeeded in joining the very top by virtue of his perseverance, stubbornness and fighting spirit. In the same way, the company broadened its scope and moved purposefully and resolutely abroad. On foreign markets, De Nul occasionally forged alliances with Ackermans & van Haaren, which had also been on its own for most of the time - unlike the teaming up between Decloedt and Algemene Baggermaatschappij. In the end, Jan-Piet De Nul became a feared competitor on world markets for all his colleagues, whose companies could look back upon much older corporate traditions than his own family business.

Consolidation and mergers (DI-DEME)

For many centuries, dredging was a trade and its traditional skills were handed down from father to son. The emergence of family businesses in the second half of the nineteenth century coincided with the need of more capital and the development of new technology and powerful plant. For the first time in history, this accumulation of capital, technology and human resources made hydraulic engineering projects feasible that until then could only be dreamed about. At the same time, this evolution caused a strong drive for tendering abroad - the only way to get a reasonable return on capital, once major infrastructure projects in their own country had come to an end.

In the same way as strong family businesses had emerged in the early days of capitalism, dredging companies merged into major groups as from the seventies of the twentieth century. The same trigger would unleash the same process. The increase of scale of projects and plant, the technological innovations, the required accumulation of capital and the growing risks led to teaming up and ultimately to the establishment of world-spanning corporate groups. Which came first, the chicken or the egg?

The increased scale of projects required bigger ships. Yet bigger ships asked for mega-reclamation projects to reach profitability. **In the seventies, fourteen dredging companies from Europe were sharing the better part of free world dredging business. By the turn of the century, only six were left, and in May 2001 further consolidation was announced which would reduce surviving groups to five**. However, it is interesting to notice that dredging capacity did not diminish during that process, and that production capacity even strongly increased.

In Belgium, Baggerwerken Decloedt had maintained a close business relation with the Brussels contractors Jean and Georges Cousin. When Algemene Baggermaatschappij (SGD) was established in 1930, Decloedt got a minority stake at the side of CFE. It would reach twenty percent in the 1970s. A decisive consolidation of Belgian dredging business happened on November, 27, 1974, when the dredging section of Ackermans & van Haaren and Algemene Baggermaatschappij (SGD) merged into Dredging International. The synergy was obvious, with Ackermans & van Haaren operating mainly cutter suction dredges for building the port of Antwerp, and Algemene Baggermaatschappij being specialised in trailing suction hopper dredges at the maintenance dredging side of the business on the river Scheldt. Ackermans & van Haaren and CFE were given equal parts in he new company while the share of Baggerwerken Decloedt in SGD was diluted to 11.5 % in Dredging International.

After the crisis in the dredging business during the 1980s, caused by a 30 to 40 percent over capacity in the sector, a second wave of consolidation happened in April 1991 when *Dredging, Environmental and Marine Engineering* (DEME) was established. It is a holding company in which both parties, Dredging International and Baggerwerken Decloedt, kept their independence as two separate operating companies. The Antwerp based Dredging International found a natural partner in Baggerwerken Decloedt, which relied on a coastal tradition. **Together they made for the number two dredging group in the world.** In the last decade, the newly established DEME-group would diversify its activities in offshore related projects, hydraulic engineering, environmental contracting and the winning and supply of marine and fluvial aggregates to the construction industry.

At the time of its incorporation, Ackermans & van Haaren and CFE each subscribed to 39.5 % of the DEME-capital. Baggerwerken Decloedt controlled 18 % and a minority stake of 3.0 % was given to a regional investment company in the Flemish region of Belgium (GIMV). After Decloedt left the group in mid 2000, Ackermans & van Haaren and CFE exercised their right of pre-emption, after which they both owned 48.5% of the capital. In the past decades, Ondernemingen Jan De Nul remained an independent, closed family business, spreading its own wings and growing on its own power.

Belgian dredging companies spanning the globe.

The several waves of consolidation in the Belgian dredging business during the last decades of the 20th century can not be described without the background of the huge international expansion of Flemish dredging contractors after the Second World

War. The establishment of Dredging International in 1974 followed the emergence of major infrastructure works in the Middle East. These were the result of a massive recycling of so-called petro-dollars after the oil crisis of November 1973 and the Yom Kippur war of October, 10, 1973. In the same way, the establishment of DEME-group in 1991, after the heavy crisis of the whole dredging sector in the 1980s, was a strong signal heralding the upcoming mega-projects in Southeast Asia. However, the phenomenal expansion abroad of the Belgian dredging business has much older roots. Already in the very first years of the 20th century, Belgian dredging companies were operating on four continents. One century later, they confirm that position spanning the globe.

Looking for opportunities in 1900

The remarkable expansion of the Ackermans & van Haaren fleet at the turn of the 19th century, forced the company to look abroad for business opportunities - once the first wave of major port infrastructure works in Antwerp was completed. It was the same motivation that had driven Dutch contractors across the border, when deepening works on the **Nieuwe Waterweg** came to an end around 1896. It happened again in the 1930s, when surplus plant became available after the enclosure of the **Zuiderzee**.

The same phenomenon was once again repeated in the last three decades of the 20th century, when the search for continuous profitability caused Belgian dredging companies to make a new run on foreign markets - at a time when huge infrastructure programs on the right bank of the river Scheldt in Antwerp, and in Zeebrugge were being completed.

As from 1903, Ackermans & van Haaren was involved in almost continuous business in several countries of **Latin America**. For ten years, the Belgian company was building and expanding the port of **Rosario de Santa Fé,** starting as a subcontractor for the Paris based companies **Hersant & Fils and Schneider & Co**. In 1912, total volume at Rosario was estimated at a tremendous 9.5M m^3 of dredging and 5.7M m^3 of reclamation. Other Argentine dredging works at that time were executed in **La Plata, Bahia Blanca and Puerto Belgrano, San Nicolas, Ensenada, Sorento, Quequen,** the **Parana delta**, the **Maschwitz canal**, etc.

From **1910 till 1913**, Ackermans & van Haaren built an **1150 m long tunnel in Buenos Aires**, supplying water from the Rio de la Plata. This contract was carried out in a joint-venture with **Société des Grands Travaux de Marseille** and the **Paris contractors Etablissements Daydé** and **Fougerolle Frères**. Other Latin American countries included **Uruguay (capital dredging in Montevideo, gravel dredging at Colon),** and Brazil where **8.5M m^3** were dredged in the port of **Rio Grande do Sul** between **1908 and 1916.**

Closer to home, Ackermans & van Haaren dominated the European dredging scene in the first half of the 20th century. In the **Bolshevik Revolution of 1917** and during the First and Second World Wars, the company lost many vessels. But at the other hand, the loss was an opportunity for building new plant and applying the latest technology and innovations. Within ten years after 1918 and 1945, Ackermans & van Haaren had regained and confirmed its pre-war leadership. After the establishment of Dredging International, this new company would take over the century-old tradition of dredging and marine engineering abroad.

The **maritime history of France** may be written by way of the involvement of Ackermans & van Haaren in building and dredging its Atlantic and Mediterranean ports. It remains largely an unwritten story, and would go well beyond the framework of this book.

There may be no one single country in Europe where Ackermans & van Haaren was not involved in dredging, from the first decades of the 20th century on. We pass over dredging assignments in the Netherlands, Germany, Portugal or Spain**.** The company played a decisive role in **Poland**, after the country had been re-created by the **Treaty of Versailles**. **Danzig** being recognized as a free port at the time, the so-called **Polish corridor** was the only outlet to the sea for the re-established nation. In quite heroic circumstances, Ackermans & van Haaren started building the new Baltic port of **Gdynia in 1921**, for which a total of **36M m^3** was dredged.

At the time of the nazi-invasion of Poland and Russia in 1941, Belgian dredging workers and their plant were blocked in Poland, where they had been building the fishing port of **Wielka-Wies**, a submarine base at the far end of the **Helska peninsula**, and the 32 km long **Warta-Goplo-canal** linking the **Vistula** and **Oder rivers**. In the years between the two world wars, plant in Poland included the ***Gelderland,*** the cutter suction dredge ***Gironde*** which became locked in the ice along inland navigation waterways, and the bucket dredge ***Gdynia*** - a scale-model of which was acclaimed as one of the technological wonders at the **Brussels World Exhibition of 1935**.

Until the Bolshevik Revolution of 1917, Ackermans & van Haaren was doing continuous business in the Russian empire. Between 1912 and 1917, the company was dredging some 2.5M m^3 for the construction of harbor basins on behalf of the Putilow-factories in St.-Petersburg. Other projects were executed for the Trouxtanny Ostrow company, the railway company Moscow-Kazan, and the electrical railway Oranienbaum. In the **Gulf of Finland, the Morskoi to Ligowa canal** was dredged for the benefit of the rubber factory Treougolniek. In a joint venture with Schneider & Co and Hersant J & G in Paris, Ackermans & van Haaren built the **military harbor *Emperor Peter the Great*** in **St.-Petersburg between 1913 and 1917**, for which proprietary quarries were operated in **Kyrslätt, Finland**, and **Wiems near Reval**. Reval is the old name for **Tallinn, the capital of Estonia** which was under Russian influence at that time. Before the First World War, Finland was a semi-autonomous grand-duchy of the czarist empire. In 1916, the Belgian company started construction of docks for the Russian navy at **Sveaborg near Helsingfors, currently named Helsinki.**

Outside Europe, Ackermans & van Haaren was also involved in dredging and hydraulic engineering in **Africa and Asia**. Assignments included the **Maghreb** countries **Morocco**, where the cutter suction dredge ***Pologne*** was dredging the port of **Safi** in 1933, and **Algeria** where the company carried out very important harbor works from 1926 to 1935. In the late twenties, Belgian dredging workers were deepening the port of

Douala, Cameroon, and later on they would become deeply involved in the Belgian colony **Congo**.

Besides a quay-wall and breakwater construction at **Now-Chahr, Persia, in 1937**, major pre-war assignments in Asia included **Pnom-Penh**, the capital of then **French Protectorate of Cambodia.** Starting in 1928, Ackermans & van Haaren assigned its flagship ***Antwerpen III*** to the Far East. For almost three years, the dredge was operating in Pnom Penh on the **Mekong river** and in **Saigon, Vietnam**.

Pooling resources for tackling foreign markets

The international expansion of Algemene Baggermaatschappij, Baggerwerken Decloedt and Ondernemingen Jan De Nul followed a somewhat different course. The impressive and rapid expansion of De Nul in Belgium, obliged the company to look abroad for optimising return on investment and for sustaining its growth. This happened from the 1970s on. Just like in its home country, De Nul remained an independent company abroad - sometimes tendering in joint-venture, sometimes allying its interests with Ackermans & van Haaren which also stood on its own feet.

When its fleet was rebuilt thanks to war-damage compensation, claims from Germany, salvage of sunken vessels, and leasing in Denmark, Baggerwerken Decloedt made an agreement with **Algemene Baggermaatschappij** for tendering together on foreign markets. There was a third partner involved, the Paris based ***Société Nationale de Travaux Publics* (SNTP).** This company had been taken over by the Belgian holding group ***Société Générale*** just after WW II, but had its roots in the early 20th century. It was the SNTP-experience on foreign markets, which acted as a catalyst for both Belgian companies to make the daring step abroad.

SNTP had been established in Paris in 1911 as an association of four civil contractors, **Wiriot, Dollfus, Sillard** and **Allard**. Its aim was common prospecting, tendering and execution of projects in other countries than France. However, SNTP lacked the resources for dredging and hydraulic engineering, which were quite often a prerequisite for successful civil contracts. The obvious solution, teaming up with Belgian dredging companies, was prompted by Louis d'Haille, the French director of Algemene Baggermaatschappij and a close business- and family friend of the De Cloedt family. During the Second World War, Louis d'Haille had left Antwerp and worked as an engineer for SNTP in Africa. After the war, he retrieved his old passion, hydraulic engineering, and forged the association between SNTP and the Belgian dredging contractors. He passed on his passion to another French engineer, Michel Chevalier, who began his career as a director at SNTP and would end as a member of the Board of Directors at Dredging International.

The contract between **Baggerwerken Decloedt, Algemene Baggermaatschappij and SNTP** was signed in Paris on **January, 21, 1954**, for a duration of twenty years. It was a framework-agreement, which would later be known as **"the pool."** The dredges ***Yperlee, Meuse, Dyle*** were brought in by Algemene Baggermaatschappij; Decloedt contributed its ***Volkracht IV and Vlaanderen IX***. The contract was a pragmatic agree ment, which had to be detailed for each specific assignment. The association was valid for the whole world, except in those countries where the partners had already dredging equipment mobilized - in practical terms Belgium, the Netherlands, Abidjan, Conakry, Douala and Madagascar.

In 1963 the pool ordered, on behalf of its partners, construction of the 4,800 m³ trailing suction hopper dredge ***Atlantique***. It also ordered the cutter suction dredge ***Tramontane***. The Pool was a very successful instrument and would exist for eighteen years, during which period of time more than twenty important international tenders were obtained. The pool petered out in 1972, at a time when Baggerwerken Decloedt felt strong enough to go its own way in Singapore, and at the eve of the merger of Algemene Baggermaatschappij with Ackermans & van Haaren.

Even before the association was formally established, the three partners had tendered together for dredging works in **Port Lyautey**, currently **Kenitra**, in what was then **French-Morocco**. The first major assignment followed the so-called ***contract no. 666* of Bluff Harbour Board** in 1956. The site manager was SNTP-director Michel Chevalier. Building a **50 ha artificial island** with two docks in Bluff, on the southern tip of **New Zealand's South Island**, was a ***Homeric*** undertaking in those days. The cutter suction dredge ***Rupel*** was towed to the antipodes in an adventurous 88 days journey. She was one of the last vessels that could safely make the passage through the Suez canal, just days before the Israeli invasion of the Sinai desert on October, 29, 1956. It took almost a week for crew, staff and workforce to reach their destination. They would stay in New Zealand, together with their families, for more than three years without a single holiday in their home country.

The years following the Bluff contract have often been called the thirty golden years of dredging for Belgian contractors. In Belgium itself, the enormous maritime infrastructure works on the right- and left bank of the river Scheldt in Antwerp, and the construction of the sea port of Zeebrugge, laid the foundation for continuous business and innovation. They were considered as a valuable reference for tendering abroad. On foreign markets, an impressive series of sometimes huge dredging and hydraulic engineering projects was carried out, amongst others in the **French ports of Boulogne-sur-Mer, Bordeaux, Dunkirk, Gravelines, Fos-sur-Mer and in the Algerian harbors of Oran and Bône (currently Annaba).**

The last contract under the terms of the so-called pool was the extension in 1969 of the first runway at **Kingsford-Smith airport in Sydney, Australia**. Besides the trailing suction hopper dredge "Atlantique" and the cutter suction dredge ***Leie,*** a workforce of 49 people with their families was sent to Sydney. The reclamation in **Botany Bay** extended Sydney's main runway to 3,962 m.

The fact that Belgian dredging groups had won this very important tender against stiff competition, meant that maturity was reached. After the Sydney contract, Decloedt went its own way and established, in a very early stage in 1972, ***Overseas Decloedt & Fils*** as a subsidiary in Singapore. The company would never ever leave Southeast-Asia and the Far East. As for Algemene Baggermaatschappij, a merger with Ackermans

& van Haaren was on hand. Apart from the crisis in the dredging business in the mid eighties, the golden sixties and seventies would extend in the booming nineties and change into even higher gear at the turn of the millennium.

Dredging: precursor of prosperity in the new millennium

The acceleration of the rate of change and the increase of scale are the determining factors in the dredging industry at the turn of the millennium. Compared to the fifties and sixties of the 20th century the market share of Belgian dredging groups in the world has significantly increased. Baggerwerken Decloedt, Dredging International and Ondernemingen Jan De Nul have taken a commanding position on the five continents and the seven seas. **The annual turnover of the three Flemish dredging groups is currently more than a third of total free world dredging. The doubling of their world market share in the last decades was mainly realized at the expense of Dutch competitors.**

The remarkable expansion of Belgian dredging business after WW II, was focused on two periods of intense acceleration: the Middle East infrastructure boom in the seventies, following the rapid increase of crude oil prices in 1973; and the even more phenomenal scope of works in the nineties, in the emerging New Tiger countries of Southeast-Asia. At both occasions, the *three Belgian D's* were prominently in play. Part of that geographic expansion, another in-depth revolution catapulted the Belgian dredging groups to the world top: an unsurpassed diversification, that would broaden their scope in the field of offshore works, environmental dredging, treatment of silt and contaminated soil, wreck removal, directional drilling, protection of shores and embankments. **As an example, environmental activities in the Belgian DEME-group currently account for almost 20 percent of total turnover. A decade ago, it was almost non-existent.**

The merger between Ackermans & van Haaren and Algemene Baggermaatschappij came at the eve of the biggest marine infrastructure projects till then. It was about the same time that Ondernemingen Jan De Nul made its decisive first moves abroad, with assignments in **Iran and Libya.** The newly established Dredging International immediately took the first prize, when it was awarded the contract for building the port of **Dammam on the Persian/Arabian Gulf in Saudi-Arabia**. In **November 1975,** the **48M m^3** contract was the biggest one that had ever been tendered. Dredging International assigned its trailing suction hopper dredge Antwerpen IV and the cutter suction dredges Tramontane and Mascaret. The company teamed up with **Hollandse Baggermaatschappij (HAM)** in a joint-venture which became known as **HAMDI**. Other contracts followed in the Arabian ports of **Jeddah, Jubail and Yanbu,** as well as in the **United Arab Emirates.**

Those were the years when globalization became a reality. At the other end of the planet, the Belgian group completed a series of major harbor projects in **Thailand**. At **Laem Chabang, Sattahip and Map-ta-Phut,** Belgian dredging workers confirmed the old saying that prosperity and development are following the trail of a dredging vessel.

Along the Suez-canal, both Dredging International and Ondernemingen Jan De Nul took part in the deepening and enlargement of this vital transport artery between East and West. **During the last forty years, Dredging International and its founding companies participated in the execution of four such Suez-campaigns totaling 30M m^3: in 1956-59, 1959-61, 1978-80 and 1992-94.**

Workforce of Ackermans & van Haaren was on site at **Port Saïd** with the cutter suction dredge ***Pax,*** when they were taken by surprise by the **Suez-crisis in 1956**. It is part of corporate history how they escaped, in the midst of British-French landing operations. They were driven by taxi to **Alexandria,** the Belgian consul in front - too late however to board an American evacuation ship. At gunpoint, they managed to catch the last train to Cairo, where they remained blocked in a hotel for three days during bombardments. In a bold and risky move to the Nile river, they found a place on a paddle steamer, overloaded with refugees. From Aswan, they continued their journey on a Sudan Railway carriage to Ouadi Alfa, where they were picked up by a charter plane that took them to Stanleyville in the Belgian Congo.

Having made their adventurous passage through the dark heart of Africa, they arrived at home in Antwerp - eighteen days after they had left Port Said. More than thirty years later, their colleagues were less happy when they got trapped in Iraq, on the eve of the Gulf War. They had been **dredging a 60 km long canal between Qur Abdallah Streets and the Iraq harbour of Umm Qasr**. On board the cutter suction dredge ***Mascaret***, they literally had a front view of the Iraq invasion of Kuwait on August, 2, 1990. However, when the job was completed on September 15th, Saddam Hussein refused to issue exit visas and kept them as ***guestages*** - a new-speak euphemism contraction at the time for "guest" and "hostage."

After a Belgian parliamentary delegation had pleaded their case in **Bagdad** and the **Belgian king** himself had taken unprecedented steps for reaching a humanitarian solution, they could finally leave the country - a full six months after some of them had arrived on the site. They received a hero's welcome in Antwerp, fully aware they had escaped from the worst. On January, 15, 1991, the allied ultimatum expired. Six weeks after their escape from Umm Qasr, all hell broke loose in the **Mother of all Wars.**

It was in those years of globalization that Belgian dredging contractors, extending a century-old tradition, earned a reputation for spanning the globe. A particularly interesting assignment was brought to a successful end at **Guayaquil, Ecuador**, in 1992. A complete suburb for more than one hundred thousand Indian people was cleaned up, after Dredging International and Baggerwerken Decloedt blew huge volumes of sand under thousands of little dwellings on stilts. People who until then had always lived in extremely unhygienic conditions, were given solid ground under their feet for the first time in their life. For this spectacular undertaking, the trailing suction hopper dredge *Vlaanderen XX* had been mobilized from Nigeria. One year later, Baggerwerken Decloedt had moved once again to the antipodes, where the trailing suction hopper dredge *Vlaanderen XI* was assigned for the dearly-won contract of building a parallel runway at Sydney's Kingsford-Smith airport.

In the seventies and eighties, the extension of the port of Antwerp on the left bank of the river Scheldt and the construction of the outport at Zeebrugge were admired as top achievements of the industry, in which Belgian dredging contractors had taken the lead. In the maritime access routes and in the fairway to Antwerp, a Belgian wreck removal industry took shape. The perilous and tricky salvage in 1984 of the French freighter ***Mont Louis***, loaded with 225 tonnes of uraniumhexafluoride, and the spectacular raising of the capsized ferryboat Herald of Free Enterprise in 1987 made the reputation and assigned lasting credentials for this new business of the Belgian dredging industry.

After the founding of the DEME holding structure in 1991, the group went ahead with a further diversification in the field of pre- and post dredging activities - establishing and developing specialized companies, such as **Silt and Soils** for environmental treatment; **Hydro Soil Services (HSS)** which specialized in quay wall renovation and directional drilling; and the expert contractor for bituminous, environment-friendly embankments and capping of landfills, **Bitumar**.

In the course of time, these specialised contractors became respected players in their own right, both in Belgium and on foreign markets. Treatment plants for contaminated soil were built and technology for sludge handling, -recycling or -disposal was developed - attracting business from all over Europe. Bitumar sent its experts to **Bangladesh, for the World Bank** sponsored shore protection works at the **Jamuna Multipurpose Bridge. HSS** realised a "world-first" directional drilling between two offshore platforms at **Manslagt in North-Germany**. The same company carried out a "world-longest", precarious directional drilling in the midst of the busy river Scheldt fairway. Its engineers received international admiration for in-house developed grouting techniques, which enabled port activity to continue during renovation of a quay-wall in the **Lithuanian port of Klaipeda**.

By the time the millennium had changed, the environmental activities of DEME had become so important that an umbrella structure was set up named ***DEME Environmental Contractors*** **(DEC),** which was the actual merger of the former DEME companies **SILT, SOILS, O.E.C. and Bitumar**. Specialized offshore contracts for the oil and gas industry became a significant activity, capitalizing on the core-business know-how in the field of pre- and post trenching, shore approach and landfall construction, pipe-pulling and installation, directional drilling, pipeline stabilization (stone dumping through **D.P. fall pipe vessels**).

Ondernemingen Jan De Nul was involved in building the ***Norfra gas pipeline*** from **North Sea platforms** to France. DEME-subsidiary **Tideway** realized the successful and environmentally sensitive landfall of ***Europipe*** in Germany; the landfall of ***Zeepipe*** at Zeebrugge, Belgium, coming from the ***Sleipner*** gas-field off **Norway**; and the ***Bacton to Zeebrugge Interconnector*** - the latter one by way of a partnering agreement with major players, such as **European Marine Contractors or Brown & Root.** In the same period Tideway in conjunction with other DEME-companies participated in the installation of the **Transmediterranean pipeline** between **Tunisia** and **Italy** and the **Gibraltar gas pipeline crossing**.

In less than a decade, Tideway gained a world-wide reputation in the field of off-shore contracting, which in this case can effectively be considered as a spin-off from the dredging industry. **Recent assignments include a landfall in Trinidad; the Magnus *Enhanced Oil Recovery Project* on the Shetland Islands; and the ambitious 504 km Malampaya gas pipeline in the Philippines**. Tideway clients for these landfalls happen to be the same oil majors that have been commissioning pipeline protection and rock dumping contracts in the North Sea - for which the DEME subsidiary built sophisticated, world-class rock fall pipe vessels such as ***Rollingstone*** and ***Seahorse***.

In the last decade of the 20th century, Belgian dredging contractors focused on Southeast-Asia where projects were tendered of a dimension, never seen before. It is fair enough to recall that Ackermans & van Haaren, Baggerwerken Decloedt and Algemene Baggermaatschappij had in fact never left the region at all. In particular, Decloedt had maintained strong business contacts in the area, after its Sydney runway extension of 1969. At a time when nobody was even aware of the emerging *Tiger* countries, Decloedt was already building ports and harbour infrastructure that would provide the basis for East-Asian prosperity some twenty years later.

Operating from its Singapore subsidiary as from 1972, the Belgian company was busy in **Pulau Sebarak, Coney Island, Pulau Seletar, Buran Darat or the Street of Johore.** Decloedt was the very first to start building Port Klang, off Malaysia's capital Kuala Lumpur. After some fifteen years of activities in Australia, a separate subsidiary ***Dredging and Development Company*** **(Dredeco)** was incorporated in Sydney on January, 24, 1973. But all this early Asian business seemed to disappear into nothingness, compared to the booming years that would follow.

Ondernemingen Jan de Nul took the lead, when the company was awarded the contract for **Hong Kong's Chek Lap Kok international airport.** In the 70s and 80s, the Belgian family business had profiled itself at the very top of the dredging industry, disputing the number one place with Dredging International for the size and the power of their dredging vessels. In the years before the hand-over of power in the former Crown colony, the **237M m^3 of dredging and reclamation at Chek Lap Kok was called the contract of the century**, the single largest land reclamation project at any one time in history.

It is saying a lot that barely **two years later, a new *contract of the century*** was awarded to the same Belgian company for deepening and maintaining **the Rio Parana** and **Rio de la Plata in Argentina**, at the other end of the world. It was a huge contract indeed, won against the competition of all marine contractors world-wide, calling for **dredging and maintenance of a 790 km waterway system, from 205 km out in the South-Atlantic to 585 km up to Santa Fé.**

At about the same time, Dredging International sent its state of the art cutter suction dredge ***Amazone*** to the **Rio de la Plata** for another ten year spanning contract: capital and maintenance dredging in the **Martin Garcia channels**, forming the so-called **Ruta-1** under the Uruguayan coast. This would create a more viable fairway to the Argentine grain ports of Rosario and Santa Fé, and boost activity in the **Uruguayan**

ports of Carmelo, Nueva Palmira and **Soriano.** If nothing else, the Hong Kong and Argentine contracts proved to the world that Belgian dredging contractors had taken the lead in the industry.

In the mid nineties, Hong Kong's Victoria Harbor was commonly called ***hopper lane***, for the more than **20-TH** dredges that were sailing through the narrow Strait at any time. **Nine out of the ten biggest dredging vessels in the world were among them, and five of these nine belonged to Belgian contractors** - **including the JFJ De Nul, the biggest of them all at that time.** Through its 50 percent subsidiary ***Far East Dredging*** **(FED)**, **Dredging International** followed another strategy in Hong Kong than De Nul - teaming up with a **Chinese partner** and diversifying its efforts in a wider series of contracts: the **Yau-ma-Tei shelter harbour, the Tai-Ho section of North-Lantau expressway, the West-Kowloon expressway, the first phase of Tsing-Yi CT 9, the Macao airport.**

Business was booming to the extent that sufficient plant was not available. In the same period Dredging International was one of the first European dredging contractors to re-enter the opening **Vietnamese market**. DI executed the extension of the access channel to **Hon Chong** as well as the first phase of the construction of the **Dinh Vu industrial peninsula near Hanoi.**

Starting in 1995 as a subcontractor to **Penta Ocean, Dredging International** was involved as a joint-venture partner in every development phase of the **Jurong** and **Tuas Extension, the biggest reclamation project ever in terms of surface area**. With several years to go, the Jurong reclamation will ultimately create over **5000 ha** of new land by amalgamating seven smaller islets off the south-western tip of Singapore. When completed this biggest-ever land reclamation contract will have added some 5% to the surface of Singapore. Five years after reclamation began Jurong is already one of the three biggest petrochemical hubs in the world.

At the start in December 1998, the **1.26B USD** ***Jurong phase 3B*** was considered the single biggest tender that had ever been awarded in Singapore. Yet two years later, the **977 ha of** ***phase 3B*** seemed unpretentious in comparison with phase 4, adding another **2500 ha**, including a second ***Tuas extension*** on the mainland. At the end of 2000, up to **12- TH** dredges at a time were taking part in this huge undertaking. Phase 4 with a contract value of **2.4B USD** is to be executed in a very short period of **65 months.**

Among its major achievements in Malaysia, the extension of Port Klang - the sea port of the Malaysian capital Kuala Lumpur - must be mentioned. It was the first assignment outside Belgium of the 17,000 m^3 trailing suction hopper dredge Pearl River, the first mega-dredge of a completely new generation that was commissioned in 1994. On an extremely successful tour of Southeast-Asia in the last years of the 20th century, the Pearl River was also the much admired giant-workhorse at **Mai Liao, Taiwan**, where Dredging International was executing a **26M m^3** dredging and reclamation project on behalf of **Formosa Plastics Group (FPG).**

For all the superlatives of contracts and plant, a particular project offers a somewhat different view of the dredging business. At one thousand km north-west of the capital of Port Moresby, in the deep inland of **Papua-New-Guinea,** Dredging International's cutter suction dredges ***Cap Martin*** and *Mudsnapper* are constructing a mine waste trap and are deepening the **Ok Tedi river**.

There is a pioneering spirit around, at this very outpost of civilization - proving that adventure is still a reality and that, even in the first years of the 21st century, the dredging industry is still a precursor of development and prosperity.

The Netherlands

The stimulus to dredge building began in early days with Netherlands' reclamation, using manual labor that built dikes to hold back the sea. Estimates hold that 37% of The Netherlands has been reclaimed from the sea.

1,000 Years of Dutch Dredging (see **Appendix C)**

"Men came, 6000 years ago, and the sea was their natural enemy. They needed the land. Forefathers of the Dutch, they began to fight back. Centuries passed. The settlers threw up fortresses against the water. They moved earth, later built defensive dikes, still later pumped dry. They chose their weapons- shovels, barrows, dredgers, (and) mills. They put their winds to work (and) they used the tides. But mostly, they pit their ingenuity against the sea. And wrested, inch by precious inch, their country from the water. –Sheer necessity got them started. And the very existence of the Netherlands is proof of Dutch qualifications. They are dredging men by nature, for Holland is the home of dredging."

The Scratcher
1435
(Vereniging "Centrale Baggerbedrift")

Early recordings of dredges included:

The Scratcher, c.1435;

Bucket dredge patent, dated c.1774. (see figures)

Nomadic tribes called Frisians appeared about 400 BC, settling in what became known as Friesland. They built mud mounds called Terpen to live on and escape the tides and storms from the sea. They spread out over more than 700 mi^2 each covering about 5 to 40 acres each. With only their hands and later shovels, the Frisians dug and transported on their backs 100M yd^3 of earth.

The first mechanization cited was the invention of the windmill, c.1600 which was used to dewater the polders. Estimates are that without constant pumping, more than half of The Netherlands would be under water.

The latter half of the 20th century has seen major changes in the merger and dissolution of dredging companies. Major projects have been conducted with Dutch leader-

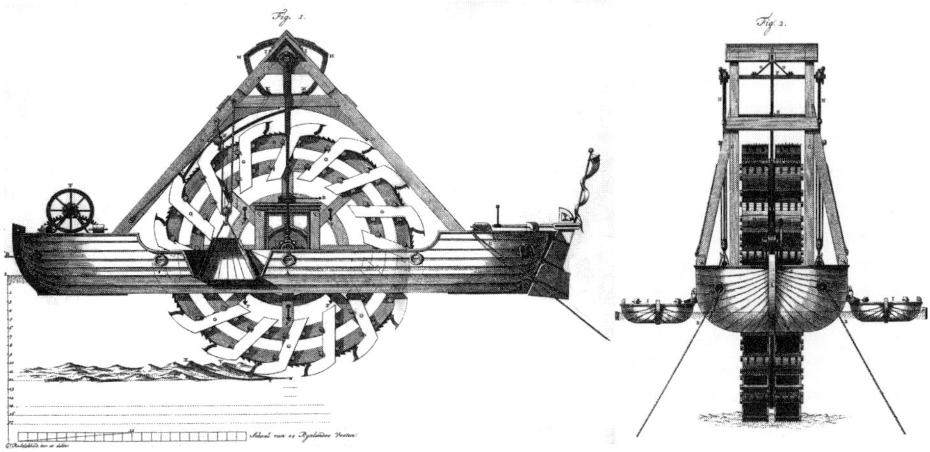

ship in dredging during this period. The companies who dominate the industry today are listed in order of number of dredges owned.

1. **Royal Boskalis Westminster N.V.**
2. **HAM, dredging and marine contractors**
3. **Ballast Nedam Dredging**
4. **Van Oord ACZ B.V.**
5. **Dutch Dredging B.V.**
6. **D. Blankevoort International Dredging Co.**
7. **Boele Dredging Contractors B.V.**

The listings of the above companies' dredges can be found in **WORLD DREDGING/Mining & Construction Annual Directory.** Further descriptions of ongoing projects particularly in SE Asia, can be found in Chapters 7 & 8, New Construction and Reclamation dredging, respectively.

France

The **maritime history of France** may be written by way of the involvement of **Ackermans & van Haaren** (Belgium), in building and dredging its Atlantic and Mediterranean ports. Just between **1905 and 1928,** total volume of dredging by Ackermans & van Haaren in France, amounted to **40M m^3**. The company can safely claim that she was continuously involved in all dredging and hydraulic engineering contracts in **Le Havre,** in the first three decades of the 20th century – including during the First World War, in quite difficult conditions. Besides dredging, quay walls were constructed, embankments were built, *wickerwork*, reclamation and shore protection works were executed. At Le Havre in 1905, Ackermans & van Haaren operated as a subcontractor to **Fougerolle**

Frères in Paris, later on for **Vigner-Smidt and Schneider** in **Paris**. In the late 1920s, the company was deepening the port entrance and the docks to minus 10 m.

At that time, Ackermans & van Haaren operated seven bucket dredges, three pumps, 30- hopper barges and 12- tugboats at the only port of Le Havre. Some of the bucket dredges were said to be the biggest in the world. Besides that plant, a quarry with an annual output of 300,000 t and a workshop were in full operation – the latter one being compared to an independent company on its own.

Other assignments by Ackermans & van Haaren in France included port and harbor projects in **Caen, Honfleur, Rouen, Graville, Bayonne** and others. In **Port de Bouc** near **Marseille** during the 1920s, the bucket dredge ***La Manche*** and others carried out dredging and extension works at the inland navigation canal from the **"Etang de Berre"** to **Martigues**. Quay wall construction, reclamation and shore protection works followed. The ancient links between Ackermans & van Haaren and French maritime ports, would be followed up after World War II when even bigger dredging contracts were executed by the Belgian company and its successors at **Sètes, Dunkerque, Le Havre** and **Gravelines.**

Port Management & Dredging

The three estuary ports: Rouen on the Seine, Nantes on the Loire and Bordeaux on the Gironde, account for most of the maintenance dredging in France. The port of LeHavre was dredged from 1964-68, excavating a new tidal basin; the south-east basin and junction canal together accounted for a volume of 25M m³. The depth of the outer channel, harbor and the old and new tidal basins were dredged to 15 m which accounted for 20M m³.The central ship canal in the alluvial plain was dredged totaling 11M m³.

Early dredging at the Bay of Fos, Marseilles, the Europoort of the south, during 1965-7 represented a total volume of 29.5M m³. The majority of that dredging was accomplished by private dredging firms. The maintenance dredging was done about 53% by the Ports' owned dredges and the rest by private contractors.

The ports of France include the following primary locations, including their annual maintenance and new work dredging c.1967:

PORT	**Mm³**
Bordeaux	12,190
Dunkirk	3,616
Le Havre	10,574
Marseilles	15,386
Nantes St.-Nazaire	4,756
Rouen	7,951
TOTAL	**54,473**

Danube River

The Danube is the second longest river in Europe, with a total length of about **2,680 km**. Navigation is possible between **Kelheim** in **Germany** and **Sulina** in **Romania**, where the Danube flows into the **Black Sea**. The seaports of the Lower Danube, lying within Romanian territory are: **Braila, Galati, Tulcea,** and **Sulina.** The commission of the Danube was established in **c.1856**. The town of Sulina on the Danube was a **Byzantine port** as early as c.**950**. By c.**1318** it was a **Genoese** trading post.

At the time of the **Danube Commission**, the navigable depth was just 9 ft, other branches of the river being 6 ft. By c.**1893**, after various techniques had been installed to increase the natural depth of the river, a decision was made to dredge the river to increase the depths to 24 ft. The first sea-going dredge, a BL dredge named *Percy Sanderson*, arrived in 1895. Then in order to increase dredging capability a second steam powered trailing suction hopper dredge, the ***Carl Kuhl,*** was built in **East Prussia** (now **Poland**). Both hopper dredges were able to dump the material through bottom doors, or, pump the material ashore. Three more hopper dredges were ordered in c.**1921**.

Between the years 1857 to 1902, the navigational branch of the river was straightened, bypassing 27-bends and reducing the length by 21.9 km. In c.1886 the Dutch shipyard Werf Conrad (later part of IHC Holland), delivered two steam BL dredges for work in Galati and Braila to a joint venture of three Dutch contractors, **Schram, Bouterse & Ozinga,** and **Galati.** These dredges had 200 ltr (7 ft^3) buckets, and production of 125 m^3/hr. The dredges continued operating until c.1975.

The largest BL dredge was delivered by Werf Conrad c.**1909**. Named the ***Braila,*** it had 1000 litre (**35.3 ft^3**) buckets, and a production rate of **600 m^3/hr.** The dredge was equipped with a centrifugal pump, capable of pumping material ashore a distance of 400 m at an elevation of 5 m. The shipyards, (later to become part of IHC Holland group), delivered a series of orders during the period of **1896-1900**. This included: two BL dredges, one hopper dredge (TH), a number of barges, and a floating elevator, destined for the construction of the ***Port of Constanta***.

Beginning c.1965, an extensive new dredge building program was initiated in Romania. The dredges built for customers in Romania until 1986, were as follows:

8-self-propelled BL dredges with capacity of 400 m^3/hr.
24-sea going motor hoppers of 500 m^3 capacity.
20-sea going, stationary BL dredges with production of 750 m^3/hr.
57-self propelled, split hopper barges with capacity of 940 m^3.
13-self propelled, split hopper barges with capacity of 400 m^3.
4-stationary BL dredges capacity of 600 m^3/hr.

Great Britain

There are some 2,000 miles of canals and waterways in Britain, which are controlled by various authorities of the government. Many of the waterways date to the time of the

Romans. The **Fossdyke Canal**, the first major canal built in England, was built for drainage by the Romans c.120 AD. Originally called the **Fossa**, it runs 11-1/4 miles from the River Trent at Torksey to Lincoln. Dredging was first budgeted c.1670.

200 Year's of Dredging in UK

The first application of steam power and centrifugal pumps to dredges, was reported in the UK as occurring in the late 1700s in canals and ports. **Riddell** reported that it was in **c.1803** that the **first steam-driven bucket ladder (BL) dredge** appeared in the UK. Production was recorded at about 120 t of sand per hour and was followed by "a vast fleet of steam-powered BL dredgers, which by the middle of the 19th century, were to be found at work in virtually every harbor in the country."

An earlier report states that **c.1797** a **John Grimshaw** used a steam engine to power his *spoon dredge.* The second application of a steam engine to a BL dredge was by Sir Samuel Bentham, in c.1802.

Few places on earth could compare with the UK as an early developer of international trade and export, dependent upon its ports and waterways. Dredging was therefore as common as farming was to the large continents. As ships increased in size and draft, dredges were improved to accomplish the maintenance and capital development for deeper channels.

The **Port of Glasgow**, Scotland was a port created entirely by dredging and increased its depth from 3.5 m in 1824, to 6.7 m in 1864, 9.0 m by 1914 and 12 m by 1936, in time for the new liner, ***Queen Mary.*** This depth was achieved using a steam powered BL dredge and included widening and straightening the 29 km channel. The economic benefits to the area were almost incalculable. "Similar benefits resulted elsewhere around the UK coast as the dredger created the water depths that allowed shipping and trade to flourish."

The steam powered BL dredge continued to dominate the port and waterway dredging activity until about the end of the 19th century, when the plain suction (S) dredge (known as the sand pump), appeared on the scene. This dredge system made possible the beneficial use of dredged material in reclamation, creating new land for port and shipyard development.

Most of the ports, seeking what they considered more convenient and economical, established their own dredge fleets. With the volume of new dredges being required, a dredge building industry emerged within the UK; in England and Scotland. The town of Renfrew, near Glasgow, became known for its dredge building industry, including such names as William Simons and Frederick Lobnitz, later as Simons-Lobnitz Co.

By the 1950s, other types of dredges including **TH,** and **BG/C** self-propelled dredges were being used. Ownership by the port authorities continued as it has to this day while private dredging firms began making some inroads into maintenance dredging that the port expansion required.

United States of America

Beginnings of Dredging

Dredging was introduced into the American Colonies from Europe in the 1700s. The Duke of Orleans solicited *by an offer of reward* c.1723 to obtain dredging of the Mississippi River entrance to a navigable depth of 15 ft at MLW. After some experimenting, a method of Agitation Dredging was introduced and was performed c.1729. A channel depth of 17 ft was obtained c.1731 and ships drawing 16 ft were passing the bar with no problem.

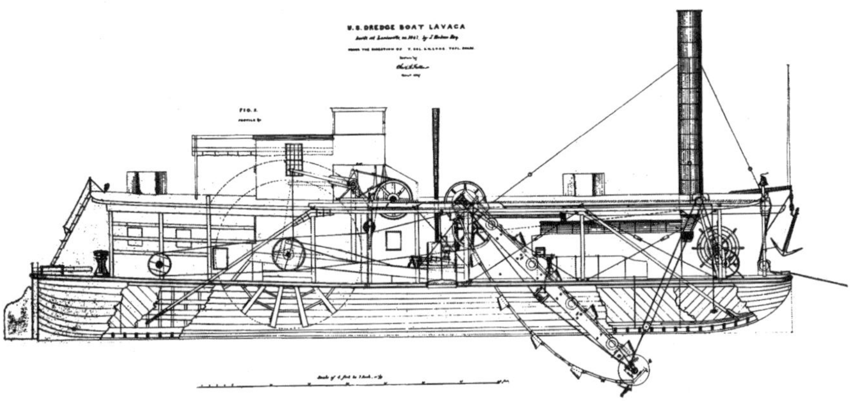

Figure 1. ***BL Dredge "Lavaca" c. 1847***

The first recorded dredge to be built in the American/British Colony was a clamshell (BC) dredge, designed by Arthur Donaldson of Philadelphia c.1774. He built the dredge to clean out slips along the waterfront. The Donaldson dredge was powered by three men and a horse. The hull was oval with a flat bottom, capable of raising 60-80 t of sediment per day and was named the ***Hippopotamus.*** This dredge was the first to be described in contemporary literature. The Pennsylvania Magazine published an article about the dredge in May 1775 which included drawings of the dredge. Based upon a report to the city of Philadelphia on the utility of the dredge, Donaldson was awarded 100 pounds for his ingenuity.

In c. 1847, the US Army COE ordered a BL dredge for use in deepening rivers in Texas during the Mexican War. It was built in Louisville, Kentucky. Later the dredge was leased to a company for contract dredging in Texas and Louisiana. (see Figure 1). This was the first case of a BL dredge being used by private contractors for harbor deepening and maintenance in the USA.

The first Steam powered dredge (USA) was built in 1804 by Oliver Evans in Phila delphia named ***Orukter Amphibilos***, or, ***Amphibious Digger*** and was self-propelled. The hull was 30 X 12 ft with the engine and long iron boiler 20" diameter set in brickwork and fired underneath. The smoke was returned through the water on its way to the chimney by means of an iron flue 9" dia.

The *Amphibious Digger* may have become the first Portable Dredge when Evans mounted it with wheels and towed it about the city (charging $0.25 per person to inspect it). He attached the drive train to a paddle wheel thus giving it propulsion. Its performance as a dredge, however, was not impressive and was several years before a subsequent dredge was able to perform successfully.

Maintenance dredging of US ports and waterways was initiated by **The Congress** in **1824**. The function was assigned to the **US Army Engineers**. With that delegation of responsibility the **Corps of Engineers** began to build its own fleet of dredges and supporting equipment. Starting with the **Mississippi** and **Ohio Rivers** navigation began to expand to the point where about **25,000 mi.** of waterways are maintained today through dredging.

Ports on the Atlantic, Pacific and Gulf coasts, were being added to the annual function of maintenance dredging. Starting with the Civil War in the 1860s private industry began to acquire its own dredges and offer competition to the Army's fleet. Harbors were in their natural state in 1861, and could not accommodate vessels coming from Europe. The controlling depth of New York Harbor at that time was 24 ft. The ports of Los Angeles and Long Beach, California, were not yet built and later were created from tidelands in the early 1900s.

Great Lakes Dredge & Dock Co. (GLDD)

The oldest, continuously operating dredging company in the US, GLDD was founded in Chicago in **c.1890** began as **Lydon & Drews Co**.

"Chicago, in the Gay '90s and early in the 20th Century, experienced rapid expansion. After the famous fire of 1871, the burgeoning city was rebuilt at a rapid pace. Completion of the Illinois and Michigan Canal, the ***Welland Canal*** and the ***Erie Canal*** connected Chicago with the East Coast and the Gulf of Mexico. Railroads bridged the Great Plains to the Pacific Coast. Chicago was at the center of the heartland, and its strategic importance was dependent on water-borne commerce. The port rapidly expanded, and Lydon & Drews, Chicago's premier marine contractor, made numerous port improvements."

The name of the company was changed to Great Lakes Dredge & Dock Co. on **May 2, 1905**. At that time GLDD owned 11-Dipper Dredges (BB), 2-Clamshell dredges (BG/C) and a large fleet of supporting equipment. This included tugs, barges, pile drivers, steam derricks, drilling scow and miscellaneous spares. The company continued to expand through acquisition of other dredging companies and equipment.

The company proceeded to grow and undertake major dredging and construction projects. The leadership that depicted each phase of development was associated with the following persons during their tenure.

1890-1905: William A. Lydon- "Launching Our First Century"
1918-1920: Roger C. Sullivan- "Under the Great Lakes Banner"
1920-1922: Harry C. Wild
1922-1935: John F. Cushing- "Growth Against the Trend"
1935-1945: MGen Edward M. Markham, USA (ret)- "Conflict within & without"
1945-1954: William P. Feeley- "Postwar Period"
1954-1966: Eugene K. Lydon- "A Period of Normalcy"
1963-1966: Langdon C. Hardwicke
1966-1978: John A. ("Jack") Downs- "Preserving the Private Dredging Sector"
1979-1986: "Domestic & International Expansion"

It was during this period that several significant acquisitions of some of the oldest US dredging companies occurred. Even more significant was GLDD taking the lead on building the first privately owned Trailing Suction Hopper (TH) dredge in the US, to begin to offer competition to the USCOE hopper dredges.

GLDD negotiated a JV with the Dutch dredging firm, **Amsterdam Ballast Dredging Co/Ballast Nedam Group**, whereby they provided their expertise on hopper dredging. This resulted in GLDD designing and building the **first Split-hull TH dredge,** the **Manhattan Island, 3,600 m^3 hopper capacity, delivered in June 1977**. This became the prototype of their TH dredge fleet.

This period also began overseas dredging projects for GLDD, including: Argentina, Saudi Arabia, Dubai, and Venezuela, to name a few. Acquisitions were made of two dredging companies' dredges, **Gahagan Dredging Co.** and **Atlantic Gulf & Pacific Co. (AG&P),** followed by the liquidation of those companies.

Weeks Marine, Inc. / American Dredging Co.

The American Dredging Co. was established in 1867, in Philadelphia; a merger of two small dredging companies. The Delaware & Schuylkill Dredging Co. became American Dredging Co. Their charter with the Commonwealth of Pennsylvania was to provide dredging of docks, canals and water-courses, and towing of vessels in the Rivers Delaware and Schuylkill.

Expansion during the reconstruction era after the Civil War resulted in dredging business in the area to accommodate export and import vessel traffic. "The answer to this need, for the Delaware River ports and others everywhere, was: **dredge or die**." Here indeed were opportunity and challenger for Philadelphia's new dredging firm, whose aggressively ambitious management lost no time in getting underway

By the year 1877 the company owned 5-Clamshell (BC/G) dredges and 8-Dipper (BB) dredges as well as supporting equipment such as tugboats, steam pile drivers, and other miscellaneous equipment. The latter part of the 1800s saw American Dredging expanding its dredging business throughout the eastern seaboard, overcoming many obstacles of weather and finance. This period was 1892-1908.

This was the period when **William J. Bradley** became president of the company. He grew up in the company and was fully aware of how to make a dredging company work, both technically and politically.

After Bradley's untimely death in 1916, **Nathan Hayward** was elected president, in **1917.** The next decade saw the boom years of the 1920s and many new dredging projects that transformed many coastal regions in the east. "As the year 1929 advanced, the Company –could well take satisfaction and a measure of pride in a decade of solid accomplishment, in the wake of a job well done in support of the war effort. Since 1919, they had obtained and fulfilled 700 contracts for dredging and related work. They had dug and moved more than 42M yd^3 of silt, sand, gravel, clay, rock and assorted name-less underwater debris."

In 1954, **Maylin H. Greaser** was elected president of the company. His background in the COE and with the company gave him excellent qualifications to take the company into a new era of expansion and projects.

After 1957, American began taking contracts in South America and in the Pacific. Later, **Dr. Gus Pistilli** became president and gave technical leadership to the company. This was followed by the acquisition of the company and its headquarters facilities in Camden, NJ by **Weeks Marine, Inc**.

In the 1990s, Weeks Marine acquired the second largest dredging firm in the US, **T.L. James & Co., Inc.** of Louisiana. They had acquired other dredging firms in the south over the years and were a leading contractor in dredging as well as land and bridge construction projects. The assets were sold to more than one other dredging firm but Weeks acquired the majority of the equipment.

Manson Construction Co.

The family-owned and originated company was founded in c.1905 as a pile-driver company by **Peter Manson,** who was a immigrant from Sweden, with other family members. As general marine contractors working in the Tacoma and Seattle area, many of the ports and waterways have been built and dredged by the firm for its approaching 100 years of operation.

The Pacific Northwest and Alaska has including **Manson-Osberg,** been the domain of Manson through various joint venture Co.'s. Known for its capable and loyal employees, no project was too difficult or large for Manson to undertake. Beginning with pile driving this brought them into bridges and docks with dredging to become a natural part of their diversification. Key members of the families who have built the legend of Manson Construction Co. are: Manson, **Haug, Paup** and **Edwards** with numerous descendants and others marrying into the group.

The Manson dredge fleet now includes: 3-TH, 6-CS, 9-BC combined with supporting tugs and barges. Their projects have taken them outside the US to Mexico, as well as outside the Pacific NW for major projects along the entire Pacific Coast, from Alaska to California and Oregon, Canada, Gulf of Mexico and Atlantic ports.

Dredges for *French* Panama project

Ferdinand de Lesseps, the French Promoter of the Suez Canal construction, undertook to finance the building of a canal across Panama. He brought two small BL dredges from Suez to Panama that began dredging **c.1882**. The company he established was named, **Compagnie Universelle du Canal Interoceanique**, that advertised for contract dredging including the furnishing of new dredges.

A San Francisco, California company, **Huerne, Slaven & Co**., won the contract to remove **7.86M yd^3** in the Atlantic entrance to Panama; at the price of **\$0.23/yd^3**. The plan was to build a scaled-up version of a dredge named **Hercules**, a **BL dredge** built by the **Golden State and Miner's Iron Works of San Francisco (patented by H.B. Angell and H.H. Lynch).** The dredge was equipped with **36-27 ft^3 buckets**, capable of producing **1440 yd^3/hr**. It was shipped to Colon in 1883, and renamed the ***Comte de Lesseps.***

All of the dredges were built in Philadelphia and transported to Colon. The buckets were fabricated out of ½ inch iron boiler plate and connected to the chain by angle iron and bolts. The chain consisted of 36-in. long links, driven by a 100 hp engine. The height above the deck of the upper tumbler was 60 ft.

The discharge chute was a funnel from 5-1/2 to 3.0 ft in dia. and extended 150 ft from the side of the dredge. The bucket ladder was 100 ft long and was hinged 40 ft from the lower end. This allowed the lower section to be raised and lowered independently, relieving the hull and machinery of the total weight of the normally constructed, one piece ladder. The steam plant was made of three tubular boilers each 60" dia., 16' long with 1,400 tubes that were 5-1/2' long each. The spuds were 60' long, 24" dia. with the pivot spud in the hull centerline and the other spud offset 18 ft from the center line.

The second dredge which had a wood hull caught fire and burned shortly after arriving in Colon and being fitted. The third dredge also of wood construction, could not be changed since it was underway. The five remaining dredges # 4 to 8 were built with iron hulls to prevent fires. As soon as they had three operating, they aligned them to dredge in single file for more efficient operation. Those dredges were so nearly automatic that in addition to shoreline handlers, the full crew of seven included the captain, engineer, fireman and four helpers.

Huerne, Slaven & Co. was reorganized in January 1884 as the **American Dredging & Contracting Co.** By the end of 1885 six dredges were working in Panama and the seventh had just arrived. The dredges had removed a total of **6,157,000 yd^3** by the end of 1885, producing a canal six miles long, 160 ft wide and over 16 ft deep.

When the French canal project went into bankruptcy, in May 1888, American Dredging had dredged a total of **22,259,254 yd³** for which it was paid **$13,643,300**. At that point an American company, the **Nicaraguan Canal and Construction Co.** was formed to attempt to build a lock canal across Nicaragua. They bought the seven remaining BL dredges from American and towed them to **San Juan del Norte, Nicaragua**. The subsequent program, however, also resulted in failure.

Dredge Mining

Gold Mining

Mining for gold with dredges began in California in the late 1800s. As placer gold miners were depleting shallow stream and river deposits, they turned to dredges to reach bedrock depths. The first bucket ladder dredge mining system (BL/M) to mine gold was built in New Zealand c. 1882, designed by a man named Scott. Two engineers who had been engaged in that design development, decided to seek their fortunes in two directions; one to England (F.W. Payne, introducing the **European-type of BL/M dredge**), the other **R.H. Postelthwaite**, to California.

The first US built BL/M gold mining dredge began mining near Bannock, Montana c.1895, based upon BL dredges being built in the Eastern US and combined with a processing plant. **Risdon Iron Works**, San Francisco (the forerunner of one of the divisions of the **Bethlehem Shipbuilding Company**), obtained the design and manufacturing rights for the New Zealand-type mining dredge and brought in Postelthwaite as their design engineer. Risdon then began building BL/M dredges for mining in California.

Postelthwaite hired **Frank Griffin**, a young mechanical engineering graduate from Harvard University who began the design configuration of the "California-type" BL/M Dredge. Griffin later started his own consultancy and joined the Board of Directors of **Placer Development Ltd.**, of Canada. In 1930 they formed **Placer Management Ltd** (forerunner to **CPD**) and **Bulolo Gold Dredging Co**. The project was to mobilize a total of eight (8) BL/M dredges in the Bulolo gold fields of New Guinea. This was one of the most celebrated placer gold mining projects of the century.

BL/M gold dredges represented the beginning of **Yuba Placer Gold Co.** by **W.P. Hammon.** Soon thereafter the type of dredge used in Montana appeared in California and developed as the **California-type**. By 1900 there were many such dredges operating in California and shortly thereafter, the dredging boom spread all over the west, north into Alaska and western Canada.

With this design guidance, **Yuba Manufacturing Co.** became the leading manufacturer of gold and tin dredges, eventually shipping and installing them worldwide. Other companies including Bethlehem Steel, Bucyrus-Erie, Link Belt Co., and Marion Steam Shovel Co., supplied many of the earlier BL/M dredges. By 1909, Yuba began designing and building its own dredges and cancelled out its Bucyrus representation by 1913.

Egypt

Suez Canal

The history of the Suez Canal is that it was first constructed during the reign of **King Senousret III** some 2000 years ago. The hand-dug canal was difficult to maintain and filled with silt. The canal was re-opened to navigation by a succession of kings; i.e., **Darius I, Ptolemy II, Emperor Tarjan**, and finally by **Khalif Omar Ebn El Khattab** after the Islamic Conquest. Remains of that Canal can still be seen.

Hand digging of the present canal started in 1859 under the direction of Ferdinand de Lesseps. After the project stumbled several times steam-driven Bucket Ladder (BL) dredges made possible effective dredging by c.1866. The completed canal was inaugurated in 1869. The dredged volume for that phase totaled 70M m^3.

Australia

The **Port of Melbourne (Harbor Trust),** was established in **c.1879**, and after planning by a leading engineer from England, dredging began in the 1880s. The Port purchased a bucket ladder (BL) dredge, but also contracted with private dredging firms to complete the deepening and channel development.

The **Yarra** and **Maribyrnong rivers,** both feed into the port and cause substantial siltation that requires regular maintenance dredging. In more than a century of dredging from 1877 to 1981, various dredges have sculptured an underwater system of thoroughfares, swinging basins and mooring berths and in so doing, dredged a total of **111.1M m^3.**

Dredging of the rivers has contributed 65.7M m^3 and Bay dredging the balance of 45.4M m^3. Of the total quantity dredged 95.0M m^3 were deposited in deep water in **Port Phillip** and 16.1M m^3 utilized in land reclamation. Dredging by contractors has contributed over 9.6M m^3 during the period.

A large BL dredge of **PMA (Port of Melbourne Authority)** is the ***A.S. Mayne.*** With **28 ft^3 buckets, 50 buckets** in the chain, it has **1440 hp** and was built in 1977 at **Vickers Cockatoo Dockyard**, Sydney. In 1979 the dredge was retro-fitted with a lubricated bucket chain which was responsible for significant reduction of noise in operation. This also led to a marked improvement in the serviceability of the chain with inherent cost savings.

CHAPTER 3

MAJOR DREDGING PROJECTS OF THE 20TH CENTURY

The following project summaries represent a selection of the largest and most important dredging projects performed worldwide during the 20th Century. They reflect the magnitude and importance of the dredging industry's contribution to worldwide development.

Japan

A period of major land reclamation in Japan performed by dredging was during 1946-85. The total volume dredged was approximately **1.8B m^3 (2.36B yd^3)** and involved a series of individual projects under a master planned program of dredging and reclamation. Over **80,000 acres** of new land was constructed for industrial development mainly in the area of **Tokyo Bay**. The government initially established an incentive plan whereby companies would provide their own financing to perform the dredging and reclamation, building new industrial land and recouping their costs from its sale.

In addition to Tokyo Bay, reclamation dredging was carried out in the areas of **Kansai, Kyushu, and Chukyo** as well. "It is no exaggeration to say that the base of the dramatic economic prosperity in Japan was supported by these reclaimed lands of the coastal areas."

In 1913 the **Tsurumi Reclamation Association (later Tsurumi Reclamation Company**) purchased the first motor-driven CS dredge (350 hp) from the U.K. This was the first dredge imported by a private company into Japan. After that land reclamation projects were undertaken in coastal industrial areas in **Keihin district (Tokyo-Yokohama area).**

In the 1970s pollution in inner bay areas became a public concern and demanded port construction technology that addressed the issue with increased attention to environmental impact assessment in port development planning.

As the program gathered momentum, competition for projects became so intense that it became difficult to decide upon bidder qualifications. It was reported that a basis for selection of bidders was that preference went to the dredging contractor who owned the largest number of dredges. This led to a major new dredge construction program so that by 1967, the fleet of dredges in Japan totaled 320 of all types.

Panama Canal, Republic of Panama

The total amount of dredging for the period **1905-70** was **376M m^3 (493M yd^3).** While the dredging included the canal and subsequent expansion, it does not include the earlier dredging by the French in the late 1800s. Much was done by the French effort but did not include the building of locks that became essential for the eventual canal. The sale of the rights and assets of the Panama Canal to the U.S. in 1904, included seven (7) bucket ladder (BL) dredges; six (6) of which were built in **France** and one in **Scotland**.

The Scottish dredge was the largest recorded BL dredge, using individual buckets of **54 ft^3** each. The BL dredges handled rock and boulders that were prevalent in that project. Beginning **c.1907** there were six (6) CS dredges (4-20", one-28", and one-36") built and shipped to Panama by **Ellicott Machine Corp**. of Baltimore, MD. This was a major development for that type of dredge in that day.

Suez Canal, Egypt

Dredging of the (107-miles) Suez Canal is marked by its modern development, covering the period of **1950-70s**. The total volume dredged was **559M m^3 (732M yd^3)**, widening and deepening of the canal to accommodate new super tankers. The Japanese

firm of **Penta-Ocean** provided up to five dredges and an Italian firm, **Vianini Dragaggi**, provided other dredges. The **Canal Authority** provided its own dredges operating along side of the private contractors.

Port of Antwerp, Belgium

Following **WWII** the total dredging conducted was **500M m^3 (655M yd^3)**. Major dredging and construction was required to restore Antwerp's port to a viable condition. Dredging was necessary to deepen and widen the channels after salvaging numerous vessels and for reclamation to expand industrial land for development.

Port of Rotterdam, The Netherlands

The total dredging conducted after WWII was **453M m^3 (593M yd^3)**. The major dredging and construction necessary to restore Rotterdam's port included expansion to accommodate super tankers and other large vessels. It became the largest port in Europe.

The Rhine and Maas Rivers discharge a great deal of sediment into the Port, which has required maintenance dredging for many years. From the years 1940-45 to 1965, the width of the entrance channel to the Port was expanded from 300 m to 650 m.

Jubail, Saudi Arabia

The main project period was from the 1970s to the early 80s **dredging** 336M m^3 (440M yd^3) **to deepen the harbor and reclamation of a new industrial complex and port.** Dredges were brought from The Netherlands, Belgium, U.S., Japan, Taiwan and South Korea in this massive program of dredging. This was the beginning of Middle East industrialization following the oil embargo of 1973.

Danube River

The canal between the Danube and the **Black Sea** to **Midia Port**, was constructed in 8 years (**1976-84**) with a total of **294M m^3**. The side canal to Midia was constructed in 3 years (**1984-7**) with a total of **87M m^3**excavated. All of this dredging was conducted with BL dredges discharging into hopper barges.

Richards Bay, South Africa

1960's-70's. 200M m^3 (262M yd^3). Dredging of a new deep-water port for large ore ships. Dredging produced one of the world's deepest ports at 100 ft depth.

St. Lawrence Seaway, Canada

1950's. 191M m^3 (250M yd^3). Dredging of a canal, by large cutter suction (CS) dredges, to provide ocean access for commerce to the Great Lakes.

Mina Jebel Ali, United Arab Emirates (UAE)

1970s. 191M m^3 (250M yd^3). Dredging and reclamation to build industrial land and a port for a new city

Steep Rock Iron Ore Mine, Canada

1944-62. 191M m^3 (250M yd^3). Dredged **300 ft of overburden** from the bottom of a Lake, to gain access to a massive ore body, permitting open pit iron ore mining. The rivers that fed into the lake were diverted to control the water level for dredging. The dredges were transported by land and assembled on site; including two 36" (CS) Dredges and four smaller CS dredges: One-30", one-28" and two-15".

Bay of Fos, Marseille, France

1960-70s. 100M m^3 (76M yd^3). This program was a major new port development, dredging and reclamation producing industrial land of approximately 14,000 acres.

Fort Peck, Montana, USA

1933-62. 100M m^3 (130M yd^3). Hydraulic fill dam provided by dredging, using four-28" CS dredges.

Port of Los Angeles, CA, USA

1980's-90s. 87M m^3 (114M yd^3). Dredging and reclamation included breakwater construction to produce approximately 700 acres of new land for container terminals. The inner harbor was deepened to over 50 ft and the entrance channel to 85 ft.

Port of Long Beach, CA, USA

1960-70s. 76M m^3 (100M yd^3). Dredging and reclamation that produced **Pier *J*,** expanded and deepened the port with channels to **69 ft**. The harbor dredging and reclamation was accelerated by the Port providing the financing instead of waiting for federal appropriations. This gave the Port of Long Beach the deepest dredged port in the US.

A companion project was funded by a consortium of major oil companies called, **THUMS Long Beach Co.;** comprising **Texaco, Humble, Union Oil, Mobil** and **Shell**. The Port dredged and built **four (4) offshore islands** within the harbor. One of the islands is 8.8 acres and the other three are 10 acres each. Over **400 wells** were drilled on each of the islands and began production of a major petroleum deposit of over **1.0B barrels**. The project required **3.8M yd^3** of dredge-fill to build the islands, plus **631 K tons** of barrier rock, quarried and barged from **Catalina Island**, 26 miles offshore.

Tenn-Tom (Tennessee-Tombigbee) Waterway, USA

The project took 12 years with a total construction cost of nearly **$2.0B** including dams and was completed on December 12, 1984. This provided the connection of the Tennessee and Tombigbee rivers, to foster commerce and trade; the largest inland water resource project ever built in the US. Total dredging for the project was: **76M m^3 (100M yd^3)** covering a distance of **232 miles** and saving shippers a total of 800 miles to reach the **Gulf of Mexico**, opening a new channel for commerce into **Mobile Bay, AL**.

CHAPTER 4

DREDGE TYPES & DESCRIPTIONS

This chapter presents an orientation to the different types of dredges and their comparisions. For a more detailed description of the dredge types, refer to Chapter 11, Technology Developments on Dredges. (Refer to Photo Section for illustrations of the different types of dredges).

Mechanical Dredge

The bucket dredge excavates material using various types and size of buckets, digging with "positive displacement." Actuation of the buckets may be powered mechanically, electrically, or by hydraulics. An advantage of the bucket dredge is that most of the water is left behind with about 90% solids in the buckets.

The different types of bucket dredges are described with a typical range of sizes of the buckets. The volume of each bucket is the dimension used in describing bucket dredges, along with its digging depth and total installed (hp or kw) power.

Bucket Ladder Dredge (BL)
Bucket size range: 1.0 to 54 ft^3
Digging Depth: >165 ft dD(50.3 m)

The BL dredge was the earliest form of mechanical dredge. The largest BL dredge (54 ft^3 bucket) was built in Scotland c.1880, for the French effort to dig the Panama Canal. The dredge was sold to the US c.1905 to help complete the canal. BL dredges continue to be used throughout the world for maintenance, construction and adaptations to clean the environment. It has also been popular in dredge mining for gold, diamond, and tin.

"In the first half of the 19th century, an unsurpassed economic boom took place in the **Liège region** of Belgium - mainly through the diversified emerging industrial empire of **John Cockerill**, the son of an English mechanic and entrepreneur who had settled in the country. His applications of the newly invented steam power seemed endless and were exported all over Europe**. The *Liégeoise* dredging vessel would remain the plant of choice and the workhorse of the dredging industry for almost 150 years.**

The *Liégeoise* was a bucket ladder dredge (BL), which has been built in several models through the years - in a towed and a sailing version. Responding to increasing demand, Cockerill moved his Liège-based shipyard along the river **Meuse** to Antwerp, where more space for expansion and better depth was available on the river Scheldt. But other companies in the Liège region continued building the famous bucket dredge. They included entrepreneurs such as **Tilkin-Mention, Bonivert, Spiroux, Jacques**

Piedboeuf and **Perin.** Their vessels would be found all over Europe, as well as on other continents."

The first steam dredges in Holland were constructed in Belgium as well. They were bought by state authorities. It was not before 1864 that a private company in Holland would acquire a steam driven bucket dredge of its own - it was a *Liégeoise* vessel. In the **mid 19th century, 22 dredges of Tilkin-Mention were operating on Dutch and Belgian rivers alone.** Even in the early thirties of the 20th century, *Liégeoise* dredges were still being used in the Netherlands, at the time of the daring enclosure of the Zuiderzee."

The BL dredge was used for civil construction in the U.S. in the 1800s, during the Mexican War. for river maintenance in Texas The dredge was built c.1847 for the US Army Corps of Engineers in Louisville, KY. Later it was leased to a company to dredge in Texas and Louisiana as a private contractor.

Bucket Grab/Clam Dredge (BG/C)
Size range: 1.0 to 200 m^3 (262 yd^3)
Digging Depth: >250 ft.(76.3 m)

The BG/C dredge is often adaptable from land-based equipment. The size of bucket seems to have no limits, only to require ever-larger crane equipment to lift the loads. Another advantage is to be less limited to digging depth than other types of dredges. It has a wide variety of designs that range from heavy construction for hard and large pieces of rock to light plate with special configurations for minimizing spillage for environmental applications.

The dual, alternating BG/C dredge is popular for sand & gravel mining. The largest is the dual 20 yd^3 bucket dredge, digging to over 100 ft. and producing 2000 yd^3/hr.

Bucket Backhoe & Dipper Dredge (BB)
Size range: 1.0 to 30m^3 (39.3 yd^3)
Digging Depth: >38m (125 ft)

This configuration as a dredge has been built for hard digging in New Construction dredging. The **Liebherr Co.** has built a number of 25 yd^3 dredges, for US and Europe harbor deepening and construction in recent years. As the size increases, the tendency has been to install it on floating barges that are sometimes self-propelled. One of the virtues of the bucket ladder (BL) dredge which is to leave a uniform bottom surface can be performed to a limited degree by the backhoe dredge.

Large dipper dredges were built by **Donge Flushing Yard**, The Netherlands, for the **Lake Nasser** pump station in Egypt. Two were built; ***Donge S-type*** dipper dredges with a digging depth of 38m it using a 4.5m^3 rock bucket and a 2.5m^3 bucket. The two hull dimensions are: 48x17.5x4.8m, and 55x19x4.8m, using Liebherr P 995 and P 996 model configurations, with digging depths of 27m and 38m respectively. Total dredge displaacements are: 1358 t and 1900 t.

Bucket Dragline Dredge (BD)
Range of sizes: 1.0 to 200m^3 (262 yd ^{3})

The BD dredge using very large buckets can excavate great volumes of material. However, studies have shown that a BD dredge is not as effective for digging and thus lives up to its name as a dragging function. In the early 20th century, BD dredges were popular in placer gold mining in California, feeding floating processing plants. This was a low-cost operation and relatively easy to mobilize. Other applications have been in phosphate, coal and iron ore mining for stripping of overburden. It is also used for cleaning shallow lakes, ponds and waterways to remove sediment accumulations.

Suction/Hydraulic Dredge

The **Cutter Suction (CS) Dredge** was "-first built by **Col. Von Schmidt, Patent No. 180-600, and operated by Schwartcopff and Hoffman, of Hamburg, Germany; c.1850."** The design and construction of a CS dredge was contracted by the **US Army Engineers** about the same time for operation in the US. Three additional patents were issued for CS dredges in the US**; by L.W. Bates, 1890; A.D. Stevens, 1896;** and **J.B. Cantwell, 1908.**

"The **Cockerill** company dredging vessels found their way to Russia. A complete dredge-building installation was sold to the port of Loban and in the sixties of the 19th century. Cockerill operated a shipyard on **Gontouriefsky island off Saint-Petersburg**. In 1897, the famous American engineer**, Lindon Bates**, concluded an agreement with Cockerill for exporting his powerful ***Bêta*** **cutter suction dredge** to Russia.

The contract had been prepared by a Belgian government mission, studying dredging techniques on the **Mississippi river**. Having seen the results of Bates' *Bêta* dredge, producing 5,946 yd^3/hour during 175 hours and a peak output of 7,789 yd^3 per hour, the Belgian government delegation concluded that this **plant was *very superior* to anything else in the industry.** The next year, Bates was invited to Brussels, where he became one of the most celebrated speakers on the **7th International Navigation Congress of PIANC**. Among the Lindon Bates dredges that Cockerill exported to Russia, one was christened ***Wolga***, and another was named ***Nicholaev*** after the reigning czar."

Cutter Suction Dredge (CS)
Discharge pipe diameter: >6.0 to 48"(1200 mm)
Power: >100 to 30,000 kW
Digging depth: >150 ft (45.8 m)

The CS dredge constitutes the largest population of dredge types in a wide range of sizes, throughout the world. Adding a cutter to the front of the ladder provided a digging function with the swinging of the dredge hull, made possible underwater excavation on a scale not previously known. The advent of the CS dredge and its development has had a profound effect on the capability of the dredging industry to handle a wide range of projects.

After WWII the demand for reconstruction and industrial development resulted in the emergence of *small dredge* builders. One of the earliest was **AMMCO** or **American Machinery Co.** in Tennessee, who began to build a simple pontoon CS dredge; low power with diesel driven cutter head and pump.

Aggressive entrepreneurs were attracted to AMMCO and to subsequently form new companies to duplicate the AMMCO type of dredge. As a result, this type of dredge was built and exported throughout the world. Meanwhile, the small, portable CS dredge continues to be built as a less expensive means of accomplishing dredging projects on a small scale.

Trailing Suction Hopper Dredge (TH)
Hopper capacity: >100 to 33,000 m³ (43,230 yd³)
Suction pipe: >12 to 56" (1422 mm)
Power: >100 to 40,000 kW
Digging depth: >210 m (689 ft)

Developed by the US Army COE in the 1800s, the TH dredge saw major developments in the 1900s. The COE developed its own fleet of TH dredges. The private dredging sector in the US began building its own TH dredges, using the innovation of split hull systems, in 1977 by GLDD.

The development of the Jumbo TH Dredge in 1994, the first being ***Pearl River*** with 17,000 m³, introduced a new dimension and capability into the dredging industry. (See Chapter 8).

An earlier limiting factor in the effectiveness of TH dredges in offshore maintenance dredging was the difficulty in navigational positioning. Navigation problems and their inaccuracies, however, were cured by the introduction of the GPS satellite positioning system.

Plain Suction Dredge (S)
Suction pipe diameter: 4.0 to 36"(914 mm)
Power: 50 to 10,000 kW
Digging depth: <400 ft (122 m)

The earliest form of suction dredge made possible by the combination of a centrifugal pump with steam power, was applied to the maintenance of canals and drainage channels. The configuration of *S* dredges can vary from a small raft with a motor/engine and pump for mining of sand or other minerals, to large systems mining in depths exceeding 400 ft (122 m).

Dustpan Suction Dredge (SD)
Discharge pipe diameter: >6.0 to 38" (965 mm)
Power: >100 to 10,000 kW
Digging depth: >65 ft (19.8 m)

The US Army Corps of Engineers was the original designer. It evolved from the needs of the Mississippi River and its demand for rapid removal of sediment to navigable depths.

The nearest type to this but on a smaller scale is the **Suction Auger Dredge**. This unique form of dredge is discussed in more detail with examples of its applications in large river systems for maintenance dredging.

Another variation to the SD dredge has been the ***Side-caster*** dredge using a pipe on an elevated swivel to discharge slurry up to 500 ft. to the side. This has been popular in **Venezuela** on the **Orinoco River** and the entrance channel to **Lake Maracaibo.** The COE uses this system on rivers and inlet channels in the US.

Bucket-Wheel Suction Dredge (BWS)
Wheel Diameter: >50 to 200" (5080 mm)
Discharge pipe diameter: >10 to 40" (1016 mm)
Power: 35-1500 kW
Digging depth: >65 ft (19.8 m)

There have been several types of BWS dredge built. A patent was filed in **1926** for a dual wheel configuration. Contemporary designs are built by several dredge builders including**: LMG, Lubeck, Germany; Ellicott International, Baltimore, MD, USA; IHC Holland;** and **Neumann Manufacturing, Australia**. Beginning as a single wheel it became evident that the lateral swinging method of dredging requires a double wheel that facilitates digging in both directions.

While attempts have been made to apply the BWS to construction dredging, its primary use has been in mining. Some of the applications have been in Brazil, Australia, Africa and US, mining **cassiterite (tin), rutile (titanium)** and **S&G.**

Bucket-Suction Dredge (BB/S)
Bucket: 0.7-3 m^3 (3.9 yd^3)
Suction pump: 4" (101.6 mm)
Power: 162-300 hp

This type of small, portable dredge is primarily manufactured in Finland called the ***Watermaster.*** Used in canals and lakes for cleaning maintenance.

CHAPTER 5

MAINTENANCE DREDGING

General

The function of Maintenance Dredging comprises the largest volume of dredging conducted throughout the world. Ports, harbors and waterways are kept viable by the regular performance of dredging which normally entails the removal of accumulated sediment. This work tends to be performed by **Trailing Suction Hopper (TH)** dredge, since projects often require transporting material to other locations for disposal or reclamation. It is preferred in busy navigation channels since anchoring and discharge pipelines of non-propelled/stationary cutter suction dredges create impediments to vessel traffic.

There are many instances where cutter suction dredges (CS) are used for pumping into disposal sites and other beneficial applications unique to the CS dredge. This includes pumping into barges for transport. Depending upon the nature of the material, bucket dredges are used and the material dumped into barges that are shuttled from the dredge to disposal sites.

United States of America

General

The largest amount of maintenance dredging in the US, is performed in the eastern and Gulf of Mexico regions of the country. Principal areas include: **Mississippi, Missouri, Ohio, Atchafalaya,** and **Tombigbee Rivers. The Gulf Intracoastal Waterway (GIWW),** over 1,300 miles from Florida to the Mexican border, carries about 1.7B tons of cargo per year and requiring regular maintenance dredging. The main ports in the Gulf include Houston/Galveston, New Orleans, Mobile and Corpus Christie.

The principal ports on the Eastern Seaboard include **New York/New Jersey, Boston, Philadelphia, Baltimore, Chesapeake Bay, Norfolk, Charleston, Savannah, Wilmington, Jacksonville, and Miami**. *(See Photo Plate 10)*

Atchafalaya River

An unusual event affecting the need for maintenance dredging occurred on the Atchafalaya River, LA, in 1952. The Mississippi River, during heavy flooding, changed its course northwest of New Orleans and began a major diversion into the Atchafalaya River. The US Corps of Engineers mobilized an emergency construction project to minimize the diversion. The result after completion of the construction of the dam, reduced the flow to about 30% of the Mississippi River into the Atchafalaya River.

Atchafalaya Bay, an area of 520 km^2 began to form into a large marine delta. ***(Figure 1. Map of Bay).*** At the time the Bay serviced a large fishing fleet and hundreds of offshore oil production platforms. The huge influx of sediment from the Mississippi River gave rise to the observation that "the new sediments are resources rather than pure cost in the traditional costs-benefit analysis." A dredging program was proposed that would build new land by sequentially controlling sub-delta formation while maintaining existing navigation. It was thought that the new land might be used initially for wetland maintenance and salinity control, followed later by low density, recreational and agricultural activity.

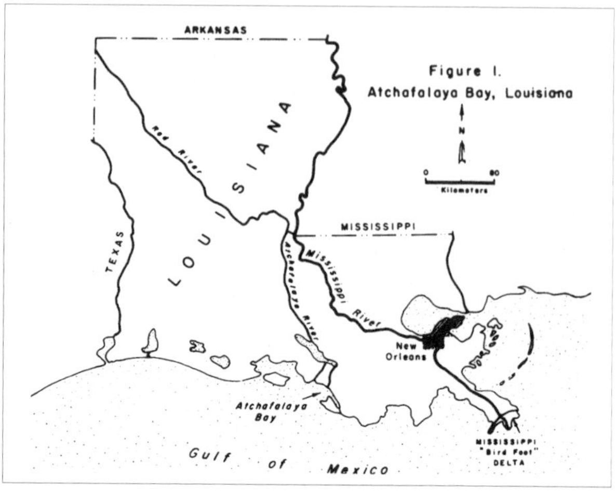

Figure I. Atchafalaya Bay, Louisiana

When the river diversion occurred it was seen that had the dam structures not been built, natural processes would have diverted most if not all of the Mississippi River. As a result, the route below Baton Rouge through New Orleans would have been abandoned. By 1972 Dr. Roy Shlemon reported that over 285 km^2 was covered by clay, silt and fine-grained sand. Since the diversion it was estimated an area of new sediment had more than doubled. He estimated that the ultimate areal expansion of the Atchafalaya Delta could create more than 950 km^2 of new land.

A proposal made by Shlemon, summarized a plan for **"A Dredging Management Program"** which was based on the following. "The concept of land building through sedimentation control is not new. It has been employed in Brazil where flood borne sediments of the **Amazon River** have been diverted into adjacent back-swamp lakes creating highly productive new grazing land (**Castro Soares, 1956**). Along the lower Mississippi River, too, (**Gagliano et al, 1970-71**) have suggested building sub-deltas to reduce land loss, enhance biological productivity, and ameliorate potential hurricane-generated surges. But sub-deltas may be perhaps better made within Atchafalaya Bay, for the beginning of the rapid-growth phase of deltaic evolution, plus the pres-

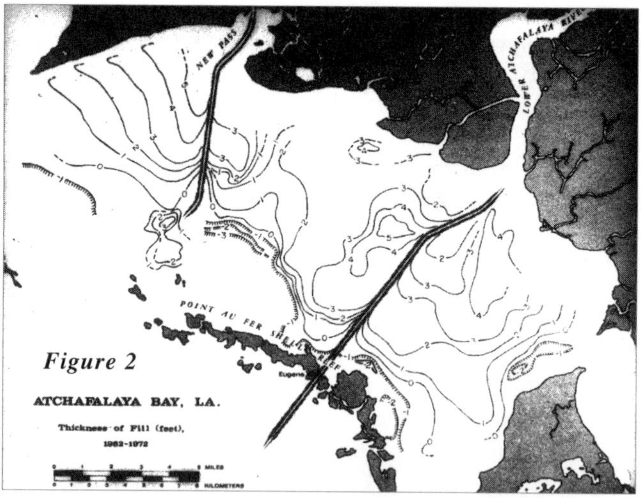

Figure 2

ATCHAFALAYA BAY, LA.

Thickness of Fill (feet), 1962-1972

ence of navigational channels, provide an ideal locality to implement a dredge management program which combines channel maintenance and sequential development of new land. This general concept is graphically portrayed in *Figure 2.*"

Europe

Major maintenance dredging in Europe is in the **Ports of Rotterdam, Hamburg** and **Antwerp**. In the latter case, the **River Scheldt** requires regular dredging to maintain the accessibility of the Port of Antwerp. Other significant dredging is conducted on the **Danube River.**

River Scheldt

"The building of the port of **Zeebrugge** was directly inspired by the calamitous experience that had once hit Antwerp and the Belgian economy as a whole. In the years before the First World War, the sheltered harbor of Zeebrugge was purposefully built on open North Sea water. Never again would siltation or closure of a river threaten the Belgian economy. Zeebrugge would develop into large scale extensions at the end of the 20th century, a dredging and hydraulic engineering project that is rated among the top achievements in the industry. Even then, maintenance dredging in the Scheldt river and in the maritime access routes currently amounted to about **20** to **40M m^3** of material per year. Total volume of dredging on the Scheldt river alone, has been estimated at **500M m^3** since the end of the Second World War."

Western Scheldt

The Western Scheldt estuary in the SW of the Netherlands has to be dredged frequently and the total volume of material dredged annually amounts to approximately **10M m^3**. The morphological processes of the estuary have to be understood in order to plan a maintenance dredging program. The estuary covers an area of 300 km^2 and is funnel shaped. *(Fig. 1)* A detailed study has been made of the tidal influences and shoaling at the five dominant nodes, or, sand bars. In this study including trends of sedimentation the dredging programs are planned and executed.

"The morphological system of a bar is relatively simple. The situation gets more complicated, in case of continuous dredging at the bar. Some complicating factors are:

The artificial interruption of the natural processes will also affect the area upstream and downstream of the bar. Due to accelerating and decelerating flows, under and over saturation of the sediment load will occur, which will influence the residual sediment transport. Therefore, bar dredging will affect the channel developments on both sides;

Dredging followed by dumping, takes place in the vicinity of the dredging areas either in eroding outer bends of ebb channels, or in secondary flood channels. The dumping increases the sediment availability and affects the natural processes in the channels.

Navigational requirements demand safe shipping lanes without sharp bends or frequent fairway changes. This makes it difficult to follow the natural development of the system and may result in a gradual shift from almost completely natural (except an over depth-ratio) to almost completely artificial channel configuration."

Conclusions

"The morphological developments of an estuary like the Western Scheldt, can be understood by studying the historical development of the estuary. Modern tools like **Geographical Information Systems (GIS),** allow easy access to and interpretation of large data bases.

Estuary channel dimensions are determined by the tidal flow volume that passes the channel during the dominant tidal phase (ebb or flood). The non-dominant tidal phase does not have a significant impact on the channel cross sections. In the Western Scheldt, the relation between dominant tidal volume and channel cross section is equal for ebb dominated and flood dominated channels.

Most maintenance dredging locations are situated at the downstream end of an ebb-dominated channel, close to a *node* where ebb dominated and flood dominated channels cross each other. The reason for bar development at these locations is a considerable deposition of sediments during flood, while the erosion during ebb is reduced or even changed into sedimentation due to short cuts or divergence of the flow. Based on the morphological analysis, a qualitative prognosis has been made of the development of the maintenance dredging volumes in future, without or with the implementation of the fairway deepening to the Port of Antwerp."

Rotterdam Fairway

This will present an overview of maintenance dredging in the ports of Rotterdam and **Ijmuiden**, the largest port in terms of vessels and cargo. The beginning of the Fairway is called the **Eurogeul**. It is located 57 km off the coast and has a depth of 25.4 m. It has a width of 1200 m and consists of a channel 600 m wide with 300 m wide banks on either side and a turning point of 2700 m diameter. The Eurogel runs into the **Maasguel** with a depth of 24.3 m.

The maintenance dredging of the access channels to the Port of Rotterdam is conducted in the parts of the Fairway shown in Fig. 3. The total quantity of dredged material in the three main areas is about 12M m^3 per year. The G-block consists of silt and sand, the E-block is mainly silt, and the F-block is a mixture of sand and silt. The cost of the maintenance dredging is between NLG15-20M yearly. The history of this dredging however shows considerable variation from year to year. For instance, 1994 was 50% higher than in 1993.

The beneficial uses of the Rotterdam Fairway maintenance dredging includes such applications as:

- To supply sand for various purposes; for landfill, replacement fill, road construction and city expansion projects;
- Beach nourishment, shore protection and offshore berms;

- Islands provide nesting sites for birds, wetland restoration and establishment;
- Land creation.

Contractual Methods

In August 1995 the **North Sea Directorate** introduced a new dredging contract. "The former contract was a charter contract with remuneration based on production. There was no fixed price per hour or cycle; the remuneration was based on the amount of dry solids and each week the principal decided on the dredging program. In the form of contract now used, the contractor takes more responsibility for the execution of the maintenance dredging. On the basis of the survey maps, he has to make the dredging program that needs the prior approval of the principal before it is carried out. The principal is the supervisor and thus the inventiveness of the contractor has to be engaged to meet the challenge of achieving more effective maintenance, leading to a reduction of the maintenance costs."

Egypt

The maintenance dredging of the **Suez Canal** continues to be a requirement of the **Suez Canal Authority**. The expansion of the Canal along with recent major construction east of Port Said and south of the entrance to the Canal on the west side of the **Red Sea** increases the demand for regular maintenance dredging.

The dredge fleet for maintenance includes: two TH dredges, 7-CS dredges including the largest, the ***Mashhour,*** with a 36" pump and 30,000 hp; one plain Suction (S) dredge, one BL dredge with buckets of 26 ft^3 and one BWS 24" dredge.

South America

Some of the largest rivers in the world are located in South America, and require regular maintenance dredging. They include the **Amazon, Parana** and **Orinoco.** In the early 1990s, a precedent-setting maintenance and management program was put out to bid by the Government of **Argentina** offering a long-term contract for maintenance and development dredging of the Parana River. The winner was **Jan de Nul NV** of Belgium, with a contract term estimated at 25 years. The project started at the **Rio de la Plata,** Buenos Aires and continues up river to Santa Fe.

The Parana project was followed by the **Hidrovia Project** which includes a 3,400 Km channel of the **Paraguay** and Parana Rivers between **Caceres, Brazil** and **Nueva Palmira, Uruguay**. The Hidrovia Project is supported by five countries of the **Rio de la Plata basin: Argentina, Bolivia, Brazil, Paraguay, and Uruguay.**

Subsequent to the study, tenders were issued and a consortium of dredging firms were awarded dredging and management contracts for the **Rosario-Victoria Fixed Link.** A concession was awarded to the **Consortium *Puentes del Litoral S.A.*,** formed by five civil contractors: **Impregilo, Iglys, Hochtief, Techint,** and **Benito Roggio**. Total

value of the contract was US$350M, of which $120M will be collected by river tolls over a period of 25 years, the term of the concession. The Dutch **JV Boskalis Intl-Ballast Nedam Baggeren** as subcontractor is responsible for the construction of 47 km of road embankments totaling **$76M,** to be completed within the concession period.

Venezuela

The two largest areas of maintenance dredging in Venezuela are the **Orinoco River**, carrying bulk cargo from iron ore mines; and **Lake Maracaibo** and its entrance channel, supporting the country's largest oil production. Large TH and side-casting dredges are used to maintain channel depths for navigation under heavy silting conditions.

Panama, Republic of

The Panama Canal requires regular maintenance dredging which is conducted using BC/G, BD, BB and CS dredges owned by the **Canal Commission**. The material varies from fines to rock which explains the varied types of dredges used.

South Africa

On South Africa's east coast innovation in maintenance dredging is illustrated by the **Portnet**. In 1990 as a result of the commercialization of the **South African Transport Services**, they organized a state-owned transportation company, called **Transnet Ltd**. Port maintenance dredging is undertaken by the division called **Portnet Dredging Services**. The main ports requiring continuous maintenance dredging include **Richards Bay, Durban** and **East London**.

The East Coast of South Africa has a net annual northbound, long shore movement of sand. The average volume of this movement is estimated at 350,000m^3 at East London, 600,000m^3 at Durban and 800,000m^3 at Richards Bay. Each of these ports has installed southern breakwaters, which if not dredged would result in build-up of a beach along the breakwaters and finally a bar across the entrance channels. It would also cause erosion of the northern beaches.

The port management has adopted dredging as the method of resolving the problem. At each port a sand trap is dredged adjacent to the southern breakwater. The size and location of the sand traps has been scientifically determined and varies at each port.

At Durban and Richards Bay, sand dredged from the sand trap is reclaimed onto the northern beaches. At Richards Bay the sand is pumped directly onto the beaches by a dredge, while in Durban some sand is pumped directly onto the beaches and the remainder into a concrete hopper. From this hopper the sand is re-handled and pumped via several pump stations to various outlets along the beach.

At the concrete hopper, re-handling pumps and the pipeline are owned and operated by the Municipality of Durban. An arrangement for the delivery of a specified amount of sand annually exists between the Municipality and Durban Port Management. When the Municipality requires additional sand for beach nourishment, they contract directly with Dredging Services.

At both Durban and Richards Bay, the beaches that are thus nourished are the major public recreational beaches. Sand trap dredging and the reclamation it provides, is the major portion of the annual, maintenance dredging requirement and accounts for 57% of the total. The remainder of the annual work load consists of dredging sand and silt from basins, channels and alongside the quay walls.

Dredge Fleet

The reorganization of the dredging fleet began in 1988. After considerable personnel changes and reduction or increases to some of the equipment, the following fleet evolved.

2- Trailing Suction Hopper (TH) dredges
1- Bucket Grab/hopper (BG/H) dredge, self-propelled
1- Bucket Ladder (BL) dredge
2- Hopper barges, self-propelled
1- Tug, bed leveling

Saldanha Bay

Located about 110 km from Cape Town on the Atlantic coast is Saldanha Bay, encompasses an area approximately four times that of the ports of Durban, Cape Town, Port Elizabeth and East London, the increase in tanker size and frequency of transit around the Cape demanded expansion of the harbor. Other demands for export of iron ore resulted in a major study in the late 1960s for increasing the port.

Delft Hydraulics Laboratory in Holland undertook studies to solve problems of mooring. Saldanha Bay had a natural potential as a harbor for 250,000 t ore carriers. In 1973 reports of dredging requirements were published and dredging plans were outlined. Estimates were that it would require $R400M to dredge the harbor.

India

Some of the major rivers and estuaries of the world are in India. The east coast of India is subject to a large littoral drift causing siltation in the approach channels to the ports, requiring major quantities of maintenance dredging. Because of the magnitude of dredging, highlights only are presented for the principal ports.

Storms and cyclones are well known in the **Bay of Bengal**. More than 200 storms have occurred during the last 100 years with two storms per year, out of which a dozen storms were of catastrophic nature. The beach material that is debouched into the Bay of Bengal emerges from the majority of the rivers of the Indian Peninsula and amounts to several million tons of sediment per year; producing the main source of the littoral drift. ***(Fig. 1)***

The East Coast of India extends from the **Hoogly Estuary** in the north to **Cape Comorin** in the south. The major rivers of the south are: **Ganga, Mahanadi, Krishna** and **Kaveri**, having their origin in the **Western Ghats** but flowing eastward into the Bay of Bengal. There are in all 225 port of different sizes and 11 major ports in India. Total traffic handled is more than 96M t/year.

COAST	TIDAL RANGE	BED SLOPE	BED MATERIAL
East Coast	1.20 m at Madras and 5.0 M at Calcutta	20 m contour at 3 to 6 kilometres from shore	medium to fine sand
West Coast	1.0 m at Cochin & 11.0 m at Gulf of Cambay	20 m contour at 12 to 20 kilometres from shore	Silt, clay and mud

Figure 1

Maintenance dredging required in the Indian ports, is necessary in order to keep channels open for the large amount of export traffic. The following is a summary of the maintenance dredging activity by port.

Calcutta Port

The **Hoogly River** channel is about 120 nautical miles in length, and is maintained navigable by dredging. Annual maintenance dredging is conducted by government and port-owned dredges. The total dredging amounts to about **20M m^3**.

Paradip Port

Two rivers, the **Mahanadi** and **Devi**, flow into the sea near the port, which is located on the east coast. The port is a lagoon-type of artificial harbor with two breakwaters. The channel is about 1850 m long and is dredged to a depth of 12.25 m. Maintenance dredging amounts to c.**3.0M m^3**. **Dredging Corporation of India** conducts the dredging using trailing suction hopper (TH) dredges.

Viskhapatnam Port

Located between Madras and Calcutta, the net littoral drift is about 0.7 Mm^3/year. Total maintenance dredging is c.**1.25M m^3/year**. The outer harbor is maintained to a depth of **10 m**.

Madras Port

Dredging has been conducted for maintenance of this port since the late 1800s. In the ensuing years, experiments with different types of dredges resulted in various acquisitions and disposals. The annual maintenance dredging amounts to c.**2.0M m^3**.

Cochin Port

Silting of Cochin port is caused by the **River Periyar,** in addition to the littoral drift from the outer approach channel during the SW monsoon. The outer channel is 8640 m long and is maintained to a depth of 12.6 m. The annual maintenance dredging is c.**5.0M m^3**.

New Mangalore Port

Located between **Mormugao** and **Cochin** on the west coast, it is an artificial or man-built harbor. Siltation is contributed by movement of sediments due to the action of the wave-induced currents coupled with tidal flow. The total annual maintenance dredging is about **4.7M m^3.**

Mormugao Port

Located on the west coast and south bank of the **River Zuari,** Mormugao Port joins the **Arabian Sea**. Siltation from the river and the sea, is dredged by the **DCI** hopper (TH) dredges. Total dredging is about **2.5-3.0M m^3 per year**.

Bombay Port

Siltation in the harbor is mainly due to the speed and depth of tidal flow, fresh water influx during monsoon, and nature of the bed turbulence caused by wave action. The resultant maintenance dredging is **4.5-5.0M m^3/annum**.

Kandla Port

With a tidal range of 7m and an approach channel of 24 km siltation is considerable. This results in annual maintenance dredging of about **2.0M m^3**.

China

Few areas of the world can compare with the volume of sediment that is transported in the rivers of China. Statistics show that 13 of the large rivers in the world carry annual sediment loads of over **5.8B tons**. The **Yellow River** is the highest and the **Bramaputra River of Bangladesh** with **499M tons** is second. The **Indus River** in **Pakistan** ranks third with 435M tons. About 30 % of the sediment load carried by the 13 rivers comes from the Yellow and Yangtze Rivers.

The effect of these rivers is to create the need for annual maintenance dredging, to keep them open for commerce. The number of dredges owned by the government agencies in China is not published but is estimated at several thousand.

Yangtze Estuary

The Yangtze River is the longest river in China. It is **6300 km** in length with a drainage area of **1.8M km^2**. Runoff is approximately **486M tons** per year. A dredging program began in 1974/5 to improve the Shanghai Port to extend the channel to 32.8 km and deepen to 7.0 m.

In order to improve the condition of the sea-route, a project of harnessing the channel of the sea-route in the Yangtze Estuary was carried out in early 1998, in three phases. During the first phase, two guide-banks, a groin field and other construction works were set up on both sides of the upper section of the North Passage. The artificial channel, which was 7.0 m deep, was further deepened, extended and widened, and the natural sea-route of the lower section of North Passage was dredged to 8.5 m in 2000. Later deepening will be to 12.5 m.

Owing to the limitations of navigation, distribution and type of sediment, only TH self-propelled dredges were used. In addition to normal dredging procedures, they also used other methods such as by-pass and side-casting. Since 1974, the total amount of dredging in the estuary was about 390M m^3, or, a yearly average of 14M m^3 maintenance dredging.

Dredging of the Grand Canal

The Grand Canal was dug in c. **486BC** starting in Beijing and ending in **Hangzhou**. It is 1,789 km long and is the earliest and longest man-made canal in the world. It is also the only transport route running north to south in China. It passes through **Beijing, Tianjin, Hebei, Shangdon, Jiangshu and Zhejiang,** connecting **Heibe**, the **Yellow River, Huaihe, the Yangtze and Qiantang Rivers**. In 1885, the bank of the Yellow River collapsed in **Henan Province**, resulting in the Grand Canal being cut into northern and southern sections. As a result North-South transportation was interrupted and the Canal became silted up year after year.

It was decided to reconstruct the Subei canal section first and from 1958-61, dredging was carried out using a labor force and excavation machinery. Curve cutting and land digging were done by hand using the large labor force. Widening and underwater excavation were done by machinery. Due to the length of the project, 15 large CS, BG and BWS dredges were used to excavate **3.73M m^3** of material.

With the rapid development of the national economy a second term of dredging was carried out from 1982-88. The width of the channel was enlarged to 60-70 m and a depth of 4.0 m with a 800 m radius of curvature. The engineering earthworks were again completed by a combination of manpower and machinery. There were 25-CS dredges; large, medium and small, and BG dredges. They were assigned from inside and outside the Province to carry out a total of 6.0M m^3 of underwater excavation work.

Sunan Section

The dredging of Subei section with the resulting large increase in transportation, resulted in greater impacts on the Sunan Canal. Serious vessel jams often occurred in the narrow and shallow sections due to lack of dredging. While dredging of the area continued from 1958-87 it only partly solved the problems of navigation of the Sunan.

As a result the National Government decided to dredge the whole canal of 208 km to a national class four standard (40 X 2.5 m and bends of 330 m). Work started in 1992 and lasted for 5 years and cost 2.7B Yuan, moving 38M m^3 by dredging and labor force. Recent construction of dredges to add to the fleet for dredging the Canal and inland rivers, have included **BC/G dredges with 0.5-1.2 m^3 buckets** along with **80- 200m^3/hr BWS** and **CS dredges**. These have dredged **4.0M m^3** of material in the Province and played a protective role for both waterways and transportation.

The **Jiangsu** section occupies one third of the length of the Grand Canal. It has improved its condition and increased its role in transportation for Jiangsu and other Provinces and cities following dredging. Vessels from 13 Provinces use the Jiangsu, which has become a main transportation artery especially for the movement of coal from north to south.

Thailand

The **Port of Bangkok** and **Laem Cha-Bang Deep Sea Port**, have been the responsibility of the **Port Authority of Thailand (PAT)**. All of the maintenance dredging is performed by PATs fleet of seven (7) dredges. The dredge types include: 4-TH, 1-BL and 2-BG/C.

Approximately **4.4M m^3** of silt is annually dredged from the access channels and the material is dumped at a deep water disposal site. The Port of Bangkok is situated on the left bank of the **Chao Phraya River**. The approach to the port is along an 18-km bar channel, 100 m wide in the reaches and 250 m wide in the bend following the bar channel. The designed depth at the bar channel is 8.5 m and the river channel varies from 8.5 to 17.8 m (MSL).

The Laem Cha-Bank Deep Sea Port is located in the **Chonburi Province**, about 130 km from the City of Bangkok. The access to the port is through a 1.5 km channel that is 500 m wide and 14 m deep (MSL).

CHAPTER 6

RESERVOIR MAINTENANCE DREDGING

(The) Sudden loss of the world's reservoir capacity would be a catastrophe of unprecedented magnitude, yet their gradual loss due to sedimentation receives little attention or corrective action.

During the 20th century, worldwide demand for potable and irrigation water resources resulted in the construction of dams worldwide. Areas such as the US, China, former USSR, Europe and other high population zones, have been engaged in massive construction projects to meet rising population needs.

The largest number of dams is in China which is reported to have c.82,000 reservoirs. The ultimate result of the dams is sedimentation accumulation, which robs capacity and thus becomes a task for the dredging industry. The average annual rate of loss of capacity is at the rate of 2.3 %. **(Zhou, 1993)**

In the US, dam construction increased after WWII so that in the decades of 1940-80, there were over 4,000 built. The largest international reservoirs in terms of volume (M m^3) are: **Bratsk/USSR- 169,000; Aswan, Egypt-162,000; Kariba, Zimbabwe-160,368; Akosambo, Ghana-160,368; Hoover Dam, US- 34,850; Glen Canyon, US-33,300.**

As of 1996 there were about **42,000 large (over 15 m tall) dams worldwide**. In the US, total sediment deposition in reservoirs reached **1.2 B tons per year**. In **Japan**, up to 1979, **425 reservoirs had lost 6.3% of capacity due to deposition**.

Dredging of Reservoirs with Limited Water

Many large dams have been built throughout the world during this century in order to create water storage reservoirs, water supplies and flood control. However, in rivers which transport considerable quantities of sediment, many of these reservoirs function as sand traps and often silt up. Their storage capacity is often reduced more rapidly than anticipated and intakes and outlets get obstructed. In many cases the need to remove these sediments has become very urgent.

Flushing and dredging are the main options to remove the sediment from the reservoirs. To date dredging has largely been limited to small reservoirs and to the clearing of the intakes. For larger scale removal of sediments, dredging has often been considered by Authorities to be uneconomic and causing too much loss of valuable water. However, in view of developments, in both excavation techniques and transportation systems, dredging can now be recognized as an economically viable solution in many cases.

This paper describes some possible combinations of traditional and new excavation and transportation techniques for rather homogeneous sediments and for moderate annual quantities of some million cubic meters, which cause only very limited or no water losses at all. Cost analysis indicates that the removal of sediments under these circumstances remains in terms of only a few dollars per cubic meter of sediment.

Since the dimensions and characteristics of reservoirs and their sediments, as well as the local circumstances differ from place to place, such an optimum dredging combination has to be determined for each particular reservoir.

Introduction

This text is intended to give a review of the considerations relating to dredging in barrage reservoirs. The initiative results partly from Boskalis' own market expectations and partly by hardware developments elsewhere. Some of the developments considered are not yet concluded, while others are established.

When hydraulic dredging, with a cutter suction dredge and boosters, is considered as a financial baseline, most likely all other systems will not produce a cheaper solution, but will either extend the technical possibilities or improve the *quality* of the dredging. In some particular situations financial gain might be expected as well.

The kind of jobs that are dealt with are those which can adequately be done at production rates from say 500 to 1000 m^3/hr per system. These boundaries originate mainly from economy of scale, and transportability aspects. Moreover an annual production level of 3.5M m^3, seems feasible related to acceptable project duration versus problematic siltation rates and considered reservoir size (10 to 100M m^3 water volume). Higher rates may also give problems when alternative transport sub systems (not the dredging installation it-self) are pursued.

Lower rates can be considered of course. In that case hydraulic dredging tends to give higher cost per m^3, but alternative technologies can become more favorable. High pressure slurry pumps, for instance, are currently built in units of about 800 m^3/hr maximum production.

There is no standard system configuration that can be nominated as the best solution for all circumstances. The contractor has to compose an *optimum system* for a particular duty, depending on client preferences, with the sub-system possibilities (dredging, transportation offshore/onshore depot location), as discussed below.

Sediment Control at Reservoirs

The Problem

During the past decades a great number of dams have been built in river systems in order to create storage capacity of water resources. These schemes have been constructed for power generation, irrigation, drinking water supply, flow regulation. Independent of the purpose of the scheme, they clearly have one thing in common, viz. the fact that they always constitute an artificial interruption in the hydro-morphological river regime. The practice of reservoir operation teaches that without an adequate control of the hydro-morphological regime such interference will be revoked by nature via sedimentation of the reservoir. This is caused by trapping of sediment conveyed by the flow entering the reservoir, due to sudden reduction in sediment transport capacity as compared to upstream.

As a result, reservoir sedimentation is one of the main threats to Reservoir Management as regards operational efficiency_and lifetime. Problems that numerous dam operators experience already are the continuous reduction of live storage capacity, the abrasion and later clogging of bottom intakes at dam as well as the horizontal forces exerted on the dam by the sediment deposited in front of the dam. Having in mind the relatively high capital costs of such reservoir building schemes, it is imperative for the reservoir management to develop and implement strategies to control reservoir sedimentation in an economically viable way.

Sediment Control Strategies

The type of year-round sediment discharge into the reservoir is the paramount, determining factor in developing remedial or controlling measures. This requires a proper insight into the hydro-morphological aspects of the upstream river regime. Such knowledge can be gained from comprehensive site measurements,_combined with indispensable numerical modeling of flow and sedimentation/erosion. Analyzing the problem via the above rational method, one can arrive at the following general measures, applicable for future and existing reservoirs:

· minimize entering of sediment into the reservoir by:
· erosion control in watersheds (terracing, ashore station, etc.)
· upstream trapping of sediment
· bypassing to prevent sediment from entering the reservoir
· minimize reservoir sedimentation by:
· passing the sediment-laden flood waters through the reservoir via bottom outlets as an 'undisturbed' stream flow
· remove accumulated sediment deposits by:
· intermittently flushing the reservoir with sediment-poor reservoir water, using bottom outlets during the pre-flood period
· applying dredging techniques
· accept reservoir sedimentation and maintain long term storage capacity by;
· increase of the dam height
· abandon the sediment filled reservoir and construct a new reservoir elsewhere in the river system.

The technical and economical feasibility of the above (combination of) measures depends on a great number of parameters such as:

· availability of capable bottom sluicing facilities
· availability of water volume for flushing
· type of year-round flow and sediment seasonal distribution
· characteristics of accumulated sediment deposits
· purpose and volume of water resources
· consequences of downstream or adjacent sediment disposal

- consequences of control measures interfering with the main reservoir operation
- environmental impact
- institutional-political limitations.

Most of the control measures mentioned have already been implemented and tested at some reservoirs. In the range between the control by dredging techniques to accepting reservoir sedimentation, the net result of these measures varied from viable to non-viable.

To date dredging has largely been limited to the (emergency) removal of sediment, which has accumulated in front of the intakes. However, in view of developments in the field of dredging and transportation, these techniques should now be recognized as an economically feasible solution to the general sedimentation problem.

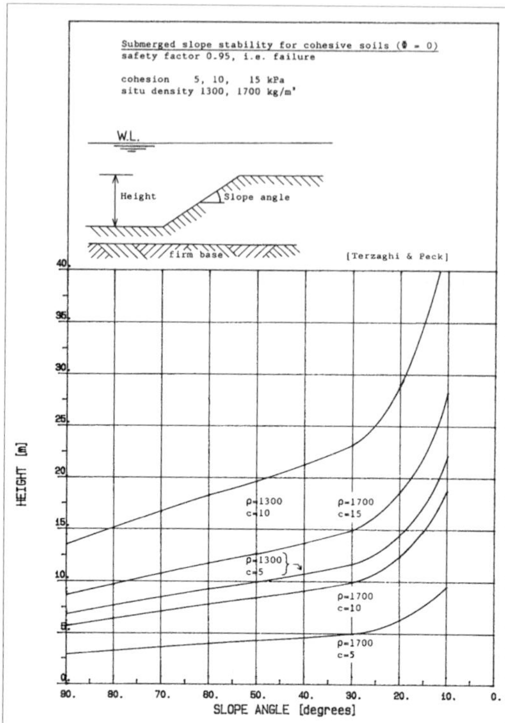

Figure 1.

On the other hand it appears that the consequences of sedimentation in numerous reservoirs have often been under estimated, resulting in the allocation of maintenance budgets which are often too low for the efficient management of the reservoir. Contrary to control by dredging techniques, the control measures making use of natural forces do not require expenditures. However, a proper cost-benefit analysis may reveal that the implementation of dredging techniques (sec or in combination with other measures), may be an economically viable solution. Such cost benefit analysis should take into account the loss in revenues due to reservoir inefficiency as well as the expenditure for downstream consequences.

Control with Dredging Techniques

Under the prevailing market situations (reservoir m³ costs versus dredging costs) current dredging and transportation techniques can provide answers to former problems of reservoir dredging. Today, higher densities are reached by hydraulic (cutter) suction dredges. In such a way the required depot volume also diminishes considerably. Otherwise mechanical excavation, which still comes nearest to the in situ density, can be combined with high density pipeline or continuous mechanical transport systems.

These transport systems do considerably augment the possibilities of choosing a fill area elsewhere than directly adjacent to the reservoir or behind the dam, for example across hill, in the next valley, in less populated or valuable areas. This might be an attractive proposal for a client. Some systems are more suitable with regard to turbidity, water consumption, excavation methods, depth and operational constraints than others.

Two types of dredging strategy can be distinguished:

1. Low capacity work: entails either a continuous program under a management contract, often with equipment owned by the Reservoir Management, or just a small contractor dredge working on a small job. Usually the dredging should just cope with the yearly siltation;

2. High capacity work: Small or large volumes at a high production rate in the case of emergencies, or regular cleaning up operations at a specified (e.g. an decade) interval, performed by a contractor.

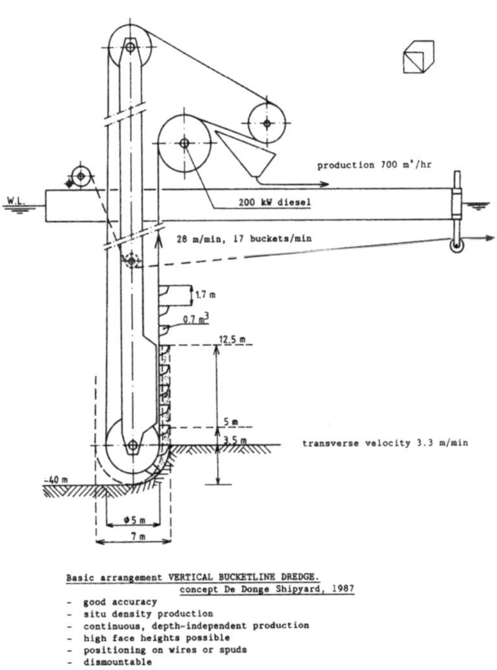

Figure 2.

History, Progress, Dredging Methods

Dredging in reservoirs has been considered earlier. Although the problem was apparently mostly approached as a "normal" job, with rather little attention to special problems as large depth, concentration, attainable, water consumption, turbidity generated, or depot allocation elsewhere than directly along the reservoir or just behind the dam. General reservoir dredging conditions are also discussed in ***[2], [6].***

The attitude of the dredging industry was partly justified because the technical means did not offer alternatives then. Dredging depth until 30m can be dealt with adequately today. For larger depths, still bearing in mind a rather high production level, mechanical disintegration and transportability constraints, suitable systems are still to be developed.

In the last 5 to 10 years however, progress has been made in both hardware, and theoretical understanding.

TABLE I RESERVOIR DREDGING

Generally: silt; 5-10 kPa, 1300-1700 kg/m³
Mobilisation: by truck

		Density water situ 20 40 60 80 100%	Equipment, method	Typical production situ m³/hr	Turbidity 20 40 60 80 100%	Investments
Dredging depth	shallow <20/30m		traditional cutter dredger	800-1100 (Beaver 1500)		p.m.
			disccutter dredger	800-1100		low (modifications)
			bucket dredger:			
			- conventional	600 (0.4m³)	screens?	
			- De Donge	700 (0.7m³) 16 buckets/min		moderate
			grab dredger, backhoe	120 (1.5m³,-5) to 200 (6m³,-20)	(open/closed grab)	moderate
	deep. 25-70m		De Donge bucket dredger	700 (0.7m³) 16 bu/min 200 kW (at -40m)		moderate
			grab dredger	60 (1.5m³) twin grab possible		moderate
		?	Pneuma system	200-800		
	deepest 70-120m		Grabs, Pneuma, Toyo, Airlift	50-100		
Transport					Spec. power [kW/ton/hr/m] $\times 10^{-3}$	Spec. investm. [$/ton/hr/m] (excl. powerplant)
	offshore		float.pipeline Ø400-Ø600	1100	0.1-0.3 at 0.1 to 0.3 kPa/m	0.1 - 0.25 low
			barges			
			special belt conveyor	250 (Telebande) to 1000 (Aerobelt)	0.25	1.8 - 0.8 high
			high pressure pipeline Ø150-Ø300	700	0.40 at 2 kPa/m	0.3 - 0.5 moderate
			cable (bucket) conveyor	(actually unsuitable offshore)	0.25	3.0 extreme
	onshore		pipeline Ø400-Ø600	similar		
			trucks (close by, flat)	similar		
			spec. belt conveyor	similar	(downhill: power recovery)	
			high press. pipeline	similar		
			cable (bucket) conveyor (sliding along slope)	600 max.	(downhill: power recovery)	
Depot circumstances			Retainer dikes required, unless natural pits exist. Return water pumping.			moderate
			No dikes necessary, no returnwater. Faster access.			low

Examples:

1. Cutter dredges with improved "disc" -cutterhead. Better concentration, up to 50%, less turbidity, less spillage, only usable in soft material. The disc cutterhead dredge concept resulted from a development made by **Technical University Delft** and **Breejenbout Dredging Company**, presently a Boskalis working company. It is extensively described in *[1]*.

2. Controlled re-fluidization of bottom sediment in order to generate dense mudflow transportation. Local mud shifting to a dredge pit, or cleaning by flushing the reservoir.

3. High pressure (100 bar) slurry pump technology exist and is suitable for these dredging purposes.

4. Conveyor belt technology. Making bends, steeper inclinations, higher throughputs, less vulnerable designs.

5. Computer orientated, mathematical modeling of silt transport and settling phenomena.

6. Knowledge of pipeline transport calculations with low and high density, silt slurries exhibiting plastic behavior.

7. Knowledge of characteristics of silt in the disposal area. Vast experience and research from harbor maintenance projects (Rotterdam). Settling, dewatering, consolidation, accessibility, retainer dike construction, can be calculated in advance.

8. Mobile, electrical, variable speed systems for high powers have become more common, cheaper, and are also in use within dredging. Power supply may be obtained direct from the high voltage, public grid ashore (probably supplied by the client).

Dredging Methods

Due to the large variety of the reservoirs with respect to depth, desired production level, turbidity and water loss requirements, geographical constraints between reservoir and dumping area, etc, it is not possible to suggest a particular dredging method for general application.

The scheme shown in ***Table I***, is intended to give some possibilities. It must be emphasized that the scheme mainly indicates relative magnitudes rather than absolute values. The scheme is not exhaustive, and not all the systems mentioned are evaluated in this paper.

Soil Constraints

It must be recognized that the material encountered in reservoirs is usually silty.

Compared to harbor silt, the density is relatively high, 1400 to 1700 kg/m^3, with cohesions up till 10 kPa. Neither plain hydraulic suction nor jetting, afford feasible means to dislodge the material, because of the resulting re-suspension, so it must be mechanically disintegrated.

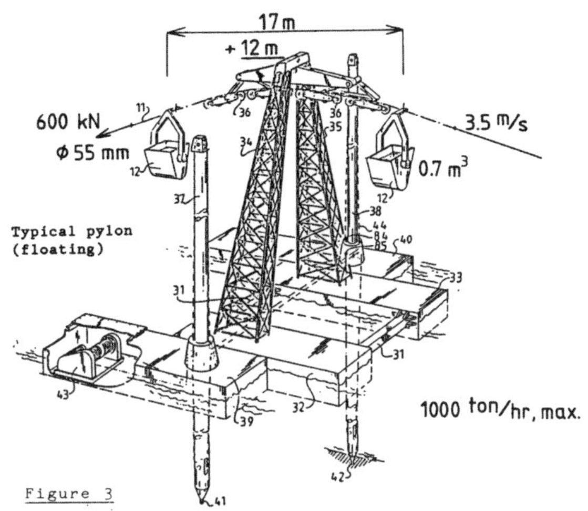

Figure 3

The typical material is expected to remain stable at steep faces, during excavation as Figure I indicates. These situations are not usually encountered in normal trailer and cutter suction dredging because far thinner layers are to be removed. With the usual cutter dredge or bucket dredge, this implies that dredging can only be done layer by layer because of danger of collapse. A vertical bucket line scrapes ***(Fig. 4.1)*** a high face, without undermining, and can therefore handle thicker layers at a time. Such a method finally results in less frequent shifting of dredge, anchors, and floating pipeline.

Various Systems

When turbidity and water loss are of minor importance and it is possible to construct a depot or fill area close by at low costs, the hydraulic cutter dredging usually provides the cheapest solution. The biggest throughputs can also be achieved with this solution. Transportation is by pipeline (400 to 700 mm) while the mixture is virtually *liquid*: 1- part soil with 1- to 2 parts water for sophisticated, improved disc cutter systems, 2 to 4 parts water for the simple, straight-forward designs.

Mechanical dredging is usually limited to lower production rates. Water loss is negligible, while turbidity can be kept low with appropriate measures as low excavation velocities, closed grabs and silt curtains.

Further transportation can either be done by high- or low- (when desired) density pipeline, barges, or even special conveyor belt systems. The **Pneuma** system (or **Oozer**), is driven by compressed air in order to bring the

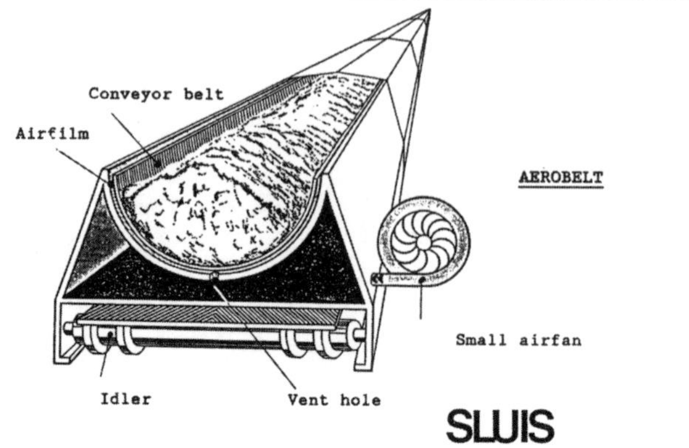

Figure 4

collected mud to the surface. It has been, or still is used in reservoirs in Japan. Efficiency is low, but medium to high densities, are said to be possible. Can dredge at largest depth, but is vulnerable to malfunction by debris such as tree trunks. Its applicability is uncertain for large-scale operation. For small scale emergency dredging, the Toyo submerged pump, fitted with a mechanical cutter has been used in Morocco (40,000 m^3), by Boskalis. Another example of this nature is described in *[3]*.

Accuracy Hydraulic Cutter Dredge

Contrary to the usual requirements, dredging accuracy is assumed to be unimportant for the client. However, in order to achieve minimal water consumption, or simply an economical pipeline density, dredging accuracy is important for the process itself.

This implies that an accurate cutting pattern has to be maintained, by the cutting device, making precise cuts adjacent to each other. At great depths (>30 m) this will pose problems when a relative small cutter head is used (crown-, but also the disc-cutter). Consequently the productions become irregular. Once the mixture has entered the suction line, further consistency change is no longer possible before discharge.

Accuracy Mechanical Excavation

Mechanical excavation by bucket dredge, backhoe, drag bucket- or grab dredge, can all provide almost situ density. But the latter two have unguided (no ladder) digging, which definitely impairs filling efficiency, especially at great depths. Nevertheless the largest depth, 70 to 100 m is only attainable with grabs.

With these mechanical systems, once the material is above water, de-watering treatment can still be done, (to get rid of water accidentally scooped). Thus high density can be maintained before transporting the dredged material further.

Vertical Bucket Line Dredge, Concept by De Donge

To comply with the special requirements of reservoir dredging, Boskalis and Shipyard De Donge, have a commercial agreement to develop an adapted design of the well-known, bucket line dredge. The objectives are:

· dredging at situ density;
· at possibly shallow as well as great depth with acceptable accuracy;
· dismountable pontoon transportable by truck;

· mainly intended for silty material with cohesion up to 10-15 kPa (weak clay) without too much wear impact;

· positioning on wires, or alternatively, when necessary, on spuds.

Description

The proposed system is basically a bucket dredge, appreciated because of its good accuracy, negligible water consumption, continuous excavation, and high digging forces, when necessary. A basic arrangement is shown in *figure 2.*

The drawback of limited dredging depth (30 m maximum for the big, conventional dredge), is overcome by arranging the ladder in a perpendicular position. This enables a lighter construction, and, most important, makes greater depths easily attainable. Also, the effective bucket volume is independent of variable ladder inclination due to depth variations, which can be large and fast in reservoirs. High vertical faces can effectively be dredged.

Unlike the clamshell or grab type dredge, with a freely suspended grab, the ladder provides sufficient guidance to the rather large bottom tumbler and bucket chain: for accurate dredging, to exert additional digging forces beyond gravity forces only.

With ample power installed, the production is independent of dredging depth; advantageous compared to the grab dredge. Nevertheless, the greatest depths (deeper than 70 m) will be beyond the bucket dredge construction limits, and can only be reached with a grab.

Dredging production of 700 m^3/hr silt from a depth of -40 m requires about 200 kW diesel power for digging and vertical transport. When heavier materials or higher excavation forces; e.g. in sand occur, a smaller bucket could be used on the same dredge. Transport to the shore can be accomplished in various ways. In the next paragraphs some consideration is given to various systems to transport the dredged material to a depot. It is possible to distinguish pipeline transport mechanical transport: barges, trucks, conveyor belts, cable bucket conveyors. Again low, and high density transport has to be considered separately.

Floating Section

Transport from the dredge to the shore makes additional demands:
- dredge maneuverability has to be maintained - shifting of the transport system itself must remain simple - workability of the transport system must be good. Severe winds, waves, and currents may occur and the water surface may be disturbed. These constraints rule out the deployment of cable conveyors and usually conveyor belts also, when used as floating installations.

The cable conveyor consists mainly of a pre-tensioned cable, carrying buckets along pylons of about 12 m height ***(figure 3)***. A pre-design indicates the following data:

cable	055	mm
tension	600	kN

bucket	0.7	m^3
bucket interval	17	m
production	700	m^3/hr
pylon distance	160	m

The main drawbacks appeared to be the continuous presence of the tension, which hampers shifting and anchoring, the absolute limitation to 1000 ton/hr (about 500 to 700 m^3/hr), the complicated loading-, linking-, and unloading stations, and the very high investments incurred.

Nowadays advances in technology have made conveyor belts somewhat more attractive for floating, and moving installation (e.g. air-film carried belt, ***figure 4***, or fully closed *tubular* belt like Sumitomo *[5]*, Telebande). Longer sections, bends, and steeper inclinations are possible. Potential productions can meet dredging requirements.

Barge transportation is a well-known traditional method. Its feasibility depends on accessibility of the reservoir, sailing distances and production level.

Pipeline systems still pose a very attractive and competitive solution, for both low and high density transport. In other ways the systems are similar to the land based ones.

Summarized:

Application	High density	Low density	Dredge freedom	Rough water, wind	Production level
floating	yes		good	good	moderate
pipeline		yes	good	good	high
barges	yes	n.a.	maximum	moderate	high, moderate
special conveyor belt	yes	no	limited	moderate	moderate
cable track	yes	no	poor	poor	n.a.

Land-based Section

Pipeline Transport, Low Concentration

When the slurry is virtually liquid (concentrations less than 20-30%), the mixture behaves like a ***Newtonian*** fluid. Due to the low grain sizes critical velocities do not have to be maintained. Generally this type of flow is well understood. Significant geodetic height differences give raise to special problems. A booster (like the common centrifugal dredge pump) is then required for about every 80 to 100 m increase in elevation. Start up and operation of such an installation becomes complicated. Non-return valves have to be installed.

In a downhill section the pipeline has to be safeguarded against vacuum surges. Snifter valves are required or, for more sophisticated systems, a high back pressure must be maintained. This implies higher strength requirements for the pipeline, which is still rather large in diameter. Back pressure can be created by an orifice (nozzle), possibly combined with a slurry turbine for power recovery.

Almost equal constraints apply for a second pumping system (when appropriate) for the return water from depot to reservoir. Installation of a pipeline in an inaccessible, mountainous landscape becomes more difficult with larger diameters.

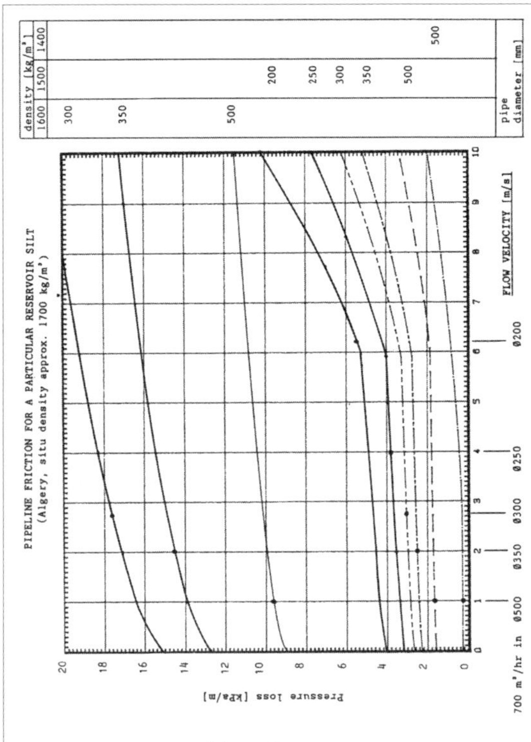

Figure 5

Pipeline Transport, High Concentration

When for obvious reasons, the silt is, or must be, dredged at high densities, other methods of transport should be considered. Apart from mechanical means, high density pipeline transport can be viable. Ability to calculate the rather high specific friction losses which occur is essential. The calculation principles have been described earlier *[4]*. When the silt is pumped in high concentrations it behaves according to ***Bingham's law.*** An example based on actual reservoir material follows.

The initial situ density averaged 1600 to 1700 kg/m³. Due to (fast) consolidation and silt and sand content, the densities in reservoirs are usually considerably higher than those found in harbor situations.

Some dilution with water will be almost inevitable:

situ density	**transport density**	
	10% diluted	**20% diluted**
1600	1540	1480
1700	1630	1560

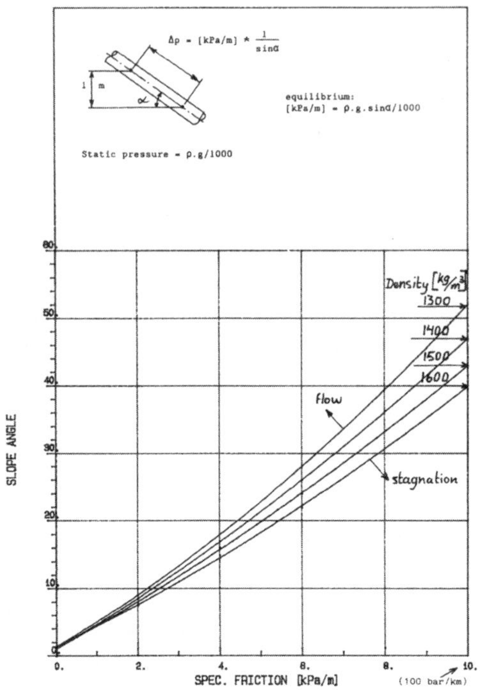

Figure 6

Some friction loss curves are given in ***figure 5***. The calculation model is based on practical experiences, and rheological tests on this particular material. It is demonstrated that dilution with 10 to 20% water (deliberate or not) causes a sharp decline in frictional losses. Furthermore, the pipeline diameter still has a considerable influence. Otherwise the flow velocity is of minor importance, contrary to Newtonian fluids.

With proper dimensioning, the friction loss level, say 2 to 3 kPa/m, is still considerably higher compared to traditional silt pipeline transport, being about 0.1 to 0.3 kPa/m.

The centrifugal dredge pump is ruled out because of low efficiencies in this type of flow. Positive displacement pumps with valves will be necessary. Wear impact of these slurries is rather low, while pump efficiency is good. Generated pressure per unit varies from 150 to 300 bar for the various designs. The main constraint for such pumps is the possible throughput, ranging from 150 to 700 m³/hr. Nevertheless it can be envisaged that only 1 or 2 such pumps can cope with a considerable transport distance, even crossing hills (geodetic height is of less significance now, compared to friction loss). Pump technology is available from the concrete and offshore industries, and coal and ore pipeline transportation.

Downward sections of a high density pipeline may be even less problematic than for liquid flows. Due to the higher friction loss, the pipe diameter may be chosen in such a way that, for a given slope angle, the static pressure more or less equals the friction loss. ***Figure 6*** gives some indications.

High pressure pipelines may be subject to safety legislation.
Mechanical Transport, High concentration

· Included in the initial De Donge concept was a transportation system by means of a cable track, as outlined in *S 5.1*. The landscape-independancy is advantageous as are the low energy consumption for vertical transport uphill, situ density transport. Gradually it became apparent however that investment costs *(see chapter 7)* were too high. Further study was abandoned, but the data derived are used as a basis for other mechanical transport. The same manufacturer revealed that a conveyor belt system was under development: the Telebande, considered to be more suitable.

· ***Telebande.*** Type of conveyor belt, supported by pylons, at regular distances (max. 100 m.). Like the cable track, it is independent of the landscape but bends are possible, and pylons may move to some extent (floating pontoons). The belt is folded together, therefore semi-liquid material cannot be lost in rising sections. Presently a prototype of 400 tons/hr is being tested.

· Another conveyor belt system could be studied as well, the Aerobelt. It is more similar to the conventional type built in rigid, constructional sections, but the numerous rotating parts are missing because the belt is carried in a semi circular trough by an airfilm. ***Figure 4***. The applicability of these systems depends heavily on capital costs, and expected lifetime. Modification of an existing built up installation, is hardly possible. Distribution at the depot requires another, following system.

· Trucks. Their use depends on distance, landscape, roads, acceptable nuisance for the community. Possibly low investment because of subcontracting.

Depot

Sometimes no actual depot is needed. When the river downstream does not dry out periodically, part of the silt can be carried away by adding it to the river. Downstream erosion is reduced simultaneously. This type of discharge is usually limited to emergency dredging in power generation schemes, when dredging production is negligible compared to the flow rate of the river.

Low Density

Large scale clean up operations, in arid climates, can only make partial use of such disposal: depots are necessary. For the usual hydraulic (cutter) suction dredging in silt, rather large depots are necessary. Due to the complete disintegration during excavation and pipeline transport of the material, initially about 2 to 3 times the situ volume is needed. After consolidation and evaporation the original volume returns, but this process may take years. Retainer dikes need to be heightened several times. Often a return water pumping system is needed, either because of water nuisance in the depot area, or water loss from the reservoir. The decantation is difficult to accomplish but necessary in order to avoid return pumping of silt *(1)*. For several occasions the total depot costs have been in the order of 20 to 40% of the total m^3 price.

High Density

Thus it must be attractive to eliminate these drawbacks by increasing the transport density to such a level that dike building, return water pumping can be omitted. Faster accessibility is another feature. Opportunities for re-handling for distribution elsewhere, by truck or rail, are better when the density is high. Pipeline transport to another valley may even be facilitated in some aspects.

Cost Comparisons

For cost comparisons between various dredging concepts many operational and technical constraints are to be considered. The feasibility of a concept is determined mainly by:

- total volume, acceptable project duration
- soil type, dredging depth
- discharge distance, discharge elevation
- water consumption
- depot location
- environmental aspects (turbidity, noise).

When investments are implied, the depreciation policy must be established. For large projects a special new-building can be justified, for smaller projects an acceptable depreciation level is only achieved when sufficient, prolonged market expectations do exist.

The results in this chapter should be regarded as tendencies, rather than absolute figures. All costs are related to the cutter suction dredge (CSD) cost, which is taken as a unit cost. The CSD costs are based on usual dredging standards. In this case no investments are involved while depreciation is based on a long, 15 year period. CSD costs are also based on silt: no wear, no critical velocity. Coarse material will increase costs (wear, boosters).

Long distance

An elaborate case study was made for a reservoir with 2 possible fill areas. Both rather far away because lack of suitable sites in the adjacent hilly landscape. Both were along the river valley and not across the hills: Conditions favored cheap conventional dredging. No exceptional dredging depth is possible. The rather long job duration (10M m^3, in about 90 to 120 weeks), permitted the use of a large diameter pipeline, 800 mm, for a 500 mm dredge, reducing the need for boosters.

Depot I: 8 to 9 km. Only modest geodetic height increase (+ 15 m).

Upstream beside the river, no return water system

Depot II: 17 km. Return water pipeline for conventional hydraulic dredging required. The discharge distance is twice as far as for depot I but there is considerable geodetic height decrease (- 30 m), which eliminates the need for a booster.

In ***figure*** *7*, the relative cost factors are shown for various cases. The mechanical dredging and transport is 2 to 4 times more expensive in this case. The systems differ in production and dredged density. When the depot should be substantially elevated however (+ 100 m), or across hill, hydraulic transportation costs increase sharply. The bucket dredge itself, and the in situ density, high pressure pump and 700 m pipeline system to the shore are only a fixed, minor part of 40% of unity.

The numerous buckets, with detachable cable clamps, are a dominant cost factor. A reduction of 50% of these costs would considerably reduce the total costs. The cable conveyor system actually consists of 6 transfer stations, each connected by a cable system. Every bucket is forwarded from one cable system to the next, because of the detachable clamp. Similar to modern alpine ski-lifts in series. Therefore, when only 1, or may be 2 cable tracks would suffice, a non-detachable bucket should still be considered. A conveyor belt system like Telebande will show expected decrease in cost of 20 to 30% (ref. manufacturer), compared to the cable bucket conveyor.

In general for the cable-track (or conveyor belt) the costs are mainly related to distance and hardly to height, while for hydraulic dredging it is the other way around (pumping only medium concentrations in silt, at low pipeline velocities). This tendency is illustrated by a next example.

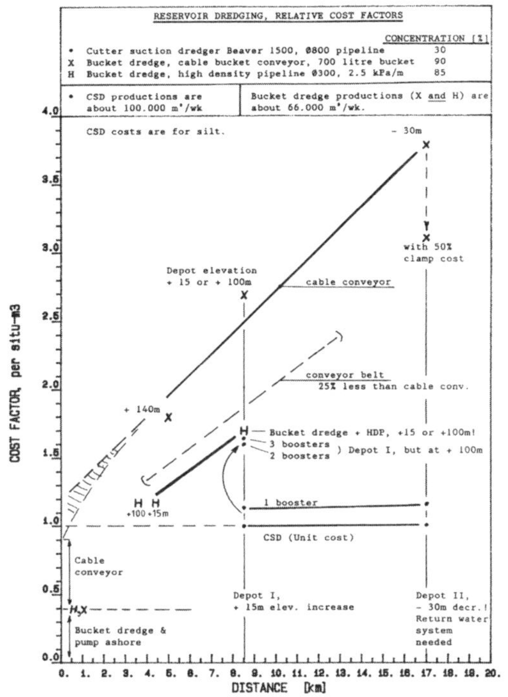

Figure 7

Shorter distance

With equal specific cost assumptions a smaller reservoir with closer fill area's, was briefly calculated (5 to 7M m³, waterdepth to 30 m, 2 to 3 year project duration).
A bucket dredge + cable conveyor solution, with a fill area supposed more remote, uphill, in isolated surroundings, the costs will approach those of the hydraulic solution. Track length 2200 + 2500 m, elevation +140m.

High density pipeline (especially at shorter distance), in conjunction with a bucket dredge, offers a competitive solution, compared to the hydraulic dredge. Much depends on the actual specific pipeline, friction loss magnitude. Geodetic heights do have marginal influence. Because of limited throughput, production increase is achieved using parallel pumps

Reservoir Dredging Projects

The dredging of reservoirs is often complicated by a number of factors, not the least of which is accessibility and where to dispose of the dredged material. The following are examples of recent US projects that illustrate those complexities and how they were solved.

Lake Carraizo Reservoir, Puerto Rico

The major reservoir of PR required dredging of sediment to recover the lost capacity. A contract was awarded to **Weeks Marine, Inc.** of New Jersey in 1997, to remove **6M m^3** of material. The project total was **$42.7M**, which included the mobilization of a brand new cutter suction (CS) dredge. The dredge was a design by IHC Holland, built in Trinity Industries Shipyard, in Mississippi, and named the ***Borinquen.*** Features included:

24-inch pump`
Hull: 200 X 42.6 X 9.7 ft.
Dredging depth was 60 ft.
Total installed power of 2800 hp
Supporting craft:
hydrographic survey/inspection boat
tender boat
cranes and earthmoving equipment

The length of the reservoir from the dam to the confluence of two tributary rivers, was 9.25 km. A requirement was to construct three separate disposal areas for the dredged material, the farthest site being 9 km upstream. This required 55,000-ft of 24-in. steel pipeline to reach the disposal sites, along with three booster pump stations, totaling 3600 hp each.

To complete the project within the two-year schedule, the dredge was shipped in pieces, trucked to the site and assembled. Windy rural roads limited the size of the loads that could be trucked. Mobilization began in February 1997 and operations commenced in September. Over 200 subcontracts, supplies and services were required totaling $15M.

White Rock Lake, Dallas, Texas

White Rock Lake was built in 1911 to supply water to the city of Dallas, TX; a 1,088-acre lake at the heart of a recreational park, home to hundreds of wildlife. Over decades the lake had been choking with vegetation, trash and silt. The problem was how to dredge the material, separate the trash for trucking to the disposal site which was a quarry located across to the other side of the City of Dallas, some 20-miles. This meant that the dredged material had to be pumped by 24-in pipeline through the middle of the city of Dallas.

The $18M contract was originally awarded to ***Soli-Flo***, a division of **Fluor Daniel**, who sold the project to its pipeline subcontractor, **Oscar Renda Contracting Co.** The cutter suction dredge, ***White Rock*** was built by **Dredging Supply Co**. The special pipeline design and pumping system was a combination of **Hagler Engineering** and **GIW Industries**.

GIW and Hagler were particularly well suited for this project because of their extensive experience pumping ore and tailings long distances, in Florida's phosphate mines and Canada's oil sands fields. Including the dredge pump, the system was composed of a total of six GIW pumps. The transport system was designed to pump 11,000 gpm at specific gravity of 1.3 or less and total dynamic head of 750 ft. It was protected by a cavitation prevention system that shuts down the entire operation if any of the pumps cavitate.

The slurry was about 30% solids pumped through a polyethylene pipe from the dredge to a 300,000 gal. sump tank where it was diluted to 80% water. It was then pumped 104,000 ft through a 24-in. steel pipeline to the disposal site and through the city streets. It passed through some of the city's highest and lowest priced neighborhoods, over a golf course and into a former gravel pit, creating considerable liability for the City. The pipeline was monitored by an electronic leak-detection system using GPS, along with daily aerial surveillance.

The system had a two-way radio telemetry signals that relayed information to and from the dredge, and telephone lines relayed information from magnetic flow meters at the beginning and end of the pipeline. This data was collected and summarized on a chart recorder that created a printout of activities on the dredge and in the transportation system.

The overall project of dredging and removing material from the reservoir, was considered highly successful and completed ahead of schedule. The city of Dallas was rewarded with a useful water supply and improved recreational lake.

Summary

Dredging of reservoirs to remove accumulated sediments and thus restore capacity so that it can perform its designed function, has become increasingly recognized as the most economical solution. While total cost of extraction and disposition of sediment is often startling to the reservoir owners, when the options are evaluated it often becomes the only practical route. Countries such as China, Egypt and Japan, as well as in North and South America, have rivers carrying heavy sediment loads and are resorting to dredging since they also have populations to sustain with periods of water shortage.

CHAPTER 7

NEW CONSTRUCTION DREDGING

The deepening, widening, or, building of new ports and waterways represents new construction dredging. It is often combined with reclamation where the resultant material is used to expand land facilities for industrial, commercial, residential sites, terminals, airports and beach restoration.

The following new construction dredging projects are described as examples to illustrate this important aspect of dredging. They are presented as examples of recent projects of significance that will present a perspective both in magnitude and complexity of engineering, planning and dredging.

Malaysia

Tanjung Pelepas Port

The first phase of one of Malaysia's most ambitious new port facilities came on stream in January 2000 and was officially opened on September 12th. Its location: Malay peninsula's most southern tip in the **State of Johor**, close to the new **Malaysia-Singapore.**

Second Crossing

Construction of a new 1800 m bridge links Singapore with Malaysia's main traffic artery. In a period of less than five years the area stretching from the state capital **Johor,** towards the west along the **Johor Straits** has rapidly changed; from a quiet oil-palm plantation area into a totally new area. It is to be developed with excellent infrastructure, housing facilities and new areas for industrial development.

Major infrastructure projects such as the Second Crossing, **Johor Airport** and the new port development at Tanjung Pelepas at the east bank of the **Sungai Pulai**, form the cornerstone of the Johor growth triangle west of the city.

Background

Since commencing operations in 1977, the present **Johor Port** at **Pasir Gudang** expanded rapidly. Growth predictions showed that the Port would suffer capacity problems by 2000. The Johor Port Authority reached maximum expansion of the Port area with the completion of ***Phase 4*** of **Pasir Gudang** and began studies in 1990 to identify a site for a new port that would have the capacity to cater to all future demands for cargo traffic. Based on a site selection study, Tanjung Pelepas was selected as the most suitable location for Johor's new port.

On 24 March 1995, the **Government of Malaysia and Seaport Terminal (Johor)** signed a 60 year concession ***Privatization Agreement*** with the Federal Government,

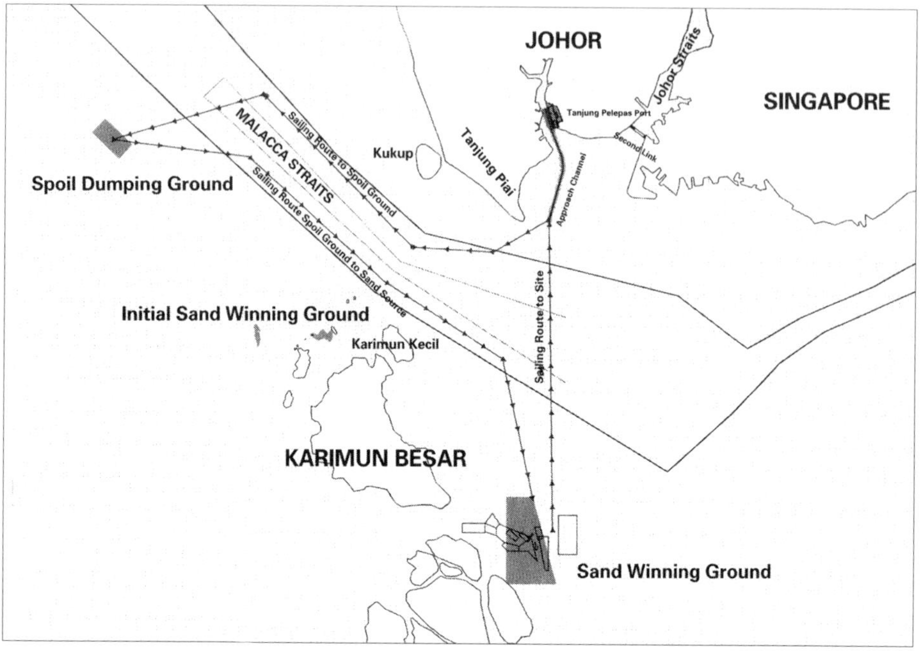

setting in motion the development of **Pelabuhan** (Port), Tanjung Pelepas.

The Port will be developed in ***Five Phases*** over a **25 year period up to 2020**. ***Phase 1***, comprises the development of six container berths, or, 2.0 km of quay together with the dredging of an access channel with a draft of 14 m enabling the Port to cater for ***Post-Panamax*** **vessels**. Preparation of a **master plan and preliminary design** were done in a very short time in the years 1995-1996, enabling the project to be started in 1997 and have its first staged completion in 1999. The studies were performed by **Moffatt & Nichol** (Long Beach, CA), in association with **Engineering and Environmental Consultants Sdn. Bnd. (EEC)** of Kuala Lumpur.

Moffatt & Nichol's schedule for submitting the master plan did not allow for extensive field measurements normally required for capital projects of this magnitude. Also, the client's commitment to have the new port on line within an accelerated timeline required assessment of financial risk and viability in an early stage.

Both technical and financial computer simulation studies were carried out based on abbreviated field data and cargo flow projections. This provided the investors with valuable insight into the viability and potential profitability of the project. Critical to this potential profitability was the government's agreement to have the existing port operations at Johor - Pasir Gudang as part of the same concession. Before reaching a confidence level sufficient to proceed with master planning and preliminary design tasks, over 30 different development and income scenarios were tested.

Computer modeling exercises were done to predict potential siltation rates under varying channel and port configurations, with a main objective to design the channel so that maintenance dredging would be minimized. Although the master plan has been changed since, the initial port and channel layout remained intact, the design detailed along these basic lines and eventually the dredging and reclamation works carried out.

Description of Dredging and Reclamation Contract

The ***Contract for Dredging and Reclamation Works*** forms part of ***Phase 1***, which comprises all works required to provide a fully operational Container Terminal by end of 1999. This included contracts for Dredging and Reclamation, Wharves, Port Infrastructure, Cranes and Equipment, as well as Port Buildings.

For the *Dredging and Reclamation Contract,* the scope comprised the removal by dredging of existing soft material to provide an approach channel, turning basin and bund foundation area.

The reclamation works under the contract comprised the construction of the Wharf Bund and filling of the terminal and infrastructure areas, to provide a stable platform for the Container Area.

The **US$158M Contract** was awarded to **Ballast Nedam Dredging**, in July 1997 with a contract period of three years. Due to the Client's fast-track programming with multiple contractors, the dredging and reclamation contract contained numerous sectional completions with heavy liquidated damages for time-overruns, to ensure critical path activities were adhered to.

Principal Contract Parties

In 1997, the **Employer - Pelabuhan Tanjung Pelepas Sdn. Bhd. (PTP),** formed a management company titled **Upper Plus Sdn. Bhd. (UPSB),** to manage and supervise the design and construction activities. That organization was replaced by **PTP Management Sdn. Bhd. (PMSB)** in October 1998, when the design tasks were completed.

Environmental and Engineering Consultants Sdn. Bhd. (EEC), was appointed by PTP to directly supervise the construction activities on Site. The construction contract was awarded to **Ballast Nedam - SKS Joint Venture**. Dredging and reclamation work was conducted by two firms, Ballast Nedam Dredging and **Dredging International Asia Pacific.**

Summary Description of Work

The Construction, Completion, and Maintenance of the Dredging and Reclamation Works, involved the dredging of **16,000,000 m^3** of soft and stiff material to form a **12 km Access Channel and Turning Basin**. This work was conducted together with the dredging of **5,500,000 m^3** of soft to medium material from a trench to form the base for the new Wharf Structure.

The initial activity concentrated on dredging an access to -12 m to allow the Jumbo Hopper (TH) dredges to reach the Site. The dredging of the trench formed a part of the soil replacement to construct a total of 2.16 linear km of Wharf and was dredged to a maximum depth of -19 m CD and executed mainly by the **750 mm, 9,870 kW Cutter Suction (CS) Dredge, *Vlaanderen XI*.** The dredged material was pumped into the deeper part of the Turning Basin and removed by the TH dredges.

The Turning Basin and Access Channel, were dredged by Hopper dredges and the end of the dredging work's high spots were removed with a **sweep bar**. The Project required the importation of some **18,000,000 m^3** of marine sand**.** This material was won from the vicinity of **Karimun (Indonesia)** from two separate locations. The sand was either dumped directly into the dredged trench, or, pumped directly into the reclamation area via floating pipeline.

Dredging Works

Trench Dredging

The trench dredging commenced from the most Northern location, working southwards. Prior to commencement the **Department of Environment (DOE)** raised concerns about the re-handling of marine clay. Modifications were implemented such as extending the discharge pipe and daily water monitoring, to ensure that release of fines was minimized.

The re-handled material placed in the Turning Basin by the CS dredges *Vlaanderen XI* and ***Hector*** (**850 mm, 10,800 kW**), was removed by hopper dredges and dumped at the designated dumping ground in the **Malacca Straits.** While removing the re-handled material, existing material in the Turning Basin was dredged at the same time.

The CS dredges suffered considerable stoppage time due to blockages. The stoppages were caused by mangrove roots, clogging the cutter head and/or dredge pump, decreasing their efficiency.

Access Channel

Initially, the **Jumbo 18,000 m^3 TH** dredge ***Amsterdam*** sailed single mud trips but after some weeks, began sailing *combi* trips taking out soft material in the Access Channel and re-handled material from the dredged Trench while bringing in sand from the concession area in NE and SW Karimun. For this activity, TH dredge *Amsterdam* was joined by **17,000 m^3 Jumbo TH dredge, *Pearl River*.** The work of the jumbo TH dredges *Amsterdam* and *Pearl River,* was supplemented by the medium **TH dredgers *Wado* (6,800 m^3), *Gogland* (7,826 m^3)** *and* ***Rigelstar* (6,073 m^3).**

The material in the shallow areas of the approach channel could not be taken away by the hoppers therefore cutter (CS) dredges were used. They dredged the material and dumped it in the deeper parts of the channel, from which the hopper dredges removed the same and dumped at the designated dumping ground in the Malacca Straits. For this purpose the CS dredges *Hector*, ***Rubens*** **(850 mm, 14,850 hp)** and ***Wombat*** **(700 mm, 4,455 kWD)** were used.

Combination Trips

Hopper dredge production is dependant on the cycle times of round trips. For the

Pelepas Project two major activities were executed by the hopper dredges:

1- removal of unsuitable dredged materials;

2- importing of suitable fill material for reclamation.

These activities were combined in round trips lasting eight to nine hours, executed mainly by Ballast Nedam's trailer (TH) dredge, *Amsterdam.* Sources for suitable sand were identified in **EEC's *EIA (Environment Impact Assessment)*** study, indicating sand winning grounds around the island of Karimun, Indonesia. Deep sea disposal areas for unsuitable dredged material were identified off **Pulau Kukup**, roughly 80 km north-west of the site in **Malaysian waters.**

Reclamation Works

Site clearance

The initial activities on site included the clearing of 210 ha of mainly mangrove vegetation. All vegetation more than 15 cm above the existing ground surface had to be removed. This caused a substantial migration of snakes, monkeys and large Monitor lizards. Mangrove root systems were left in place to serve as a natural reinforcement of the top layer of the very soft marine clay, on which the first layer of hydraulic fill would be placed. The mangrove roots made installation of the vertical drains more difficult but there were clear advantages to leaving them in place.

Trench filling

Although suitable sand sources were known in Indonesia, a large sand search investigation was initiated which was concentrated around the designated dumping ground in the Malacca Straits, northwest of Karimun and along the coast of **Pontian,** up to **120 km** from the site. Some small pockets of sand were found but were exhausted by the end of 1997. From January 1998 the sand was won from Karimun southeast.

Filling of the trench was done partly by direct dumping. First, the jumbo TH dredges dumped as much as possible until limited by their draft. Then the medium-sized trailers dumped their loads until they were also limited. After this, the remaining filling of the bund was achieved by hydraulic filling via pipelines. The final slope and berth pocket were trimmed by a CS dredge to the required profile. Excess material was pumped into the reclamation. The TH dredge *Wado,* proved to be a very efficient medium-sized hopper with limited draft to dump the top layers of the bund, achieving a much higher density compared to hydraulically placed fill via pipe lines.

Reclamation Filling

Reclamation was done in Four (4) layers.

1- **First layer** up to +3.5 -4.0 m CD; with a thickness of between 0.5 and 1.0 m was to facilitate the installation of the **vertical drains**. The drains were installed in nearly all areas with a spacing of 1.25 m. The total length for each drain was between 20-30 m .

Output was max. of **100,000 m^3** per day with **17 machines**. The total length of the drains exceeded **20,000,000 m**. Because of the relatively thin first layer of sand, this work could not be done by jumbo trailers owing to their high pumping output. Therefore, the first layer was done mainly by the medium trailer *Wado* and in the beginning also by the CS dredge *Vlaanderen XI*, which used the trench as a rehandling pit.

2- **Second layer** by Jumbo trailers up to +5.5-6.0 m CD that provided sufficient sand to construct dikes.

3- **Third Layer** up to +7.5 - 8.0 m CD.

4- After a waiting period of three weeks the **last layer** up to +10.5 m CD was placed.

To avoid any delay by the jumbo trailers, two floating pipelines with two connection points were used. In addition, there was also the opportunity to dump the load in the trench if any problem on the reclamation area or with the pump-ashore facilities of the dredgers arose.

Soil Improvement

The consolidation of the wharf bund and reclamation area, was done by a combination of vertical drainage (**wick drains**) and the placing of **sand-surcharge**.

The over 20M m of wick drains, were installed by a specialist subcontractor as soon as the area was accessible for crawler cranes. The drains were installed with a spacing of 1.25 m and the total length for each drain was between 20-30 m. Output was a maximum of 1000 m per day. Placing of the vertical drains started within five months after contract award. At its peak, 21 rigs were operational. Within a seven (7) month period some 13M m of drains were installed over **600,000 drainage points**.

Surcharge Removal

After a waiting period of six months, surcharged areas could be excavated to a level of +5.50 m CD. The material went to inland areas and to the future disri-park area. The total quantity excavated and transported was approx. **4.0M m^3**. Output was more than **100,000 m^3** per week. For this work hydraulic excavators and articulated dump trucks were used.

Conclusion

The *Pelabuhan Tanjung Pelepas Dredging and Reclamation Works* were executed successfully, not-with-standing adverse site conditions and strict contractual requirements. The timely mobilization of sufficient dredging plant and competent local subcontractors greatly reduced time overrun risks. Changes of work were handled with flexibility from all sides. Technical problems were solved with the intention to complete the works in the shortest possible time.

One of the main success factors was the excellent performance of the large fleet of **seven (7) trailing suction hopper (TH) dredges**. They achieved an overall efficiency of nearly 90 % working continuously around the clock. The lion's share was undertaken by Ballast Nedam's new 18,000 m^3 jumbo TH dredge, *Amsterdam* and its crew, working nearly **70 weeks** on this project with weekly sand deliveries in excess of **200,000 m^3** per week.

USA

Port of Los Angeles, Deepening and Reclamation

The ***Los Angeles Pier 400 Project*** started construction in June '94 with the contract award to the **Pier 400 Constructors**; a joint venture between **Connolly-Pacific Co.** and **Great Lakes Dredge & Dock Co.** The partnership was a "unit price" type of JV, whereby each partner is responsible for its own unit prices to the owner.

The initial contract was the *cornerstone* of the Port's *Pier 300/400 Implementation Program*, an expansion project to help accommodate an expected doubling of cargo by the year 2020. The project included the development of 3-miles of new, wide navigation channels, a turning basis and 5,000 ft of berthing area south of Pier 300 on

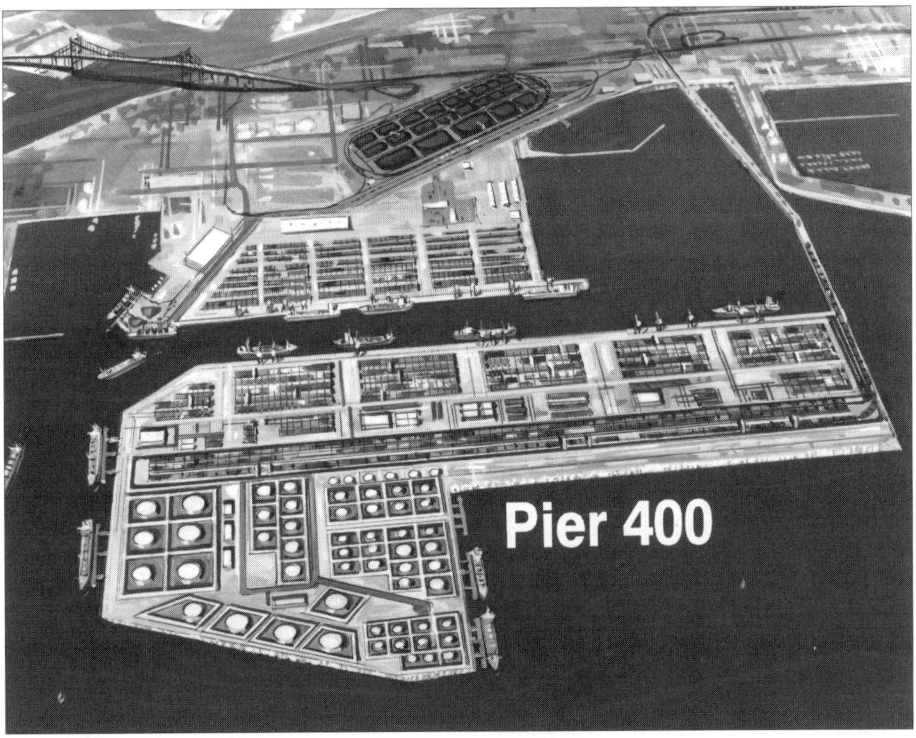

Terminal Island. *(See Photo Plate 6).*

The dredging of some 30M yd^3 of material produced water depths of 45-63 ft to accommodate deep-draft container ships and dry bulk vessels. The dredged material is contained behind rock dikes, creating a new 265-ac land mass called Pier 400.

Connolly-Pacific was the JV sponsor project manager, responsible for furnishing and placing rock dikes; removal of on-site sunken wrecks; removal of the Terminal

Port of Los Angeles deepening and reclamation GLDD's dredge "Florida" (34" CS, 16,000 hp) working at Pier 400 with Manson Construction's "Mr. Manson" (26" CS, 6,000 hp) and Crane Barge "Viking"

Island treatment plant outfall; furnishing and placing filter cloth; installing aids to navigation structures and geotechnical instrumentation. GLDD was responsible for all dredging work items.

Manson Construction Co. served as a subcontractor for rock transportation, placement and dredging. **CRSS Constructors, Inc.** had an overall management contract with the Port of Los Angeles to supervise and direct the execution of the project.

One of the subcontractors providing daily survey reports of the dredging progress (i.e., depths and widths of cuts), was **Gahagan & Bryant**, Tampa, FL. With a project completion date for that phase of May 1997, Pier 400 became the largest harbor development project ever built in the US.

The 2nd Stage of the *Pier 400 Project* was completed in August 1997. Almost 600 acres of new land were constructed from the dredged material on which new container terminals, liquid bulk and rail facilities were erected. The ***1st Stage*** contract was **$148.6M** followed by ***Stage 2*** for **$150M**, completed in January 2000.

The *2nd Stage* contract deepened the 63 ft channel to **81 ft**, lengthening a 50 ft channel and constructed a **75 ft** channel to the east side of Pier 400. The dredging and landfill project is the cornerstone of the Port's *Pier 300/400 Implementation Program,* **$650M** encompassing 24 separate but related projects.

The project makes up part of the southern terminus of the Alameda Corridor, a 20 mi rail and road transportation route, connecting the harbor with the rest of the US by truck and rail.

New York / New Jersey Port
Comprehensive Study of Dredging Process

The Port Authority Board of Commissioners, authorized a study of harbor dredging in the Port of New York and New Jersey to identify, develop and explore innovative and cost-effective ways to remove and dispose of sediment from the port's channels and berths.

The goal of the $4 million study is to find the most environmentally sensitive and the most effective way to deepen the port's channels to 45 feet, and then to 50 feet. The deeper channels are needed to accommodate a new generation of cargo mega-ships that can handle over 6,000 truck-size containers and require deep water to operate. The study is being done in two phases. The first, which will take approximately two years, will examine what the Port Authority can do to help private industry develop new technologies to process dredged material more efficiently and reduce the cost of recycling and reusing it. The second phase will explore long-term methods to improve the entire dredging and disposal process, unchanged for decades.

"The Port is a primary cog for our regional economy, and we want to ensure that it remains the economic engine that continues to create jobs and opportunities for New Yorkers," said New York Governor George E. Pataki. "To sustain this level of activity, it is imperative that we upgrade our port infrastructure to ensure that we meet the for growth that will be placed on our maritime transportation system, while protecting and enhancing the environment."

New Jersey Governor Christie Whitman said, "We must continue the positioning of our Port as a hub for maritime activity on the East Coast if we are to sustain the economic benefits that it provides. We also are keenly aware in New Jersey of the environmental sensitivities associated with dredging. This action will address the need to upgrade our port facilities to accommodate modern, larger vessels in an environmentally sound manner."

"The vitality of our Port and the economic prosperity it brings to the region hinges on our ability to provide the infrastructure needed for modern maritime operations, including accommodating a new generation of immense deep-draft ships," said Port Authority Chairman Lewis M. Eisenberg. "We must ensure that we are prepared to handle the projected increases in maritime cargo, which will double in 10 years and could more than quadruple in 40 years, creating 238,000 additional jobs and $65 billion in economic activity. And we must continue to do everything possible to protect the environment."

"Much of the dredging technology we currently use in the Port has been in place for decades and is not as cost-efficient or practical as we believe it can and should be," said Port Authority Executive Director Robert E. Boyle. "Working with the two states - and the dredging industry, this study will help identify how we can enhance the performance of dredging operations and control disposal costs to make sure our Port remains economically competitive with others around the country. Any changes that are made must be done in an environmentally sensitive manner."

"An important component of this study will be an extensive effort to educate the public on innovative methods we can use to create various products from dredged materials and use it in beneficial ways," said Richard M. Larrabee, the Port Authority's

Dredging in the Port of New York and New Jersey with GLDD's backhoe (BB) dredge "New York."

Director of Port Commerce.

Larrabee said a demonstration project is exploring the possible use of dredged material to build highway embankments. Research also has been conducted on the use of dredged material to make products such as construction aggregate, concrete and bricks.

In addition, dredged materials have been used to cover former landfills and to reclaim abandoned industrial land or *brownfields*. Several such projects have either been completed or are under way in the region and others are planned.

During the next 12 years, channel deepening will require the excavation and disposal of approximately 59M yd^3 of sediment and rock. In addition, annual maintenance dredging results in an additional 2M yd^3 of sediment in need of disposal.

Dredging in the Port of New York and New Jersey is done with a crane and a bucket, just as it has for decades. Dredged material is then put in an ocean-going scow and transported for disposal or processing. The study will explore whether new, modern equipment can be used to make the process quicker and more efficient.

Regulatory changes also have made dredging and disposal of sediment a major challenge for marine terminal operators. In 1997, the federally mandated ocean disposal site, commonly called the Mud Dump, was closed and re-designated as the Historic Area Remediation Site (HARS). Currently, only clean dredged material - the definition of which has recently been modified and is under further review - can be taken to the site.

"It is essential that the Port Authority do the studies to develop new technologies to improve the dredging process and to use dredged material as a resource," Larrabee said. "These studies will lead to innovative approaches to make the dredging and disposal process more efficient, predictable and cost effective."

Dredging of the Houston-Galveston Navigation Channels

In 1837, the steamship ***Laura*** traveled from Galveston Bay up Buffalo Bayou to what is now Houston. The trip, in water no deeper than six feet, proved the bayou was navigable by sizable vessels and established a commercial link between Houston and the rest of the world. Through that link, the City of Houston was born. And through that link, the Port of Houston has become one of the world's largest ports, with more than 7,000 ships calling annually bringing goods ranging from iron to sugar for U.S. consumers.

Over the years, the growth of the Port — and of Houston — has been facilitated by local contributions and federal funds to gradually deepen and widen the Houston Ship Channel. In 1914, the waterway officially opened as a deep-water channel, at a depth of 25 feet. Since then it has been expanded to the depth of 40 feet and width of 400 feet.

The largest port in the United States, by cargo volume, the Port of Houston brings $7.7 billion annually to the region's economy and creates more than 200,000 employment opportunities. In 1998, 169.1 million tons of cargo moved through the Port. A total of 7,093 ships were recorded at the Port of Houston during 1998.

In the late 1980's, the Port of Houston Authority presented a proposal to deepen and widen the Houston Ship Channel. In order for the Port to continue to grow, the Ship Channel needed to be expanded to better accommodate the variety of vessels calling at the Port. In 1987 the U.S. Army Corps of Engineers released its Feasibility Report and Environmental Impact Statement (EIS) addressing the environmental impacts of the channel expansion project. The report raised concerns with environmental groups who adamantly opposed the project.

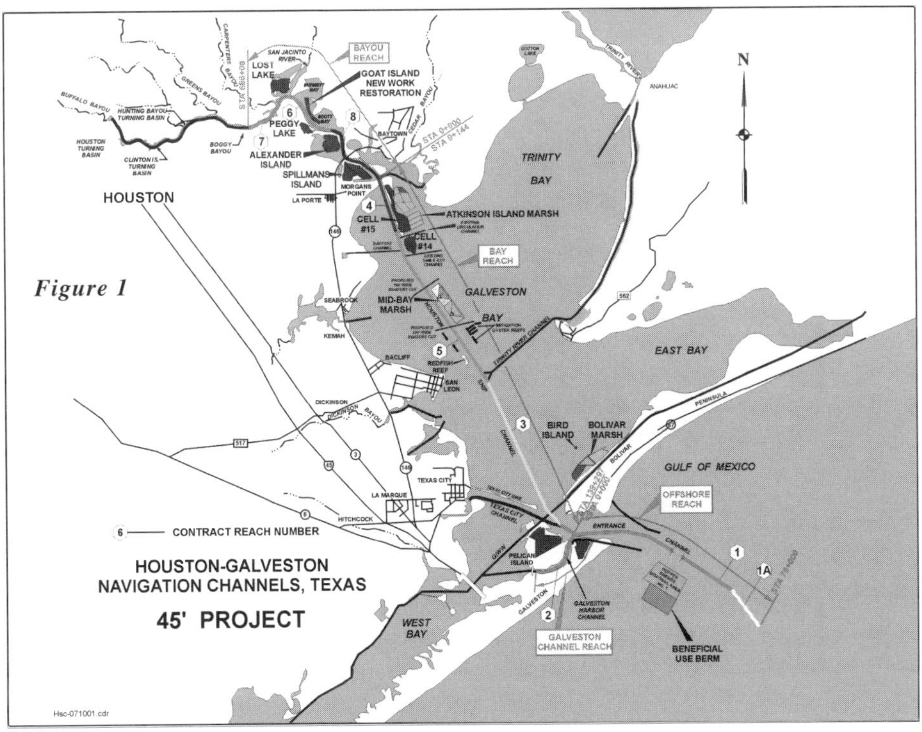

Figure 1

The opposition's concerns fell into three categories: environmental, economic, and procedural. The procedural concerns included whether the Corps of Engineers was in compliance with the *National Environmental Policy Act* and whether they had prepared a comprehensive Environmental Impact Statement. The economic concerns included whether the channel depth and width were chosen correctly, better alternatives should have been identified, benefits developed were too high, and cost estimated were to low. The environmental concerns included the project's impact on turbidity, circulation, cumulative effects of dredging and salinity and contaminants.

Figure 2

After the project was on the road to completely being stopped, the Interagency Coordination Team (ICT) was formed by the Galveston District of the Corps of Engineers. The ICT is a problem-solving group, created in late 1989 to address issues and concerns arising from the Feasibility Report and the EIS. Its goal is to attain through the project, a net benefit to the Bay. Members of the ICT include: U.S. Army Corps of Engineers (Galveston District), US Environmental Protection Agency, U.S. National Marine Fisheries Service, U.S. Fish and Wildlife Service, Natural Resource Conservation Service, Texas Parks and Wildlife Department, Texas General Land Office, Texas Water Development Board, Texas Natural Resources Conservation Commission, Port of Houston Authority, and the City of Galveston.

As part of the ICT, the Beneficial Uses Group (BUG), a collaboration of agencies - five federal, two state and one local - is finding environmentally responsible ways to utilize the dredged materials. This subcommittee of the ICT serves as a problem-solving group involved in the planning, engineering, and post-project monitoring of dredged material from the ship channel and bay. Today, the ICT meets bi-monthly to discuss the progress of the project. They continue to guide the HGNC Project toward its goal.

In 1989, Harris County voters approved the bond funding to deepen the Houston Ship Channel from 40 feet to 45 feet and widen it from 400 feet to 530 feet. Currently, the U.S. Army Corps of Engineers, with the Port of Houston Authority as a non-federal sponsor, are working to deepen the channel. This effort covers a distance of about 14.4 miles out into the Gulf of Mexico from Galveston, a distance of about 26.0 miles across Galveston Bay to Morgans Point and then 13.9 miles from Morgans Point into the Land lock reach to Boggy Bayou. In addition to this, 3.8 miles of the Galveston Channel will be deepened. With construction starting in October 1998, it is anticipated that this phase will continue until a new channel has been completed. It is anticipated that over the economic life of the project (50-years), 352,384,300 yd^3 of material will be dredged. Of this amount 63,046,700 will be dredged initially and 289,337,600 will be dredged over the remaining life of the project.

The current expansion project - which is being paid for by local bonds and federal funds - is considered vital to the economy of the area and the continued success of the Port, which contributes $7.7B in revenues to Houston's economy each year. Today, the Port, as one of the largest ports in the United States, needs the larger channel to meet the changing demands of shippers and accompanied by additional safety and efficiency requirements.

Once all the environmental concerns were worked out through completion of the studies agreed to in the ICT, the project proceeded to construction following the signing of the Project Cooperation Agreement with the Port of Houston Authority on June 10, 1998. Construction contracts for dredging the Houston Ship Channel were divided into 9 dredging contracts, two offshore (jetty and entrance channel and the entrance channel extension) for a distance of 13.9 miles, three in Galveston Bay (lower bay, mid bay and upper bay) for a distance of 26 miles, three in the Bayou Reach (lower bayou, goat island, and upper bayou) for a distance of 13.9 miles *(Figure 1)* and the Galveston Channel.

Hopper dredges or scows placed the material dredged from the entire Entrance Channel in mounds or rows near the existing underwater offshore disposal area. The irregular shape will provide better habitat to begin the food chain. The offshore beneficial use berm is being constructed in an area 14,100 ft by 11,100 ft and is located approximately 7.8 miles out into the Gulf of Mexico from Galveston following the Entrance Channel. The dredged material was placed in rows by **Bean Stuyvesant LLC** approximately 300 ft apart and at a height of 4 to 6 ft above the Gulf bottom. The underwater berm is located in approximately 40 feet of water and is approximately 23,000 ft southwest of the Entrance Channel. The irregular shape will always remain in a similar depth of water and will provide better habitat to begin the food chain for the aquatic species living in the Gulf of Mexico and using these mounds as a food source. These mounds can be seen in *Figure 2*.

The material dredged from the length of channel through Galveston Bay will be used to construct saltwater marsh and habitat for birds. Over the life of the project approximately 4,250 acres of saltwater marsh will be constructed as well as a six-acre Bird Island. Please refer to *Figure 1*. As the marsh cells are constructed, initially one will be filled with the material dredged from the widening and deepening. When the channel shoals and requires further dredging to assure the channel remains at its authorized depth, maintenance dredging will be performed and the material dredged will be placed in one of the cells constructed. Each cell has been designed to hold the material dredged during one maintenance cycle. The amount ofshoaling that occurs in a reach of a channel depends on where it is located within the system. Some parts of the channel are

Figure 3 Marsh

dredged as frequent as once per year and others are never dredged. In the Bay Reach *(See Figure 1)* capacity in the three newly constructed disposal areas will have capacity sufficient for 20 years of maintenance material placement with the remainder of the needed capacity to be constructed later in the project life. As previously indicated, the disposal areas and cells are sized to handle one maintenance cycle and there are sufficient cells to contain the maintenance material that will be dredged during the next 20 years. The number of cells and the size constructed is based on the historical dredging records.

The land-lock reach material, *(See Figure 1),* the material dredged from Morgans Point upstream and is bound by land on both sides of the channel, will be placed in existing disposal areas and used to restore Goat Island. Goat Island existed on the eastside of the Houston Ship Channel between the San Jacinto Monument and Baytown but was lost due to erosion and subsidence. At one time the piece of land called Goat Island was not an island and was actually connected to land. Over time it became an island and was approximately 320 acres in size at that point in time that it is being restored to.

After extensive investigation, the BUG determined that the best way to help Houston, Galveston, and surrounding communities is to restore wetlands in Galveston Bay. Because of the large amounts of wetland lost over the years this project will make a start to replace those lost. Wetlands, or marshes, are areas where a significant amount of water saturation in the soil determines how the soil will develop and what it will support. The saturation level affects the types of plants and animals that live in thesoil and on its surface. There are many varieties of wetlands, depending on the climate, topography, landscape, and proximity to an open body of water. For instance, Galveston Bay has *inter-tidal salt marshes,* dominated by plants that have adapted to the high level of salt in the environment. Similar wetlands can be found along the Gulf of Mexico and portions of the North American coastline.

Wetlands provide a number of benefits to many species:

* Wetlands aid wildlife by providing food, protection from predators, and spawning and nursery habitats for many fish and wildlife species.

* Wetlands improve water quality by filtering out pollutants in the water, helping to produce high-quality fish populations.

* Wetlands prevent erosion by absorbing wave energy in areas prone to losing land to the ocean.

* Wetlands form buffer zones during storms and protect lands from flooding by intercepting storm runoff and gradually releasing floodwaters to downstream systems.

* Wetlands provide many recreational activities, such as opportunities for bird and wildlife observation, hiking, hunting and fishing.

Despite these benefits, wetlands throughout the United States are being destroyed by either natural threats or human impact. Wetlands restoration is essential to maintaining the delicate ecological balance along the coastlines in the United States and around the world. Organizations such as the ICT and BUG are working to combat the destruction of wetlands and restore those that have been lost.

Engineering challenges accompanying the project are many and varied. Marsh cells construction and the pumping distances involved in the evaluation of material to be dredged had to be explored. As with all channels in the vicinity, the material dredged from the widening and deepening is made up of a variety of materials that range from very soft silty material to hard stiff clays. The levees that are hydraulically pumped to contain the soft material are best constructed out of the hard clays. The composition of the existing material varies considerably. How the dredged material would react to the dredging process and what amount of material will show up at the dredged material disposal site were very important factors to consider during the design process. Would the material become too fine (silt) and flow out of the disposal site or would clay balls

Figure 4 Bolivar Island

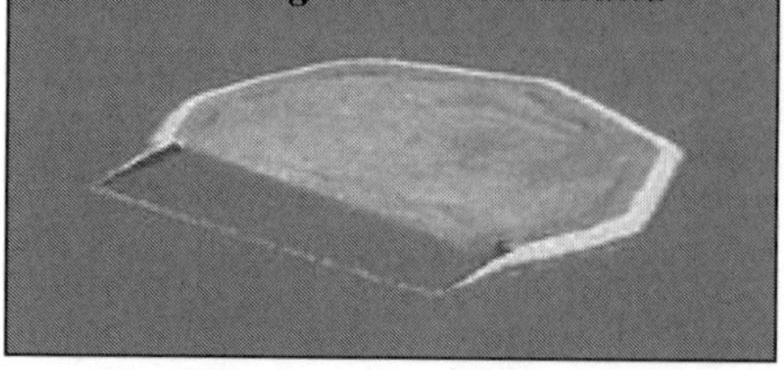
Figure 5 Bird Island

roll stay intact and clog the pipes?

Because of the long pumping distances, in some cases up to 9 miles, contractors had to be careful to know what type of material they were dredging and in turn how much water to keep in the system to get dredged material flowing to the disposal site. Should a pipeline become plugged at this length, a considerable amount of effort and time is required to clear the line. Because of the variation in material being dredged additional effort had to be spent to determine how much the material would settle This process is critical because an estimate has to be made to predict the resulting elevation of the dredged material. This, in turn, determined where the marsh plants would or would not grow.

In order to help with this prediction a model developed by the **Corps of Engineers Waterways Experiment Station** located in Vicksburg, Mississippi was used. This *Primary Consolidation and Desiccation of Dredge Fill (PCDDF) Model* predicts the resulting elevation of the material which must be the elevation that the marsh grass grows. In Galveston Bay the optimum range for plant growth is approximately 0.6-foot. Below this the lower elevation range is water and no plants grow; above the upper elevation different types of vegetation can be found. These elevations vary as you change locations in the Bay. In the Lower Bay Contract (Bolivar Marsh) the target range for the final elevation of the dredged material ranges from 0.6 to 1.4 feet Mean Low Tide (MLT). In the Upper Bay Contract (Atkinson Island Marsh) and Mid Bay Contract the final elevation varies from 1.9 to 2.4 feet MLT. The target elevation was developed with the help of the BUG. This ultimate elevation is reached only after the material has been allowed to dewater for approximately two years.

Construction in the Bay began with the Lower Bay Contract that was awarded to **Great Lakes Dredge and Dock Co**. This construction reach covered dredging the Houston Ship Channel for a distance of 11.1 miles and included construction of three new cells for marsh construction *(Figure 4)* and a 6-acre Bird Island *(Figure 5)*. Two of these cells are 270 acres in size and one is 300 acres in size. Once these have been filled another cell will be constructed that will be approximately 460 acres in size. When construction started there were no disposal areas in this vicinity that were used by the project. All disposal areas used in the Bay were on the east side of the channel and were open bay underwater disposal areas. The locations of newly constructed cells for marsh development and the Bird Island can be seen on *Figure 1*.

This Bird Island was constructed to address a need for colonial water bird habitat in Galveston Bay. The Port of Houston Authority (PHA) and the U.S. Army Corps of Engineers, following the suggestions of the ICT and in particular the BUG plan, built the bird sanctuary in the form of an isolated island, one mile north of the Bolivar Peninsula and close to the newly constructed array of marsh cells awaiting filling and planting. The island was built entirely using materials dredged from the expansion of Houston

Ship channel.

This overall contract was started on 13 October 1998 and all dredging was completed in July 2000 after nearly 10.5 million cubic yards were dredged at a cost of approximately $56 million. The Bird Island is already home to a large variety of birds. The island's peak elevation is 10 feet above mean tide and descends to mean sea level, creating a habitat for many bird species *(Figure 6).*

It is expected that during the spring of 2001, the vegetation planting will be started on the island. When completed, the island will support several habitats. The higher elevation will support a small grove of native trees and brush. The lower elevation will provide grassy areas and a sandy beach. Roseate Spoonbills, Reddish Egrets, Snowy Egrets and Great Egrets, among others prefer to nest in the brush and trees. The grassy zone is expected to attract nesting Brown Pelicans, Laughing Gulls and Forster's Terns. Beach that is bare of vegetation is the nesting area preferred by Black Skimmers, Royal Terns, Caspian Terns and Sandwich Terns. A small lagoon on the beach will provide young birds with a safe place to swim. The entire shoreline is protected from erosion by stone erosion protection. The erosion protection consists of limestone ranging in size from 70 to 1,500 lbs. Around most of the island this erosion protection was place in a 3-ft thick layer, 1-ft vertical on 2.5-ft horizontal side slope between elevations -2.0 ft to +8.0 ft MLT. On the side where a beach was constructed, a row of erosion protection was placed onto the Bay bottom to elevation 0.0 MLT and sand was pumped to transition from the top of the erosion protection (0.0 MLT) to the top of the island that was at approximately +10.0 MLT. The island will be monitored and maintained throughout the 50-year life.

Figure 6 Rip-Rap, Bird Island

Also included in the project is the construction of 118 acres of oyster reef habitat as mitigation for those oysters destroyed along the channel and at the Mid Bay disposal site The mitigation contract began on April 13, 1999 by **Luhr Brothers, Inc.** The reefs were constructed in 20-acre pads with 413,000 yd^3 of crushed limestone at a cost of approximately $14M. The pads were generally constructed with limestone placed either 2 or 3 ft above the Bay bottom, depending on the location in the Bay. This location can be seen in *Figure 1.* All pads were complete on July 27, 2000. Inspection of the pads by Texas Parks and Wildlife in August 2000, showed oysters the size of quarters already living on the newly constructed reef.

Since the construction plan was to stagger the dredges working in the channel, the Upper Bayou Contract was started on October 13, 1998. **Renda Marine Inc.** is performed the contract that was scheduled for completion in February 2000. Approximately 2.98M yd^3 of material has been dredged at a cost of $15.8M. The dredged material is placed in an existing upland disposal facility, called Lost Lake Disposal Area that can be seen in the land lock reach on *Figure 1* and is located north of the Houston Ship Channel just as it turns to the west. Lost Lake is approximately 675 acres in size and using disposal area management will be of sufficient size to contain all the maintenance material dredged from this reach during the 50-year economic life of the project.

Another contract is the Upper Bay contract performed by **Weeks Marine, Inc. (WMI).** The WMI contract started February 15, 2000 and all dredging was completed February 22, 2001. Approximately 8.0M yd^3 have been dredged at a cost to date of $21M. The only remaining item is the completion of the rock erosion protection for the marsh cell levees. Four cells (180 acres, 190 acres, 175 acres, and 230 acres) were initially constructed to contain the first twenty years of material. These cells were formed by hydraulically pumping a levee with a 15-ft top width at elevation +10.0 MLT. In future years, 5 additional cells (145 acres, 180 acres and 3 at 170 acres) will be constructed. The four newly constructed cells are being protected from erosion by limestone riprap. This riprap is 3 ft thick and is being placed on geotextile fabric. The riprap ranges in size from 26 lbs to 750 lbs and will be or has been placed around the exterior of the newly constructed levees between -2.0 MLT to +4.0 MLT.

The final contract is the Lower Bayou Contract. This contract was awarded to a joint venture of **Bean Stuyvesant LLC.** The contract was awarded in May 2000. The dredge performing the work, the ***Meridian*,** moved to work on the East Coast and is scheduled to be back by November 2001 to complete the ship channel work. The contractor estimates that the dredging will be complete within 2 months after return. Once dredging is complete approximately 7.4M yd^3 will have been dredged and placed in the Atkinson Island *(Figure 7)* and Cell 14 and 15 disposal facilities.

Two new contracts are scheduled to be advertised shortly. These are the Mid Bay Contract and the Goat Island (Mid Bayou) Contracts. It is anticipated that the Mid Bayou Contract will be the most expensive of all contracts of this project. The Mid Bayou Contract will contain approximately 10M yd^3 of material to be dredged. Because of the foundation conditions and lack of suitable material to build the marsh cell containment levees, about 5.0M yd^3 will be taken offshore and placed in the existing offshore beneficial use berm. When design for the plans and specifications, was started, probings of the Bay foundation were made and it was shown that the Bay foundation was very soft in some areas. In some cases the probe went down 40 ft without hitting anything that provided resistance to the weight. These areas and others that show very soft material on the bottom will be avoided. The remaining 5.0M yd^3 will be used to construct a marsh site at Mid Bay. Mid Bay Marsh will contain 3 large cells to plant marsh, 2 are 150 acres in size and the last is 203 acres. Also included are 2 small marsh areas of 20 and 24 acres and 2 upland areas of 20 acres each. All material from the Mid Bayou Reach, approximately 5.6DM yd^3 will be used to re-construct Goat Island. This island previously existed and protected the area east of theship channel from wave

Figure 7 Atkinson Island

attack. This island will be approximately 320 acres in size and will be protected from erosion from the ships and wind.

One possible feature that has excited the boating community is the possible restoration of Redfish Island. See *Figure 1* for the location of Redfish Reef on which a 10-acre island will be constructed on the northerly third of the reef. This once-popular boating destination island disappeared due to erosion and subsidence. The Port of Houston Authority and the U.S. Army Corps of Engineers plan to partially restore the island. Using a design developed with help from the BUG after consultation with boating groups, the Texas Department of Health and oystermen, the restoration will employ a limestone rock and breakwaters in the shape of a hook to shelter boats. The island will be approximately 10 acres and will function as bird and oyster habitat while maintaining calm water for a safe boater anchorage. This island is presently scheduled for construction in the fall and will take approximately six months to complete, subject to approval of the construction by the Corps of Engineers Division Office located in Dallas,Texas.

To add to the safety of the ship channel, barge lane construction is being added to the project. As authorized by Congress, the Port of Houston Authority and the U.S. Army Corps of Engineers are incorporating the construction of barge lanes into the expansion plans. Upon completion, barges will have a lane on each side of the channel allowing the larger ships to have the channel without competition from the shallow-draft vessels. The barge lanes are being created from the mouth of Galveston Bay up to Morgan's Point, near the entrance to Buffalo Bayou. The barge lanes will be constructed to a 12-ft depth and will extend from the side slope of the ship channel template along the 12-ft depth for a distance of 500 ft from the centerline of the channel on both sides of the channel.

Also involved in the project construction are the removal of 90 pipelines that transverse underneath the channel. These pipeline contain all kinds of products and in some instances have more than one product in a line. At this time, in most cases, only the portion of the pipeline that impacts the channel construction is being removed and usually replace. Removal of the pipelines was authorized by the Congress as an owner cost. This removal effort by the pipeline companies is currently underway. A number of the owners of these pipelines have filed a lawsuit against the Port of Houston Authority and the Corps of Engineers in an attempt to recover some of the cost of the removal and replacement costs.

Egypt

Port Said East, Dredging & Coastal Engineering Works
Egypt's Mega-Development

Where the Suez Canal connects to the Mediterranean Sea, the Egyptian government decided, after careful consideration in various studies, to construct a new port. It was initially to be used for container terminals, but with options for extension for other port uses. *(See Photo Plate 16).*

A reference design was prepared by an Egyptian engineering company, **Hamza Associates** of Cairo. The design specified the main functions the port has to fulfill, and gave a possible lay-out with construction requirements. This included dredging of an access channel, a berthing channel, and a turning basin. Construction of a breakwater and rock protection works along the coast and dredged slopes was included.

The final joint venture called **Port Said Dredging Contractors (PSDC),** included **Jan de Nul (Belgium), Boskalis Intl (Netherlands), Ballast Nedam Dredging (Netherlands), and Hyundai Engineering & Construction (Korea).** A new contracting procedure was utilized which combined engineering analysis, to a **Design & Construct.**

Lump Sum Contract

For the dredging works, a fleet of 4-CS dredges and 6-TH dredges were mobilized; all completed within 1-1/2 year. In the access channel a heavy CS dredge was used, working inland to create depth for other CS dredges to commence dredging of the Berthing Channel from shore line inwards. Dredges pumped the material into land disposal areas. Sand and silt was separated by selective dredging, to provide the best material for the future container terminal area. Excavating the Berthing Channel in a series of parallel cuts and consecutive layers and systematically cutting the Turning Basin down to the required dimensions, CS dredges continued until the project boundaries. There they met the Suez Canal dredges, working from the By-pass Channel, to make the southern connection.

Once sufficient depth was created in the outer Access Channel, TH dredges entered the area to further deepen the channel to the design depth. Initially small TH dredges were deployed, while with increasing depth larger vessels took over, offering more economic dredging performance. Due to the fine material, hopper loading was performed without overflowing. The TH dredges discharged their loads at the offshore disposal site.

At certain stages of the project, more than 20 hydraulic excavators were in operation with a fleet of more than 40- 20 m^3 lorries transporting material to the required locations. To achieve maximum capture of sand in the terminal area and maximum containment of the remaining fine material, a sequence of diked disposal ponds with weirboxes was installed. Effluent water was returned to the sea or to the channel, where it was regularly checked for sediment concentration.

Along the Suez-canal, both **Dredging International** and **Ondernemingen Jan De Nul,** took part in the deepening and enlargement of this vital transport artery between East and West. During the last forty years, Dredging International and its founding companies participated in the execution of four such Suez-campaigns totaling 30M m^3: in 1956-59, 1959-61, 1978-80 and 1992-94.

Workforce of **Ackermans & van Haaren** was on site at **Port Saïd** with the cutter suction dredge ***Pax,*** when they were taken by surprise by the **Suez-crisis in 1956**. It is part of corporate history how they escaped, in the midst of British-French landing operations. They were driven by taxi to **Alexandria,** the Belgian consul in front - too late however to board an American evacuation ship. At gunpoint, they managed to catch the last train to Cairo, where they remained blocked in a hotel for three days during bombardments. In a bold and risky move to the Nile river, they found a place on a paddle steamer, overloaded with refugees. From Aswan, they continued their journey on a Sudan Railway carriage to Ouadi Alfa, where they were picked up by a charter plane that took them to Stanleyville in the Belgian Congo.

Having made their adventurous passage through the dark heart of Africa, they arrived at home in Antwerp - eighteen days after they had left Port Said. More than thirty years later, their colleagues were less happy when they got trapped in **Iraq, on the eve of the Gulf War**. They had been **dredging a 60 km long canal between Qur Abdallah Streets and the Iraq harbor of Umm Qasr**. On board the cutter suction dredge ***Mascaret***, they literally had a front view of the Iraq invasion of Kuwait on August 2nd, 1990. However, when the job was completed on September 15th, **Saddam Hussein** refused to issue exit visas and kept them as ***guestages*** - a new-speak euphemism contraction at the time for *guest* and *hostage.* After a Belgian parliamentary delegation had pleaded their case in **Bagdad** and the **Belgian king** himself had taken unprecedented steps for reaching a humanitarian solution, they could finally leave the country - a full six months after some of them had arrived on the site. They received a hero's welcome in Antwerp, fully aware they had escaped from the worst. On January, 15th, 1991, the allied ultimatum expired. Six weeks after their escape from Umm Qasr, all hell broke loose in the ***Mother of all Wars.***

Egypt's Mega-Development at Port Said East

There are half a dozen projects throughout Egypt, which qualify for the mega-projects. Most of them are located in heretofore, isolated desert environments. The total Port Said East project area is immense: over 500 km² in the northwest Sinai, bordered on the west by the Mediterranean Sea.

The project consists of four distinct but interrelated components:

1. Road and rail crossings of the Suez Canal;
2. Agricultural and fisheries development of approximately 400 km²;
3. An international port of about 22 km² of land and water including a container transshipment center to serve the eastern Mediterranean;
4. Extensive town-ship and industrial zone development covering some +/- 100 km².

The seaport site at Port Said East extends 11 km south from the Mediterranean coast and is 2.0 km in width on the east side of the Suez Canal's second channel. It is a ***Greenfield*** development completely independent of the existing Port Said, which is located on the west bank of the old channel.

The port is being developed surrounding a man-made basin currently being dredged. Entry channels link the port to the Mediterranean at the north and to the Suez Canal on the west. This latter opening was the subject of intensive study and debate as the Suez Canal Authority took pains to ensure there would be no negative impacts on the Canal itself or on passing vessel convoys.

Dredging and breakwater construction (for entry channel protection) are being built by Ballast Nedam at an estimated cost of US$250M. Ultimately more than 8000 m of berth length will be available with water depths to 16 m. Construction of the planned 2400 m container trans-shipment terminal is underway.

The principal engineering challenge at the port as well as on the adjoining industrial land, is geotechnical. Prior to construction of the Suez Canal this was the location of the third major deltaic branch of the Nile. It consists of poorly consolidated silt and Canal dredged material some 6.0 m deep. Preparing the site for use is no easy task.

The solution adopted has been to place a surcharge layer of sand to a height of CD +8.5 m for a period of not less than 12 months. It was determined that wick drains would not be required.

The most interesting and ambitious development program for the new port is that of the tenant **Al-Farama Group,** which envisages a multimodal transport and logistics hub.

This project is intended to exploit opportunities created by:

- The adjacent transshipment seaport;
- An air cargo airport proposed for development on-site;
- The rail connection (adjoining the western boundary of the Al-Farama site), linking the transshipment port with Suez to the south, Cairo to the SW and **Al-Arish** to the east; and
- Road transportation to Cairo, Port Said, Damietta, Suez and Al-Arish.

Conclusion

The challenges at Port Said are real, as is the promise. Over the next 10 years a dramatic flowering of the desert will occur in the NW Sinai thanks to imaginative Government initiative, to bold business investments and to the engineering expertise essential to make it happen.

CHAPTER 8

RECLAMATION DREDGING

There are two main categories of dredging reclamation with several subdivisions or applications for dredged material. The first is for the construction of new land. The other is for beach or coastal restoration to correct erosion, provide protection of real estate and for recreational use. In describing Reclamation Dredging there is an overlap with New Construction and Maintenance Dredging. Reclamation is one of the beneficial uses of dredged material.

Creation of new land by dredging has proven itself to be a cost-effective real estate investment. The problem of sand shortages in SE Asia, has given rise to a major vessel construction program, building of Jumbo Hopper Dredges (TH). The larger size makes the economics of dredging at a long distance and transporting the material for *sale* of the mined product feasible. This can be defined as ***Dredge Mining***, with delivery of the ore to market for reclamation.

Hong Kong

The new international airport constructed on the site called, **Chek Lap Kok**, was master planned by **Bechtel Corp.** and aided by several European engineering firms. Total dredges involved in the project included: 14-Trailing Suction Hopper (TH) dredges; 4-Cutter Suction (CS) dredges, and several grab bucket dredges. [7] The project was considered at the time, to be the largest single contract for dredging in the 20th century. Dredges were brought from The Netherlands, Belgium, UK, USA, Germany, Japan, China and Korea. Some reports stated that there were up to **sixty dredges** operating on the project at the peak of construction. *(See Photo Plate 13).*

Production and Quality Control During Dredging Works At New Hong Kong Airport

The Site Preparation Contract (SPC) of Hong Kong new airport, was a lump sum contract of more than **HK$ 9 billion**. The Marine Works Joint Venture had to dredge and rehandle a total amount of more than 238M m^3. This huge volume was dredged in the relatively short time of two and a half years and required a very strong planning, production control and quality control program.

Production control was based on hydrographic surveys once or twice a week. For the control of fill volumes at **Chek Lap Kok Site** and the volumes and quality of the sand in the borrow areas, extensive use was made of the geostatistical visualization 3D software package ***BLUEPACK.***

The site preparation contract was one of the first contracts in Hong Kong in which the responsibility of the achievement of quality and compliance with the works specifi-

Boskalis' 8,000 m³ TH dredge "Cornelis Zanen" reclaiming for railway and station construction, as part of the Chek Lap Kok Airport Project, Hong Kong, China.

cation was put fully to the contractor. Quality control of the de-mudding operations was very important, due to very irregular dredge level.

Before and during fill operation, extensive seabed monitoring was executed to monitor and identify possible build-up of unsuitable material. Although the quality of the marine fill was checked before placement in the platform area, an extensive Cone Penetration Test Program was executed to approve the quality of the fill. During this program, calculated Relative Density Values from the CPT results were compared with the 95% Proctor Density Criterion.

Introduction

For the construction of the platform area almost **10M m² (2470 ac**.) of land was reclaimed west of the existing island Chek Lap Kok, preceded by dredging some 70M m³ of unsuitable material by cutter dredges and trailing suction hopper dredges (THSD). The total reclaimed volume was 196M m³, from which 120M m³ came from the land operation of the original Chek Lap Kok Island and 76M m³ was introduced from marine sources by TH dredge. In order to extract the sand an additional net quantity of about 30M m³ of overburden was removed from the marine borrow areas.

Looking at the size of the dredging area and the quantities of dredged material involved, it will be clear that control of the dredged levels and volumes was essential on this job. As is known, every 10 cm extra overdredging required 1.0M m³ extra fill and the possibility of additional removal of overburden. This was realized from the early tender stage.

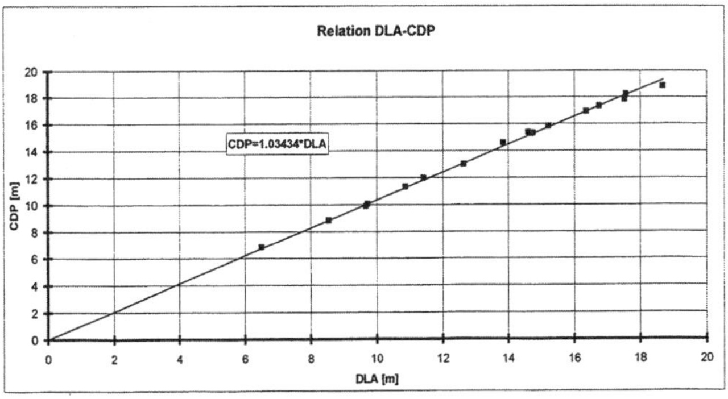

Figure 3. Realized Depth and Cutter Head Depth

Production Control

Dredged volumes and levels

Control of the dredged volumes and levels was reached by using the geo-statistical visualization software package BLUEPACK (Best Linear Unbiased Estimator), developed by the **Centre de Geostatistique** at Fontainebleau, France. This software package was used extensively on this project. In the tender phase much work was done using BLUEPACK, so it was evident to continue with the models created with BLUEPACK from the platform area as well as from the borrow pits.

The Client provided this CPT data in the tender phase as levels picked from some 3500 CPT tests. These levels were used to krig (some kind of statistical interpolation), a regular grid and was accepted by the Client. This grid combined with grids derived from land and marine surveys as well as the supplied formation levels gave the needed quantity information.

The overall marine project progress as well as individual performance of each major production unit was controlled by hydrographic surveys once or twice a week and by Tons Dry Solid (TDS) measurements on board of the trailers.

Quality Control

The theoretical dredge level of the Chek Lap Kok project was in general, the top of an alluvial clay layer situated under a layer of unsuitable material (500 kPa level). This level varied substantially between 0.00 and -28.00 related to Principal Datum (PD). The unsuitable material was removed by till 2.5 above dredge level by THSD, while dredging till dredge level was executed by cutter dredges and rehandled by THSDs.

The Cutter Head Display Programs (CDP)

The Cutter Display Program (CDP) was developed to visualize the required dredge level as an aid for the dredge master and as a monitoring tool for the superintendent/ surveyor, especially meant for areas where the theoretical dredge level varied from place to place. In particular areas, steep slopes as 1:4 were encountered; making dredging

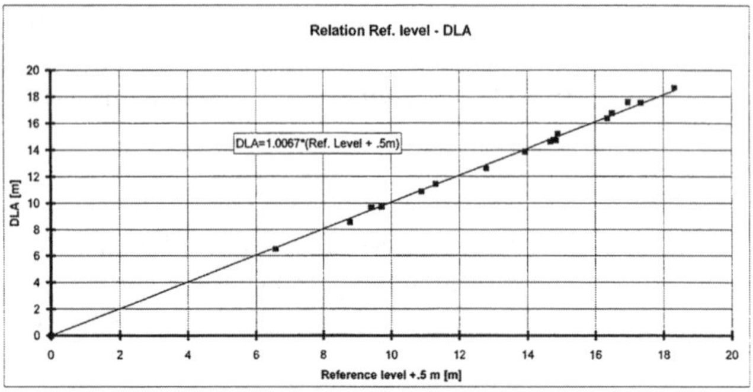

Figure 4. Target Depth and Realized Depth

without a computer equipped with adequate software impossibility.

At the cutter dredge cutter head depths were recorded and stored on the hard disk. At regular intervals these data was sent to the survey department in order to check these values with data from hydrographic surveys. In *Figure 3* the average Dredged Levels Achieved (DLA) determined by hydrographic surveys are plotted against the average values of the cutter -head depth for the areas where the cutter dredges worked.

The difference between CDP-value and the Dredged Level Achieved (DLA) was 3.5%. *(Figure 3)*. Data from the CDP together with leadline soundings was used occasionally to get approval of the areas by the Client. All calculated volumes were based on the so-called Reference Level + V2m, which means V2 m above the 500 kPa Level. This level is realized within 2/3%! *(Figure 4)*. This close ratio proves that quality control on dredged levels paid off and that accurate dredging is possible.

The Draghead Logging System (DLS)

The DLS program was designed as an instrument to check the draghead logs from the hopper dredges. The draghead logging system recorded data on floppy discs with a distance interval of 10 meters, or, if earlier reached, at a time interval of 10 seconds. The data was recorded with the criteria that the draghead(s) must have reached a pre-defined depth level prior to recording, thus avoiding tracks being stored during sailing (no dredging).

The position as well as the depth of the draghead(s) was stored. From these logs, conclusions may be drawn about the accuracy of the dredging operations, the malfunctioning of the draghead sensors, etc.

Draghead Logging was basically carried out for two reasons:

1. Monitoring of the dredging process to avoid overdredging. The software reads the draghead discs as they come from the various dredges (once a week), compares the information with a bottom layer and calculates the difference between draghead depth and the requested bottom layer.

2. Updating data of the bottom layers with the recorded information. In certain cases determining the bottom of some of the borrow areas, in particular the East Sha Chau pits, was very difficult with the echo sounder, because of the dirty water and suspended silt layers whirling through the deep borrow areas. In these cases the draghead

logging discs were used to produce dredged levels and were cross-checked by hand soundings. Accuracy of these data was less than depths determined by bathymetric surveys under normal conditions, but the best available under the given circumstances.

Barge Monitoring System (BMS)

As in many other countries, the **Hong Kong Environmental Protection Department (EPD)** requires any barge operating in Hong Kong waters to be equipped with a position and status logging system. Without this system a dumping license is not granted. The reason for such a system is to register and control dumping activities in Hong Kong waters.

EPD prescribes the logging signals, data format and interval time. The system has to be in a closed environment accessible for EPD officers only. For hopper dredges it was agreed to exchange the draft signals for indication of the dump status for a signal from the bottom doors.

In case it was necessary to review a complete track of the hopper, including the sailing and dumping cycle, this could be done by using the BMS (Barge Monitoring System) data, where the position of the dredger and the status of the bottom doors (open or closed) was recorded continuously. However, these data were exclusively and compulsory recorded for the EPD and were in principle, not available for the project. Calibration of the dragheads was done regularly and in the peak season (12 dredges) this required quite some time.

Report on Seabed Monitoring

The purpose of seabed monitoring is to identify the possible build-up of unsuitable fill material in de-mudded areas and to take preventative and/or corrective actions.

The Survey Department utilized two types of echo sounders from the same manufacturer. Both the *Atlas Deso 22* and the *Atlas Deso 25* have two measuring frequencies, 210 Kc and 33 Kc. Although the contractual frequency to be used, 33 Kc, is the same for both types of echo sounders, the Deso 25 analogue record, generally gives a much better view of possible undefined floating layers of different densities. This can be ascribed to the use of advanced electronic techniques and recording on thermal paper.

Theoretically, 33 Kc should penetrate through silt layers easily, however in practice this frequency also reflects from layers with different densities, probably due to the fines in these floating layers. Digitally, only the first echo return from both frequencies is logged onto floppy disc and accordingly presented on a bathymetric plan. Due to the higher resolution, the more sensitive areas (Seawall areas and those areas with deep CPT's) prone to siltation were surveyed by the survey launch fitted with this *Atlas Deso 25* echo sounder.

Upon completion of a bathymetric survey the raw data was processed. In principle, the first echo return was presented on the bathymetric plan, thus obscuring the possible true bottom (top of approved fill material or Reference Level). Any build-up of unsuitable material visible on the analogue record of the echo sounder, which could have caused entrapment, was highlighted on the bathymetry. This was done manually by transforming the images from the echo trace onto the bathymetric plan.

Due to its complexity this exercise was not performed on each survey, but reserved for suitable surveys, where the coverage and the echo trace were of importance. However, a lot of bathymetry surveys showed return echoes, which frequently penetrated through the first layer of a different density. In these cases the second echo was processed and/or digitized and the Operations Department was informed accordingly.

These *Seabed Monitoring Plans* were forwarded to the Works Manager, who took the appropriate action. If considered necessary, the Head of the Survey Department initiated additional leadline soundings. Bottom samples could be taken if required or further geotechnical tests initiated. Earlier bathymetric plans covering the same areas could be used for determining the possible cause and more importantly the preventative and/or remedial actions.

Quality Control on Marine Fill

In the tender stage the design of the borrow pits was based on the volumes available and the quality of the sand. The only requirement for the use of sand from marine sources was the percentage of fines content which may not exceed the 20% limit.
The testing of the marine sand was done in the hopper and when placed in the reclamation area. From each hopper load samples were taken. Every week two samples from each Trailing Hopper Suction Dredge were selected at random and for sieve analyses sent to a laboratory.

The Planning and Production Department evaluated the results in order to verify the compliance with the specifications. If a trend became visible in deviation with the specifications, the necessity of corrective action was considered. It will be clear that, considering the huge hopper quantities involved, ad hoc corrective actions could not be based on small deviations of two samples from one hopper load; a statistically significant deviating trend had to become apparent to justify the corrective action. Besides that, determining sieve analyses was time consuming with the result that the hopper load had already been placed in the reclamation area before the data were available for eventual corrective actions. Such a quality control is a waste of time and money.

The sampling of the marine fill in the reclamation area was originally planned to consist of continuous vibrocores down to dredge levels in a 50 m grid. This would mean that many thousands of vibro cores (land and marine-!) would be needed. All cores had to be tested (sieve analysis) and kept for farther reference. This would have resulted in severe logistical difficulties with several rigs on site, on site logging, sample storage, and analysis and management of test results.

Cone penetration Tests (CPTs) can provide the same information (and more) in a faster, cleaner and more manageable way. Cone resistance, friction ratio and pore pressure, give adequate information on the nature, density, and fines content of the fill material. Although the contractor was aware that he had to reinforce his attention to the fill quality, the switch over from vibrocores to CPTs, the reduction of the logistical

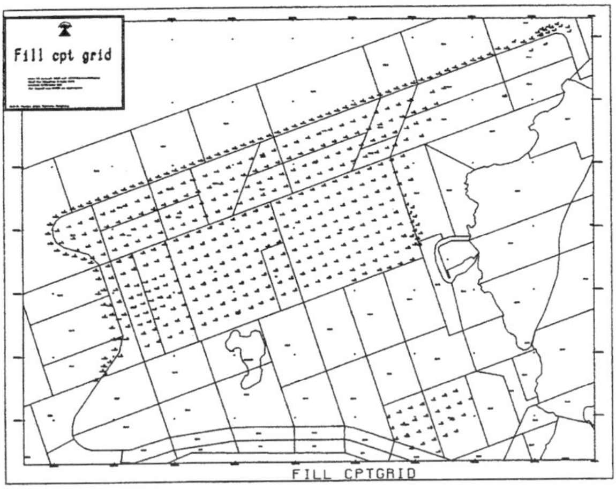

Figure 5. Fill CPT Grid

problems was more important.

In order to convince the Client, a trial area was chosen (B1/3) where fourteen vibrocores and CPT's were carried out at the same location. The results showed that CPT's can be used and are even preferable, as CPT's have no ambiguity because of core losses and bad recovery as vibrocores can have. To approve the fill material, however, the contract requirements had to be changed. The requirement applicable to vibrocore samples was a fine content of less than 20%. This requirement had to be translated to CPT-values. The following agreement was made on required cone and sleeve resistance and pore pressure to approve the fill. The fill material complied with the specifications if:

- the friction ratio was less than 0.45 % and
- the deviation between pore pressure and hydrostatic pressure was no more than 0. 2 bar, or
- the cone resistance was not less than 4 MPa, or
- the fines content in the gradings were less than 20%, when vibrocore samples were taken.

Non-compliance of any test was based on a continuous length of 1.00 m or 10% of the total test length, should more than one non-complying interval of less than one meter have occurred.

These Specifications were applied to every CPT. If a CPT appeared to be out of those specifications, a vibrocore could be taken and if the samples did not meet the requirements, the extent of the non-complying area had to be determined by additional CPT's.

The CPT testing grid had 100 m spacing and the marine CPT's in the seawall areas had 75 m spacing. It was also agreed to carry out a limited number of vibrocores on land in a grid with 200 m spacing, to supply additional samples to the Client. This also provided a means of continuously comparing the CPT results with physical samples obtained from the vibrocores because the vibrocores were taken at the same location as the CPT tests. This resulted in more than 600 CPT tests and more than 40 vibrocores. *Figure 5* shows a map of these CPT locations as used on site. The contractor's attention beforehand for the quality of the fill, during the design of the borrow pits paid off. None of these tests failed!

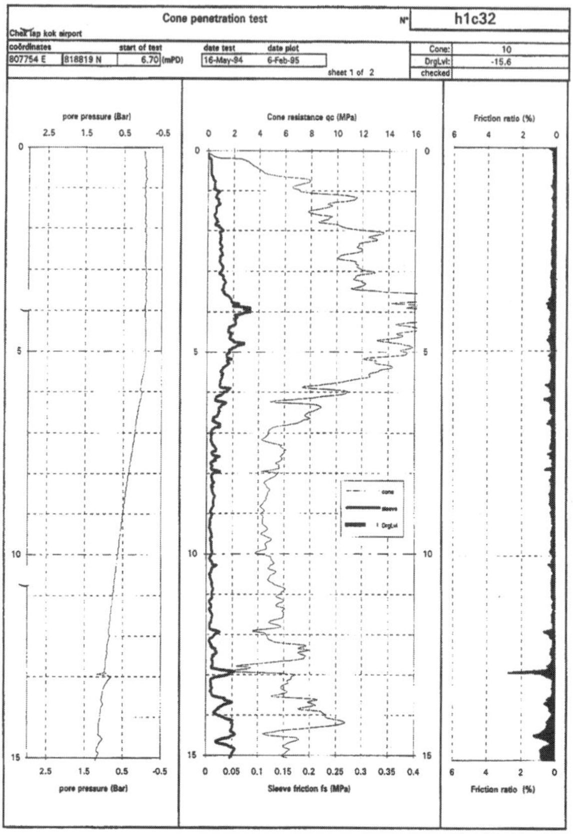

Figure 6. CPT test H1c32

The Airport Authority also obtained important and relevant geotechnical information regarding the quality of the sandfill, from the extensive series of CPT's. CPT results can be fed directly into many geotechnical calculations like settlements, bearing capacity, liquefaction potential and relative density. In new contracts it is advisable to express the quality requirements of the fill directly in CPT results and not in SPT blow count or Proctor density.

Furthermore, one has to realize that a direct correlation with the visual results of the vibrocore tests can never be 100 % exact, as with so many other tests.

Interpretation of study results

CPT test hlc32 *(Figure 6)* shows some typical results on land through the sand fill. Ground elevation was 6.70 m PD and according to the measured pore pressures this corresponded with the water table at about 5m depth (at 1.70 m PD). The fill material that was hydraulically pumped above the water table had a high cone resistance, indicating high relative densities. (Confirmed by *in situ* density tests). Below the water table the cone resistance dropped off dramatically to below 4 MPa.

At about 22 m deep, a sharp increase in friction ratio and pore pressure could be observed, indicating that the original base material (alluvial clay or marine mud q > 500 kPa) had been reached. Taken samples showed that the CPT is very sensitive to silty-clay layers. Mostly the thickness of these layers was exaggerated. Probably the cone resistance was already decreasing in the sand just above the silty layer because of a 'Punching effect'. It should also be mentioned that small clay nodules (1-2 cm) could be present in the sand, which had the same effect on the CPT as thin clay layers. The CPT results could also be very useful as a tool for establishing the actual fill levels and the

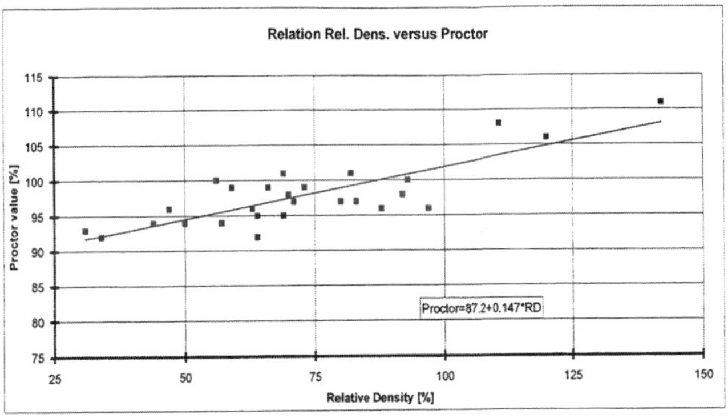

Figure 7

amount of over-dredging in a certain area.

In Situ Density Testing

According to the contract, the upper 4 m of the platform should be compacted to a Proctor Density of more than 95%. The fill material had to be tested for every 50 cm layer. Because the Type C material (marine sand) was placed hydraulically, the material could only be tested at the surface.

It was agreed that Test 15B of BS 1377 (sand replacement method with large pouring cylinder) would be used to determine the density at the surface, in the areas where only capping was carried out with marine sand. More than 50 of these tests were carried out in the capping layer. All tests showed Proctor Densities of more than 95%. It is important, however, that the surface of the test area is not disturbed and that the test is carried out with great care.

In areas where sand was placed in a much thicker layer this in situ test was not sufficient. It was suggested to use the CPT results from the quality control program to determine whether the required density was obtained. Before this proposal was accepted it was necessary to prove the suitability of CPT's and to determine a criterion applicable to CPT's to replace the 95% Proctor Density criterion. This resulted in some extensive testing (22 CPT's & 27 sand replacement tests) in a trial area situated in area H1-6.

In Figure 7, the relative density is plotted against the measured proctor density. In this case the 95% Proctor Density corresponds with 53% Relative Density. This is in accordance with literature values.

Baldi: $$D_r = \frac{1}{2.38} \ln\left[\frac{q_c}{248(\sigma'_v)^{0.55}}\right]$$ [Baldi, 1982]

D_r = Relative Density [-]
q_c = Cone Resistance [kPa]
σ'_v = Vertical effective Stress [kPa]

These CPT results can also be used to compare the Relative Density with empirical relationships, such as from:
(insert formula)

Using Baldi's equation and comparing the calculated values with the measured Relative Densities shows a reasonable correlation *(Figure 8)*.

In *Figure 9* the measured cone resistance q/c is plotted against depth, together with the corresponding 53% D/r lines according to the above mentioned relationships.

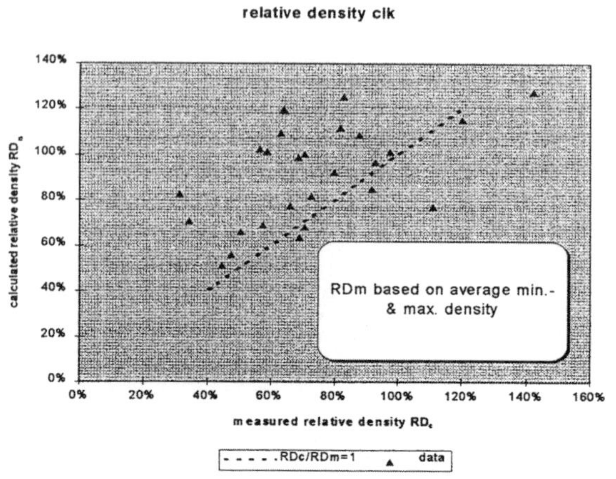

Figure 8. Calculated (Baldi) and Measured Relative Density of CLK

This figure shows clearly that all q/c values are equal or greater than those corresponding with the 53% D/r and so fulfilling the requirement of at least the 95% Proctor Density.

The Client accepted this method.

At one point in the project it was decided by the Client that at certain places in the platform area, the marine fill had to be compacted (This was probably a result of the Quality Control by CPT's). In itself it had nothing to do with Site preparation Contract 201. Under this Contract there was only a requirement about Proctor Density of the upper 4 m of the sandfill.

A separate contract was awarded to ***Bachy Soletanche***, to compact all the sand in certain areas up to final dredged level. Under this contract it was required to carry out a substantial amount of CPT tests before and after compaction had taken place. Chek Lap Kok Marine JV executed these CPT tests as a subcontractor to Bachy Soletanche.

The CPT tests turned out to be a good instrument to monitor compaction. It was observed in general that after compaction the naturally dense upper four to six meter were reduced in density while the lower looser material had considerably gained density. Also after compaction, the density was not only varying considerably with depth but there was only a high variability in the lateral sense. *See also [G.W. Plant, 1998, Chapter 18].*

Settlement of Fill Areas

Areas filled with material both from land and marine sources and subjected to settlement, could be classified into three major groups:

- Filled from Dredge Level to Formation Level with marine sand only, total fill height varied between 18 and 24m.
- Filled from Dredge Level to approximately -10 m PD in the Northern Runway area, total fill height varied between 2 and 15 m.
- Capping layer on top of (partially) rock filled areas (run ways), layer thickness approximately 2 m.

The Contractor was obliged to ensure that when handing over areas to the Client (sectional completion formation level was achieved within a tolerance of +1-150 mm). The time period between completion of an area and handing over to the Client showed large variations, from weeks to over a year.

Over-heights, installed to allow for settlement and based on the expected rate of settlement, were strongly influenced by the duration of such a period. In the tender stage, soil experts from the different companies made settlement calculations. The calculations showed variations in settlements after two years between 0.20 m and 0.80 m, depending on the type of drainage. After final discussions, it was recommended to take an average settlement of 60 cm into account for all areas, based on settlement calculations using the ***Terzaghi/Koppejan*** theory. A possible diversification of this estimate towards the Northern and Southern half of the platform could be made, i.e. 75 cm for the Northern half and 45 cm for the Southern half.

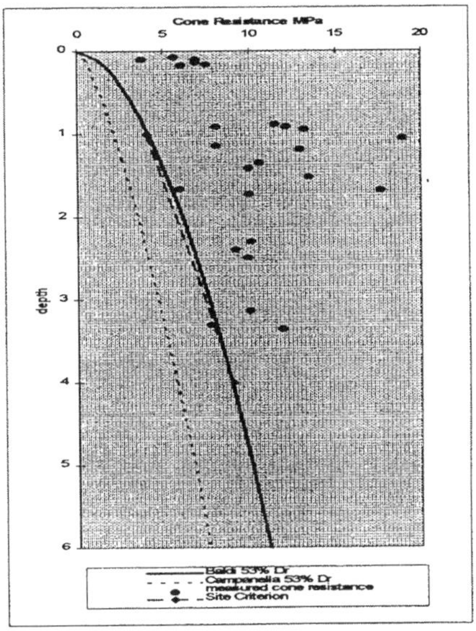

Figure 9. Measured and Calculated D_r (Baldi) as Function of Depth

Settlement Control

Settlement control in sand filled areas was in principle carried out by the Survey Department, measuring the fill levels in a two to four weekly cycle, starting within one to two weeks after the final level was achieved. Monitoring could only be carried out roughly, as surveying based on a logarithmic time scale was not possible due to time constraints.

Recommendations

Calculations and presentation of settlement predictions were of limited use on site for prediction of the extra fill heights. Although it is known that settlement predictions can vary significantly, more detailed information would have been very useful.

Instead of presenting one figure for a two year period and two settlement curves (Terzaghi/Koppejan) for one fill height, curves could have been made for three or four different fill heights.

Presenting a figure for the overall, theoretical settlement is obviously a requirement for fill volume calculations. However to determine over-height, the setting out levels on site, figure or curves are required from initial settlement of the fill during construction. Presenting predictions this way will also allow for closer monitoring of actual settlement, providing that surveys are carried out at time intervals based on a logarithmic scale.

Conclusions

1. For the necessary control over volumes and levels, modem 3D visualization software packages are indispensable at the site office as well on board of the dredges.
2. Production and quality control are in the interest of both the Client and the contractor.
3. Cone Penetration Tests are much preferable above vibrocores for fill quality control. CPT's provide much more information in a more efficient and unbiased way than vibrocores.
4. For hydraulically placed fill material CPT's are preferable above proctor tests for in situ density testing
5. The method of placement and placed below or above water is essential for the *in situ* relative density.
6. To determine accurate extra fill heights to allow for settlements, detailed settlement prediction calculations are required and a close monitoring system on site.

SINGAPORE

The 1980s were the years of recession, the 1990s brought the recovery, but the booming South East Asian economies of today will bring unprecedented economic growth within the next five years. Singapore, barely affected by the Asian crisis of the late 1990s, can already boast a healthy growth figure of around 7%.

But Singapore is full; the miniature state, no bigger than 42 by 23 km (total surface area 64,600 ha), is bursting at the rims. Apart from a population of almost 3.5M, the city also knows a growing manufacturing industry. That is why Singapore wants to let whole new patches of land rise out of the sea, some thousands of hectares (ha) of land. Soil that is now still 25 m below sea level will have to be heightened to 5.0 m above the water.

This enormous land reclamation project will require around **800M m^3** of sand. With a total value of some **Euro 3.7B,** it is the largest dredging job in the history of the dredging industry. The new reclamation project will provide further space for the booming **Oil & Gas industry** where factories, storage tanks and terminals are competing for space.

Trailing suction hopper (TH) dredge ***HAM 310,*** "is on its way with a full load of sand destined for the new reclamation area. A computer screen on the bridge shows the area, subdivided into square sections. To the left is Jumbo Trailer ***WD Fairway***, a dredge from **Boskalis**, pumping sand onshore and on the right, the dredge Jumbo Trailer ***Lange Wapper*** of **Dredging International,** is rain-bowing sand from its bow discharge on the shoreline.

A computer screen shows that the ship rises from 11 m to 5.0 m draught. A quick

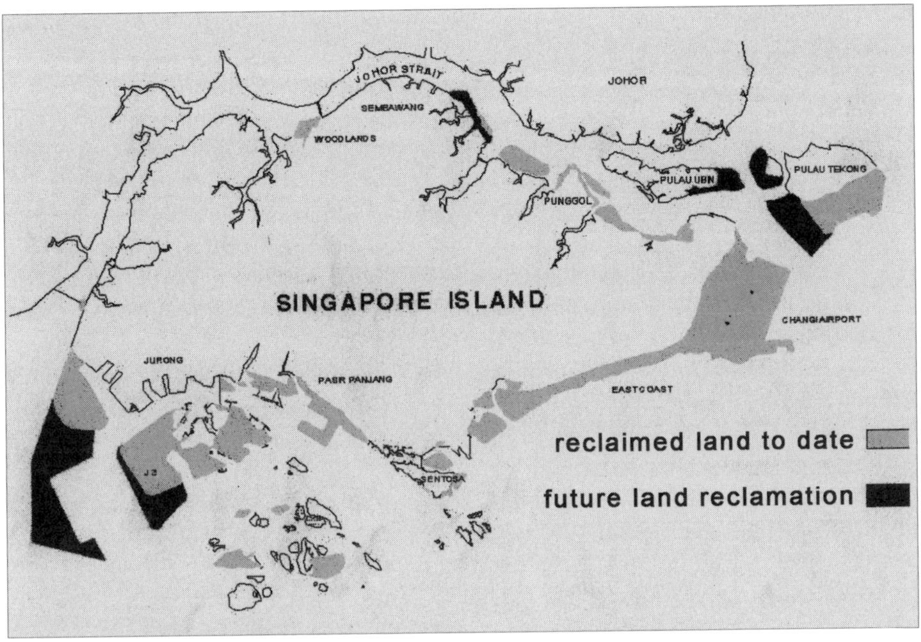

calculation leads to the conclusion that with the nine large ships and six barges at work, some **300,000 m³** of sand is moved into the reclamation area every 24 hrs. In the month of May, the production peaked at the gigantic amount of **500,000 m³** per day."
(See Photo Plate's 1,2,4,5 and 7).

Mega Operation

The project was contracted by a consortium of five dredging companies: **HAM, dredging and marine contractors, Boskalis International** (both from the Netherlands), **Penta Ocean Construction** (Japan), **Koon Construction** (Singapore) and **Dredging International** (Belgium). The scale of the works was so large that parties had to join forces to be able to bid for a project of this size.

Land reclamation in Singapore, started in the 1970s as seven little islands were enlarged to create more land. The next step was to consolidate the islands into one single large island, of some **1500 ha: Jurong Island.** The project Jurong 2 adds another 700 ha of land to the existing island. Today's project under execution, ***Jurong 3***, is enlarging Jurong Island with a further 1000 ha, a project requiring an additional **218M m³** of soil. The extension involves a **$750M** investment.

Final delivery of the area takes place sometime next year, but already large chunks of *Jurong 3* have been leased out for 99 years to chemical and petrochemical companies. It says something about the lack of space and the pressure on our project.

The demand for new land is so large that developer and client, **JTC (Jurong Town Corporation)** wants to let another patch of land rise from the sea: ***Jurong 4.*** Work on *Phase 4* has also started. Straight across from **Jurong, *Tuas View A*** and ***Tuas View B*** are also under construction. These giant projects have recently been awarded to two different consortia after a public tender issued by the Client. They will add another 2500 ha to the Singapore surface.

The works for *Tuas View B* and the extension of Jurong, were awarded to the same consortium that is executing the *Jurong 3B* project: **Dredging International, Bos Kalis, HAM, Dredging and Marine Contractors; Koon Construction; and Penta Ocean Construction.** The *Tuas View A* project, has been awarded to a consortium of **Van Oord ACZ; Ballast Nedam Dredging;** and **Hyundai** (Korea).

In parallel to the works in Jurong, currently a fifth project has been awarded: above **Changi Airport** in **eastern Singapore**, the islands **Pulau Ubin** and **Pulau Tekong,** will be enlarged with several hundreds of ha. This new land is not destined for industrial development, but will be used for the expansion of housing facilities in Singapore.

"In order to create all this land within the time frame set by the **Singapore Government,** a large part of the world's dredging resources are needed. Given the large transport distances and the competitive price level on which the works have been let to the consortia, only large trailer dredges and the so-called Jumbo trailer dredges can do the job efficiently. By consequence all *jumbos* of all the large dredging companies are engaged.

Over the years, competition among the dredging companies has triggered a drive for cost efficiency, which has led to an increase in the capacity of dredging units. With a price tag for a *jumbo* trailer dredge in excess of US$100M, these units are not only very costly but also require the world as market, given their appetite for cubic meters.

Reclamation Areas

Bulldozers are driving to and fro, on the sheer endless sandy plain of the reclamation area. Looking out over the sea, hundreds of ships can be seen anchored just off the coast. That picture will be history soon; factories and container terminals will take their place. From the air the reclamation looks like a huge patchwork blanket in various shades of brown. The different sand borrow areas each produce their own color of sand. The sand is of superior quality, one of the demands made by the client. Each hopper takes samples of its load. The client analyses grain size and percentage of silt on a daily basis.

The subsoil off the Singapore coast appeared to be of varying quality. Layers of clay have been removed and replaced by sand. On top of this sand, another layer of some 30 m high is placed. They measure the amounts of sand that are deposited, but also look at how the sand spreads out.

We use the satellite navigation systems, to check depth levels at the various positions and relay this information to the dredges. The measurements form the daily basis for our work which are used to determine where the Jumbo's have to dump, where the larger trailer hoppers have to rainbow or pump, and where the barges are to dump."

Perfect Sand

Sand is what these reclamation projects are all about; good sand makes good money. For this project, most hoppers have to sail for hours before they can dredge up sand. For instance, a trailer has to sail to a borrow area in the **South China Sea**, some four hours sailing. Upon arrival, it is the computer screens on the bridge that indicate where the hoppers are allowed to dredge up sand. White lines on the screens for areas where the hoppers have been before, black areas where there is still sand to be dredged.

The mate and pipe operator look for a black area and slowly lower the suction pipes when they arrive. When they reach a depth of **25 m,** the ship starts trembling and shaking as the sand is brought on board.

Jumbo Trailers in Action

Returning to the Jurong reclamation area, the Jumbo Trailer has several options for discharge and placement:

1. Dump in nearby deeper waters for initial land building; or,
2. Re-dredging the dump site by cutter suction (CS) dredge, pump ing into the designated area; or,
3. *Rainbowing* by spraying slurry out the bow discharge pipe into the reclamation zone; or,
4. Connecting the bow discharge to a floating pipeline and pumping ashore.

The jumbo trailer has a computer automated dredging and discharge system that controls all functions; from its diesel engines to hydraulic drives, pumps, and navigation. The hopper discharge is divided into six separate doors that can either be opened in the conventional way for direct drop of the material, or, sluiced through a channel located at each door. The channels are interconnected and used for sequential unloading, pumping via the bow discharge pipe; either *rainbow* spraying, or, connecting and pumping via floating pipeline ashore.

Reclamation programs designed by the Singapore government.
Jurong Phase 4 & Tuas View, Extension B - Sand Supply

Scope of Works:

- Reclamation Filling:
  Jurong Phase 4: ca. 155M m^3
  Tuas View Ext. B: ca. 395M m^3
  Total: ca. 550M m^3
- Dredging of Waterways around Jurong: c.25M m^3
- Design and construction of two VLCC jetties
- Sandkey Construction: c.24 km
- Land Reclamation Surface: ca. 1400 ha (3,500 ac)

Project Value: SG$ 3.5B (US$ 2.06B)
Contract Start Date: Jun. 30, 2000
Contract Completion Date: Nov. 29, 2005
Joint Venture: Penta Ocean, Koon Construction, Dredging International, BosKalis, HAM
Client: Jurong Town Corporation

Tuas View Extension A - Sand Supply
Scope of Works:

- Reclamation Filling: ca. 260M m^3
- Sand key Construction: ca. 9.0 km
- Land Reclamation Surface: ca. 950 ha (2,375 ac)

Project Value: SG$ 1.6B (US$ 941B)
Contract Start Date: Jun. 30, 2000
Contract Completion Date: Nov. 30, 2005
Joint Venture: Hyundai, Van Oord, Ballast Nedam
Client: Jurong Town Corporation

Jurong 3B - Sand Supply
Scope of Works:- Reclamation Filling: ca. 200M m^3

- Sandkey Construction: ca. 11 km- Land Reclamation Surface: c. 977 ha Project Value: SG$ 1.2B(US$ 706M)
- Total Value of Executed Works (Aug. 31, 2000): SG$ 840M (70%) Contract Start Date: Dec. 19, 1998

Contract Completion Date: Dec. 18, 2001
Joint Venture: Penta Ocean, Koon Construction, HAM, Dredging International, BosKalis, HAM
Client: Jurong Town Corporation
Total Labor Force Employed: ca. 600

West Natuna Transportation System - Subcontractor
Scope of Works:

- Engineering works for seabed preparation and rock armor works
- Seabed preparation works consisting of trenching, pre-sweeping and placing of pre-lay over a distance of 107 km
- Placing of a rock protection over the first 21 km of the pipeline between Pulau Sakra (Singapore) and the south border of the Phillip Channel (Indo nesia), after all major shipping lanes are crossed (750,000 t in Singapore and 1,350,000 t in Indonesia totaling 2,100,000 t).

Subcontract Value: SG$ 115M (95%) (US$68M)
Total Value of Executed Works (Aug. 31, 2000): SG$ 110M 95%) (US$ 65M)
Contract Start Date: Aug. 27, 1999
Contract Completion Date: Sept. 14, 2000

Main Contractor: McDermott South East Asia Pte Ltd, PT McDermott Indonesia
Client: West Natuna Transportation JV

- **Conoco Indonesia Inc.**
- **Premier Oil Natuna Sea Ltd**
- **Gulf Resources (Kapak) Ltd**

Pulau Ubin / Pulau Tekong
Scope of Works:

- Reclamation Filling: 190M m^3
- Dredging of Waterways: ca. 3.0M m^3 of capital dredging
- Construction of an Offshore Dumping Ground: capacity 24M m^3 of soft clays
- Managing the Offshore Dumping Ground
- Construction of a staging ground at the Mainland: 6.0M m^3 of good earth, 3.5M m^3 of clay
- Managing a staging ground at the Mainland
- Land Reclamation Surface: ca. 1400 ha (3500 ac)
- Sandkey Construction: c.15 km

Contract Duration: 60 months

Client: Housing and Development Board

Dredging Volume: 25M m^3 capital dredging

North Eastern Coast Reclamation
Phase 4 **- Areas at Punggol and Coney Island - Main Contractor**
Scope of Works:

- Reclamation Filling: ca. 10M m^3
- Trench Dredging and Filling: ca. 3.5M m^3
- Retaining Wall and Revetment Works: ca. 8.0 km
- Land Reclamation Surface: ca. 200 ha

Project Value: SG$ 256M (US$ 151M)
Total Value of Executed Works (Aug. 31, 2000): SG$ 200M (78%) (US$118M)
Contract Start Date: Nov. 6, 1997
Contract Completion Date: Nov. 5, 2001
Joint Venture: TOA Corporation, Jan De Nul NV
Client: Housing and Development Board
Total Labor Force Employed: ca. 450

Petroleum Industry Dredging Projects

Pertamina (Indonesia's National Oil Company) and Sembawang Gas (Singapore) have entered into a gas sales agreement for the supply of natural gas from the West Natuna Offshore gas fields for power stations and oil/petrochemical complexes on Jurong Island. The contract involves a daily supply of 326 ft^3 of gas for a period of 22 years, due for April 2001. For this purpose, 640 km gas pipelines have to be installed.

Project: West Natuna Transportation System (WNTS)
Client: West Natuna Transportation JV comprising of the operators of the gasfields:

Conoco Indonesia Inc., Premier Oil Natuna Sea Ltd, Gulf Resources (Kapak) Ltd.
Main Contractor: McDermott South East Asia Pte. Ltd (Singapore part of the work), P.T. McDermott Indonesia (Indonesian part of the works).
Subcontractor: Jan De Nul Group of Companies.
The Works: Engineering, Procurement, Construction, Installation and Testing of a 640 km gas pipelines between the West Natuna Gasfields, in Indonesia and Pulau Sakra in Singapore.
The Subcontract Works:

1. Engineering works for seabed preparation and rock armor works;
2. Seabed preparation works consisting of trenching, pre-sweeping and placing of prelay over a distance of 107 km i.e. from Tkp 0 -Tkp 107;
3. Placing of a rock protection over the first 21 km of the pipeline between Pulau Sakra (Singapore) and the south border of the Phillip Channel (Indonesia) after all major shipping lanes are crossed. (750,000 t in Singapore and 1,350,000 t in Indonesia, totaling 2,100,000 t) Date of Subcontract: Aug. 27, 1999.

Contract value: approx. US$ 70M

36" Yung-An to Tung-Hsiao Offshore Pipeline Project
Client: Chinese Petroleum Company
Main Contractor: Hyundai Heavy Industries
Subcontractor: Jan De Nul Group
The Works: - Engineering, Procurement, Construction, Installation and Testing of a 236 km, 36" gas pipeline between YUNG-AN and TUNG-HSIAO (TAIWAN).
The Subcontract Works:
Trenching and Backfilling of entire pipeline route including one conventional landfall.
Contract value: approx. US$ 45M
Equipment:
Trailing Hopper (TH) Dredges:
Gerardus Mercator (18,000 m^3), ***JFJ De Nul*** (11,750 m^3) for trenching/pre-sweeping and backfilling, Cutter Suction (CS) Dredge ***Vesalius*** (34"- 12,600 hp) for excavation of shallow landfall section.
Achievements:
Dredging at 112 m (367.4 ft) deep (*first ever* - May '99).
Innovative use of dredging equipment:

- **Special designed jet head for post-lowering of pipeline**

Baie du Tombeau Sewage Outfall Project
Client: Waste Water Authorities
Subcontractor: Jan De Nul Group
The Works:

- Installation of 1.2 m dia. sewerage pipeline (1,900 m onshore/1,400 m offshore);
- Procurement, Construction, Trenching, Installation, Backfilling.

Contract value: approx. US$ 15M

Equipment: - CS dredge ***Marco Polo*** (35", 21,913 hp) for dredging trench (5.5-33 m) (Coral, Basalt).

Land based equipment (onshore/nearshore),

- **SDV *Pompei*** for installation of pipeline and back-filling.

Australia

Sydney Airport-Beneficial Use of Sand Dredged in Botany Bay

"The construction of Sydney (**Kingsford Smith**) Airport's new parallel runway, involved the reclamation of some **170 ha of Botany Bay,** using approximately **15M m^3** of sand dredged from within the bay. The sand is retained by 7 km of vertical seawall, constructed from precast concrete panels, using the reinforced earth principle. The 130 week contract, **AUS$ 210M (US$ 150M)** for the design and construction of the new runway was awarded to a consortium of **Baulderstone Horn-ibrook Engineering Pty Ltd, Dredeco Pty Ltd, Dredging International NV,** and **NV Baggerwerken Decloedt & Zoon.**"

Sources of fill material were considered from six different sources and three methods of transportation. "Of these options, the use of sand was considered to be the most suitable, as adequate reserves were available close to the site and because of the advantageous characteristics of sand as a medium for filling, requiring a high compaction."

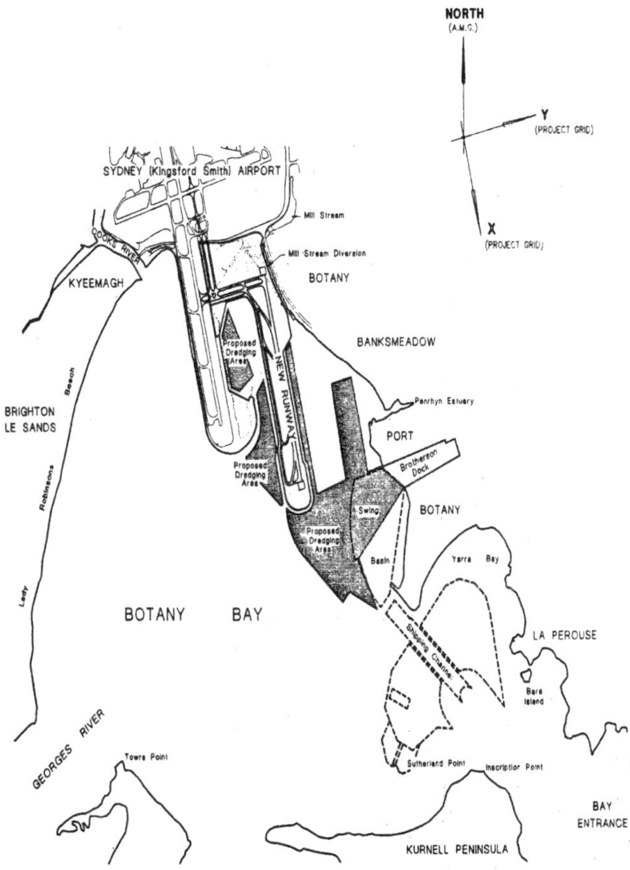

Figure 1: Runway location and proposed dredging areas

It was considered that the choice of sand for reclamation was largely responsible for the improved construction efficiency, resulting in completing the project six months ahead of the schedule of 130 weeks completion. The CS dredge ***Vlaanderen XI*** (13,400 hp) was used to increase the dredging production rate. In addition, two smaller CS dredges were used to carry out detail work, the dredge ***Blilba*** (1,566 hp) and ***Mudsnapper***

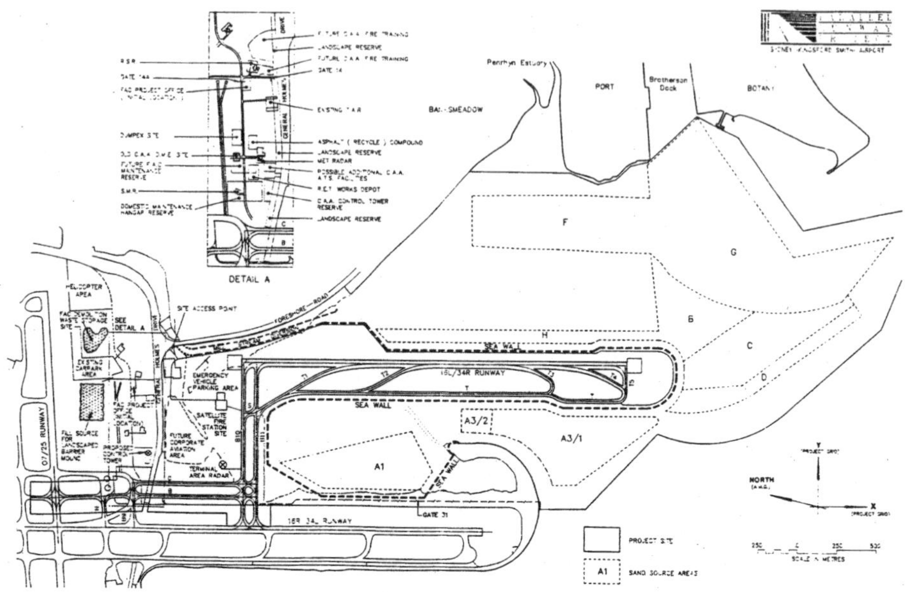

Figure 2 **PROJECT SITE** (ILLUSTRATIVE ONLY)

(690 hp).

The seawall was constructed by forming a bund of dredged sand along the line of the wall, to the finished height of the bulk fill. A trench was excavated in the center of the bund and de-watering spears installed along each side to allow excavation to the required foundation level of the wall, 3.0 m below low water level.

There was an environmental restriction placed on the project, to prevent any solid particles from being dispersed throughout the rest of Botany Bay by the action of the tides. Turbidity screens were purchased from Japan at a cost of **$5.0M,** covering the area at the seawall to a maximum length of almost 5.0km. The skirt of the silt screen was anchored to the seabed and the top was attached to flotation cells on the surface.

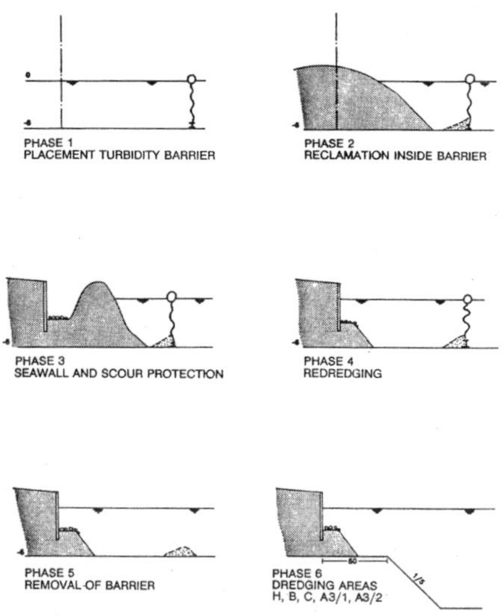

Figure 3: Dredging Sequence

The screen was considered successful in containing the turbidity generated by the dredging, and ensured that the suspended solids outside the curtain were below the specified maximum of 20 mg/litre.

Beach and Coastal Restoration

Introduction

Figure 1

Coastal erosion is endemic. For example, 80 - 90 % of sandy beaches in the United States have net erosion. In many cases, erosion is exacerbated by man's activities including damming of river sediments that provide sand to beaches and building structures and interrupt longshore sand flow.

Beaches and coasts around the world are restored by periodic beach nourishment involving moving sand from offshore or terrestrial sources to the beach or littoral zone. *Figure 1* shows a typical beach-nourishment project, often called a beach-fill, employing a dredge to pump sand from offshore to the beach and littoral zone. Beach nourishment adds sand back into the littoral system thereby restoring beaches and coasts, but it does not stop the erosion. If the sand flow has been interrupted, it remains interrupted. Therefore, as with other infrastructure such as roads, beach nourishment requires maintenance through a long-term commitment to restoring the sand balance.

Beach nourishment has become increasingly popular because it avoids possible problems of downdrift erosion that structures can produce by interrupting the longshore sand flow. As a beachfill erodes, sand moves to downdrift beaches providing additional benefits. This is the source of the Dutch observation that one has to conclude that artificial beach nourishment has from a financial point of view, a very big advantage. It is in fact an investment on which one has hardly any risk. Every grain of nourished sand is effective.

Artificial beach nourishment may prove to be somewhat cheaper or more expensive than anticipated, but what has been invested is not lost. Structures can benefit beach restoration by extending the time between renourishments but possible impacts on adjacent beaches must be evaluated. The National Research Council in the United States specifically recommends that agencies modify regulations to allow structures to be built in conjunction with beach nourishment provided effects on adjacent areas are acceptably small or mitigated. *(See Photo Plate's 3 and 15).*

History

Beach and coastal nourishment has been practiced in the United States for over 50 years. Almost 85M m³ of sand were placed on beaches in southern California, United States, in about 60 projects between 1919 and 1978. Almost all of these projects were associated with development of harbors or excavation for coastal construction with the surplus sediment disposed on nearby beaches.

This disposal transformed rocky and sand-starved beaches to wide beaches renowned throughout the world and seen on the television program, ***BAYWATCH***. The U. S. Army Corps of Engineers documents 56 major beach-nourishment projects in the U.S., for about 130M m³ of sand were placed from 1959 to 1993. Considerable sand quantities from navigation-channel dredging operations also are disposed on beaches thereby keeping sand in the littoral system.

Beach nourishment is a major tool for beach and coastal restoration throughout the world. The **Gold Coast, Queensland, Australia**, has been nourished with about **5.0M m³ of sediment. 8.5M m³ of sediment have been placed near Zeebruge, Belgium.** Between 1981 and 1987 over **9.0M m³ of sediment was placed along 50 km of coastline along the Black Sea.** Beach nourishment is the preferred approach for shore protection and restoration in Denmark with about 2.5M m³ of sediment being placed yearly.

Germany has spent over $3.0B since 1955 for coastal protection of the German North Sea coast with beach nourishment, the common approach for addressing coastal-protection problems since 1972. Spain placed over 100M m³ of sand between 1983 and 1997. The Netherlands placed about 60 M m³ of sediment in 50 projects from 1952 to 1989. The Dutch adopted a coastline preservation policy in 1990 that calls for annual beach nourishment of about 6.0M m³. The Dutch policy "... favors a periodic sacrificial nourishment because of its cost-efficiency, flexibility, and minimal environmental impact." Dutch dredging contractors had over 200 beach-nourishment projects world wide between 1983 and 1992.

Summary

	Years	M m³
USA-56 major beaches	1959-93	130
Southern California	1919-78	85
Black Sea	1981-87	9.0
Denmark	Annually	2.5
Germany-North Sea Coast	1955-	3,000
Spain	1983-97	100
The Netherlands-50 projects	1952-89	60

Benefits

Protection of land-ward development from flooding and storm damage, is a key benefit provided by beach nourishment. There is considerable infrastructure along most coasts within the United States, for example, having $3 trillion in development along its coasts. The protection provided by wide beaches was documented by damage data for over 500 structures compiled following Hurricane Eloise in the United States.

The data showed average damage costs suffered by structures increased rapidly with structure proximity to the shoreline. Beach nourishment can significantly reduce damage by increasing structure distance from the shoreline. For example, protection provided by a beach-fill placed in 1986-87 at Myrtle Beach, South Carolina, United States, was evident in damage surveys following Hurricane Hugo in 1989, where less damage occurred in the fill-protected area than in areas adjacent to the fill.

The second benefit of beach nourishment is enhanced recreation, afforded by wide beaches and accompanying tourism. Miami Beach, is an example of economic benefits of beach nourishment. Its infrastructure was markedly deteriorated in the late 1970s because the narrow beach was not attractive to recreational beach users and tourism had declined.

After the beach was nourished in 1978, a large increase in beach users provided the economic incentive to restore infrastructure and completely rejuvenate Miami Beach. Beach attendance after beach restoration almost tripled in five years. Miami Beach has two million foreign tourists a year who spend $2B and this compares with the $2.5M annual capitalized cost of the beach renourishment. Federal tax revenues from foreign tourists visiting Miami Beach, are over $130M a year, greater than the total annual Federal expenditures for beach nourishment in the United States.

Beach tourism contributes significantly not just to local communities, but the national economy. Beaches are the leading destination of tourists in the United States with 40% of people answering a major survey listing beaches as their favorite vacation destination. A recent study showed that California beaches generated $14B in direct revenue and $73B when indirect and induced benefits were included. California beach tourism generated direct Federal taxes of $2.6B and $14B when indirect benefits were included. Similarly, beach tourism in Florida creates $15.4B in gross regional product, $5.0B in wages, and $1.4B in tax revenues that are collected just from foreign tourists. Foreign tourists spend over $80B annually in the United States, and tourism generates a rare trade-surplus component to the overall trade deficit.

Environmental Effects

Beach-nourishment projects must consider environmental effects. Effects may occur at the fill or the sand source, called the *borrow* site, and can include turbidity generated at the fill or borrow site, direct impact on organisms through dredging operations or burial at the beach, and impacts on nesting sea turtles and birds. Generally, borrow sites are not used as a sand source if the sediment contains a significant quantity of fine-grain sediment, since the fine material may be suspended in the water column at the borrow site or beach as it is reworked by waves.

This turbidity may impact marine flora and fauna, including coral reefs and sea grasses by covering them with fine sediments. An analysis of ambient currents can be made to determine if suspended fine sediments will drift to sensitive areas. It may be possible to choose a time of year when currents are such that the sediment does not drift to sensitive areas. If this cannot be done, it is necessary to search for borrow material with less accompanying fine-grained sediment.

The beach-fill operation will directly affect organisms at the beach and borrow site, however, studies have shown that flora and fauna often recover relatively rapidly after a beach nourishment operation. Direct impact to mobile species such as fish is relatively small.

Beach nourishment is usually not conducted during sea turtle nesting season to avoid disturbing them. Fill sediment may also be too compact and difficult for sea turtles to dig nests, but the beach can be tilled to loosen the sand to desired levels. Prior to beach restoration at Miami Beach there were no nesting turtles because the beach was too narrow. Since beach nourishment, the annual number of nests has increased from zero to over 300, and there have been 300,000 hatchlings. Similarly, beach nourishment provides habitat for nesting birds. At West Hampton Dunes, New York, a beach-nourishment project about 3000 m in length has resulted in the return of an endangered shore bird, the piping plover, producing about 20% of the total plover fledglings in the State of New York.

Design

A major deficiency in beach-fill design in the past, was failure to realize that the subaerial beach was part of a larger beach system and the entire profile down to an approximate closure depth had to be nourished. The idea of nourishing the entire active profile is based on equilibrium beach concepts and the ***Bruun rule.***

The *Brunn rule* assumes that beaches erode such that they maintain an approximate equilibrium-profile shape, and there is a simple lateral displacement of the profile as erosion occurs. A logical extension of this concept is that the erosion can be countered by building the profile back out uniformly, and this requires fill volumes to include sufficient sand to build out the sub-aqueous portion of the active profile. However, fills placed in the United States before the late 1970s, commonly did not consider the requirement for nourishment of the sub-aqueous portion of the profile and, 'therefore, fill volumes per unit length of coast often were too low.

If the fill's sand size is finer than the natural sand size and the volume per unit length of fill is too low, the dry-beach width after adjustment to an equilibrium can be zero, with all the sand moving out to nourish the profile's sub-aqueous portion.

The typical beach-nourishment project in the United States involves *stacking* sand primarily on the sub-aerial portion of the profile during fill construction, because this usually is the least expensive method to place sand and assure it is being placed uniformly long the fill length.

The initial width of the sand wedge *stacked* on the sub-aerial beach is not the designed beach width. The designed width is a lesser width that occurs after nature does the work of redistributing the initial *stacked* volume along the active profile usually during the first winter season. Normally, less than half the *stacked* sand remains on the sub-aerial beach after this adjustment with the remainder depositing on the sub-aqueous active beach. This expected behavior of a beach fill is often not communicated effectively to the public, and this has led to perceptions that beach nourishment is not effective.

Beach-fill designers and public officials responsible for funding the beach restoration project know that the beach's initial construction width is a temporary condition, and there will be an adjustment to equilibrium that will result in a design width narrower than the construction width. However, the public may not know or even if told the design width is 30 meters, the average person cannot judge a 30-meter width. Therefore, when the initial construction width is 60 meters, most people assume this is the design width and are distressed when the fill seems to lose half its width very rapidly. Public education is important if people are to understand the true value of beach restoration.

The importance of fill length on longevity has drawn increasing attention in recent years. Fill longevity depends on the square of fill length. If a fill is too short, sand will rapidly move laterally. Beach-nourishment projects in the United States are generally sponsored by local communities, and are combined with Federal and state contributions. As a result, sometimes fill lengths are determined by the lengths of administrative and political boundaries rather than optimally planned to provide maximum longevity. Again, further public education is needed so that communities can see the advantages of working together and maximizing fill length.

Properly designed fills have performed well in the United States and throughout the world, and as a result beach nourishment has become the method of choice for beach preservation in many countries. In addition, sand dredged from navigation channels has been increasingly viewed as a resource, and rather than disposing beach-quality sand in deeper ocean depths after it is dredged, it is now often placed on the beach or in the near-shore littoral zone.

Future

The primary life-cycle cost of beach nourishment is not the initial placement, but periodic re-nourishment. Reducing beach nourishment costs will depend on developing new technologies:

- To transport sand at a lower cost to projects;
- Identify locations of beach-quality sand closer to projects; recycling sand in littoral cells;
- Extending the time-frame of beachfills, they remain at desired widths through use of structures or coarser sand (since a lower volume of coarse than fine sand, is required to achieve a specified beachfill width).

CHAPTER 9

ENVIRONMENTAL DREDGING

Background

Few events have impacted the dredging industry as much as the Environmental Movement beginning in the late 1960s. Today, most dredging projects in the US, Europe and SE Asia are subjected to rigorous evaluation of the environmental impact; modifying its methods of disposal, timing for implementation or simply preventing its performance. However, many dredging projects have been initiated that might never have been done except for the environmental demand.

The beginning of the environmental movement in the US, as it affected dredging, was in 1968. The event that sparked attention to dredging, was a cursory inspection of the Great Lakes by Department of Interior engineers. The purpose was to identify those industrial activities on or bordering the Great Lakes that could have a potential environmental impact.

The report was adopted and issued by the EPA called the *Jensen Memorandum,* suggesting a chemical and biological definition and later referred to as the Jensen or Great Lakes Criteria. The magazine, Business Week decided to publish the report and listed the targeted industries with Dredging at the top. This started the actions of halting dredging due to what was identified as pollution resulting from dredging up material *in situ.* The new chairman of WODA (World Dredging Assn), W.H. *Bill* Bauer at his first meeting with WODA in San Francisco, Dec. 1968, appointed a new committee: *The Environmental Effects of Dredging.* This committee began a program of networking with Universities and agencies including the Corps of Engineers, to stimulate interest in addressing the program.

One of our first actions was a contact with a research scientist, working on dredging-related problems in San Francisco Bay; Dr. Joel Gustafson, Professor of marine biology at SF State University. What turned out to be what I consider one of the most significant technical papers of the environmental movement related to dredging, based upon Gustafson's own research, was presented by him at a meeting of the WODA Pacific Chapter, in San Francisco. The publishing of his paper in World Dredging magazine, caused great consternation among members of EPA and other govt. agencies, who were maintaining the idea of dredging turbidity being a great evil. Gustafson's paper is included in **Appendix A:** ***Beneficial Effects of Dredging Turbidity, Supposed Toxicities of Marine Sediments,*** **Gustafson, J.F., PhD.**

Following in the footsteps of Environmental Effects of Dredging was the **Endangered Species movement.** The regulation of industrial and recreational activity became as onerous as the various forms of environmental restrictions. Examples include sea turtles, birds, small fish and it continues to grow as liberalization of species designations has grown. The cost to industry and government in direct and indirect expense is estimated to be in $billions.

Over the past 30 years the demand for dredging to cope with the removal of accumulated pollutants, has dominated research related to dredging. The principal research facility on dredging in the US is the Waterways Experiment Station (WES), COE, Vicksburg, MS. Beginning in early 1973 the Dredged Material Research Program (DMRP), was established at WES and funded at $35M. It continues today in the form of applied research in the Dredge Research Program (DRP) and accumulated expenditures are now well over $100M. Meanwhile, the data has resulted in beneficial uses in cleaning up former industrial sites as well as harbors and waterways.

Environmental Dredging is playing an increasing role in the cleanup of polluted waterways, ports and industrial sites throughout the world. The heavily industrialized areas of Europe, the former Soviet Union, Asia and the United States, are experiencing an accelerated activity of dredging to remediate conditions resulting from industrial expansion as well as military activity during the *cold war* period (i.e., 1947-89).

Full Scale Sediment Remediation

A Survey of Recently Completed Projects

Introduction

Over the past several years, the field of contaminated sediment remediation has advanced rapidly with respect to the number of sites being evaluated under various federal, state and local regulatory authorities, but progress, through actual execution of remedial construction and completion of projects, has been extremely limited. Several authors have presented summaries, overviews and case studies on the various demonstration and limited-scale contaminated sediment remediation projects that have been completed in North America (Averett and Francingues 1996, Garbaciak and Miller 1995, Stern, et al 1994, Wible, et al 1994, Otis 1994, Miles and Marr 1994, Murphy et al 1994, and Wardlaw et al 1995).

The purpose of this paper is to provide a brief summary on three full-scale sediment remediation projects that were completed in the past two years. The projects represent an interesting cross-section of the variety of remedial alternatives that are *surviving* the remedial investigation, feasibility study and final negotiation/settlement, sediment stages of project development to be selected for final, detailed design and construction. The availability of appropriately detailed information also varied between these projects, and the authors do not represent these summaries as being the most complete discussions of the individual projects. Any factual errors are unintentional, and the reader is encouraged to contact the original project proponents for further information.

Additionally, this paper is by no means a comprehensive listing of every sediment remediation project completed in North America in the past few years. Although the authors endeavor to keep track of all such similar efforts, it has become increasingly difficult to do so as the number of projects being evaluated and considered for remediation has grown beyond the handful of well-known sites that characterized this field up until the past few years. We invite any additional information the reader may wish to provide to augment future efforts such as this one.

Marathon Battery, New York

The *Marathon Battery Project* involved cleanup of a Superfund site located in the Village of Cold Spring, New York, located upstream of New York City on the Hudson River. Marathon Battery produced nickel-cadmium batteries for the military and for commercial use from 1952 until 1979 (Simmons, Matthews, and Multari 1994). During its operation, wastewater was discharged into the Hudson River and into cove and marsh areas hydraulically connected to the Hudson contaminating sediments in these areas with cadmium, nickel, and cobalt. The site was placed on the National Priorities List in 1981.

Remedial investigations and feasibility studies culminated in a plan to dredge the top one foot of sediment to achieve 10 ppm residual cadmium in sediments in the cove and achieving an action level of 100 ppm cadmium in the marsh. Attaining the marsh action level required excavation of 12 to 42 inches of sediment (Nocera and Simmons 1994). After dredging and excavation, the ROD specified that sediments would be dewatered, chemically fixed, and transported to an offsite disposal area.

Remedial design for the Marathon Battery remedial action was originally performed by the Corps of Engineers through an interagency agreement with the U.S. Environmental Protection Agency. After the project had been advertised for a construction contract, the principal responsible parties (PRPs) agreed with the U.S. District Court to perform the remedial action in accordance with the RODS and the construction plans and specifications developed by the Corps of Engineers. Cleanup activities began in 1993 and were substantially complete in 1995 **(Taylor, Tames, and Elia 1994).** Value engineering proposals were successfully made by the PRP contractor modifying some of the original design concepts.

Sediments in shallow, open-water areas (coves/ponds), were removed using a small hydraulic, horizontal auger dredge operated by **Aqua Dredge, Inc.** The 8-ft wide auger was reported to be capable of a vertical precision control of 0.1 inch (Taylor Tames, and Elia 1994). Low tides limited daily production times for the dredge, but during optimum conditions, a production rate of 1000 yd^3 per day was reported. Dredged material from the hydraulic dredge was piped to the pre-treatment/dewatering process. Some areas of the site were obstructed with rocks, debris, and other physical limitations to the small hydraulic dredge. These areas were excavated with a clamshell bucket and transported by scow to an off-loading area prior to treatment. Marsh areas of the site were excavated using low ground pressure amphibious excavating equipment with hoppers for transport to the treatment facilities. Project plans called for removing approximately 52,000 yd^3 from the cove and pond areas, 10,000 yd^3 from pier areas, and 14,000 yd^3 from the marsh (Taylor, Tames, and Elia 1994).

The treatment and disposal process for the dredged material consisted of dewatering, solidification/stabilization, and transport by rail to a landfill in Michigan. The PRP contractor originally chose a mechanical dewatering system consisting of screens and centrifuges. The variability of the physical characteristics of the dredged material, rocks, wood, and vegetation, plugged the coarser screens, and the fine silts and clays blinded the finer screens and overloaded the centrifuges (**Logigian, Dudek, and Palermo 1994).**

To overcome these problems, the dewatering system was modified to include settling ponds prior to the mechanical dewatering system to equalize the feed and remove debris. Settled solids were excavated mechanically from the ponds and stockpiled into a paved staging area to await further dewatering using belt filter presses. Excess water from the dredging operation was treated with a polymer and filtered through a sand filter prior to release to the waterway. Dewatered solids were stabilized using the proprietary Maectite process to pass leachate testing, loaded onto rail cars, and transported to an off-site landfill in Michigan (Taylor, Tames, and Elia 1994).

Orion Project, New Jersey

The beneficial reuse of dredged material has been accomplished on a large scale at several sites in the U.S., with beach nourishment or offshore island creation, both using uncontaminated sediments, representing nearly all of the completed projects to date. At the Orion Project, in Elizabeth, NJ, contaminated dredged materials, unsuitable for open-ocean disposal, are being stabilized and used as structural fill to aid in a landfill closure and redevelopment project (Walsh 1997). This project, when completed, is believed to be the largest single usage of stabilized contaminated dredged material as a structural fill in the U.S.

The Orion Project, a 166-acre privately-owned waterfront property in Elizabeth, NJ, is owned by **OENJ Corporation**. The site is a former municipal solid waste landfill, and is being redeveloped by OENJ in preparation for the construction of a 1.2M ft^2 retail mall facility. A total of 2.2M yd^3 of structural and peripheral fill is needed to complete the project, which includes landfill closure and retail mall site preparation.

A joint venture of OENJ, **Walsh Remedial Construction Services and Great Lakes Dredge and Dock Company** was formed to execute the project, including dredging, amendment of the dredged material to improve physical stability and chemical leaching properties, and placement of the material on site. To date, a 100,000 yd^3 demonstration project has been completed, under contract to the Port Authority of New York and New Jersey, while additional dredged materials are currently being processed at the site.

The dredged material processed during the demonstration project originated in Reach A of Port Newark, NJ. The material was dredged with a closed-bucket clamshell, transported to the disposal site via scow, where it is mechanically unloaded into a vibratory grizzly feed hopper. After debris screening the material is pumped from the offloading facility to the processing plant via a positive displacement pump (Walsh 1997). Difficulties with debris handling in the early stages of the project were anecdotally reported to cause excessive pump failures resulting in extensive operational downtime and subsequently greatly reduced processing rates. Refinements to the debris handling operations have improved pumping operations and speeded processing throughput times.

The dredged material was subjected to a battery of tests, as prescribed by the NJ Department of Environmental Protection, focusing on the TCLP test as an indicator of leaching characteristics. An unspecified mixture of pozzolanic-based materials is added to the dredged material in a pug-mill centered mixing plant. The amendment mixture varies according to the feed material characteristics, and the final product is suitable for use as structural fill on the site.

The site will be brought to final grade through the placement of up to 15 feet of additional recyclable and clean fill materials and will eventually be covered by an asphalt parking lot (Walsh 1997).

Manistique Harbor, Michigan

Manistique Harbor is located in the Upper Peninsula of the state of Michigan, at the mouth of the Manistique River, a tributary to Lake Michigan. It was the location of lumber mills and logging operations during the heyday of the Midwest logging industry, and continues to be the location of a pulp mill operated by **Manistique Papers** and a coal fired power plant operated by **Edison Sault Electric Company**. **USEPA** had identified **PCB** contamination in the sediments of the river and harbor exceeding 1,000 ppm. Through a series of negotiations, the Federal government and the PRPs were able to reach a settlement which involved the dredging, treatment and removal of all sediments contaminated with greater than 10 ppm total PCBs from the system.

The specifics of the negotiations between the parties, and the extensive debate surrounding the need for removal of the sediments vs. the effectiveness (both technical and economic) of capping the sediment *in situ*, while of great interest to the readers of this paper, are not the authors' focus. Readers are encouraged to contact the USEPA project manager, James Hahnenberg, for additional details. This paper will focus on the removal actions completed to date (April 1997).

Three areas of Manistique River and Harbor are being remediated, Areas B, C and D. Diver-assisted dredging was used beginning in the summer of 1995, to remove soft, silty sediment deposits in Area B. To minimize PCB losses due to sediment re-suspension and transport away from the dredging site, the area was surrounded by a sheet pile steel cutoff wall, with floating booms and silt curtains augmenting the sheet pile. USEPA reported that "no re-suspension of sediments occurred" during the dredging activities, determined through visual observations, surface water analysis and other measurements (USEPA March 1996). Approximately 10,000 yd^3 of sediments were removed prior to the shutdown of operations in November 1995.

The materials removed in 1995 were processed through a series of settling, particle separation and dewatering steps. An unusually high amount of wood chip and sawdust type material was present in the dredged material, which in turn served as a strong concentrator of the PCB contamination. Through the screening and separation processes (including hydrocyclones, plate and frame and belt presses), the volume of contaminated material exceeding 50 ppm total PCBs, and therefore requiring disposal in a TSCA-permitted facility, was reduced to 3 percent of the total dredged mass (USEPA

March 1996).

Dredging operations, this time using a *floating hydraulic auger dredge,* were recommenced in *Area B* in summer 1996 (USEPA September 1996). An additional 8,000 yd^3 of sediments were dredged and processed as discussed above. In total, about 3,520 tons of sediment and other waste materials (wood chips) was shipped off site for disposal (USEPA April 1997). These volumes and masses represent clean up of all contaminated sediments in *Area B* that exceed 10 ppm total PCBs. Confirmatory sampling concluded, to USEPA's satisfaction, that no residual PCB concentrations above 10 ppm remained in the sediments of *Area B.* After the completion of dredging activities a layer of gravel was placed in *Area B* to improve the fish habitat characteristics of the newly exposed bedrock bottom.

Additional work in 1997 will include the removal of a temporary **geotextile** cap previously placed over *Area C*, the removal of sediments from the capped area, and the removal of all sediment contaminated with greater than 10 ppm PCBs from *Area D*. Any areas with residual PCB concentrations over 10 ppm following dredging, will be covered with clean sand.

Total costs for the project are estimated to range from **$6M to $11M,** with the PRPs providing a cash payment of $6.4M and additional in-kind services as a final settlement of their liability. USEPA will cover the balance, if any, of the costs incurred in completion of the site remediation.

Summary

The three projects discussed above are representative of the still unsettled nature of the contaminated sediment, remediation field. Although more and more projects are moving into the design and construction stages, no consistent pattern has yet emerged concerning removal, processing or disposal techniques. Site-specificity remains the name of the game, and demonstration projects are still the main avenue through which the application of treatment technologies are being considered. Our ability to adequately capture true costs remains a significant challenge, as the few projects that are completed are often done under unusual funding circumstances that leave details either sketchy or protected by confidentiality agreements.

The next few years promise to continue the rapid expansion in the database of completed sediment remediation projects, through both the advancement of work on several Superfund sites and through the development of large-scale operations for the processing of dredged materials from navigation projects in such areas as New York and Oakland.

Cable Arm Bucket

Remedial dredging introduces constraints and sensitivities not present in maintenance or other dredging projects and have resulted in innovations in equipment design to cope with the unique needs. Those needs are summarized as follows.

1. The need to minimize resuspension and downstream transport of contaminated sediment.
2. The need for near 100% removal of contaminated material.
3. The need to minimize the volume of water removed.
4. The need for precise removal of the targeted material with a minimum over dredging.

Manson Construction Co. BG/C dredge "Viking," maintenance dredging Newport Beach, CA, Bay Channel with a Cable Arm 36 yd³ clamshell bucket.

The **Cable Arm Bucket (CAB)**, was invented and patented in **1989** by **Ray Bergeron**; his special environmental bucket design. The opening, cut and closing, of the Cable Arm clamshell bucket is controlled by two steel cables. One cable controls the opening of the bucket using four spreader chains to guide the two side plates as they open. The second cable controls bucket closure, lifting and operates through a series of cantilevered sheaves to allow closure of the two side plates at a slow, controlled rate. A tag line (chain) connected between the bucket and the crane maintains bucket alignment; i.e., prevents the bucket from spinning.

Donjon Marine Co., Inc.'s BC dredge "Michigan" environmental dredging with a 16 yd³ Cable Arm clamshell at Port Newark, NJ.

To minimize downward pressure during descent of the open bucket through the water column, vents in each side plate open allow water and trapped air to escape. The side plates open to nearly 180 degrees and, with the bucket's *floating pivot* action, produce a larger, level, rectangular cut; eliminating the pothole effect produced by a conventional bucket.

For example, a conventional 6 yd^3 capacity clamshell would produce a footprint of 60 ft^2 (10' X 6') while a similar CAB would produce a 144 ft^2 (18' X 8') footprint. (see illustrations).

"Seven case histories have shown that the CAB successfully met the dredging criteria on four of the projects, despite the presence of typical environmental dredging constraints. Cycle times on completed projects ranged from 3-8 min., longer than for standard navigational dredging, but necessary to ensure proper operation and to manage the constraints and sensitivities unique to environmental dredging."

Environmental Benefits of the Bucket Ladder Dredge (BL)

Dredge Selection for Environmental Cleanup

The emphasis on dredging of polluted aquatic environments in the 1960s, focused upon the **cutter suction dredge (CS)**, the principal type of dredge in the US. The fact that this type of dredge usually discharges at **10-20% solids**, became a major concern

Figure 1 *Typical European BL dredge used for harbor maintenance, offloading dredged material into barge.*

Figure 2 ***BL Dredge "Ecodrag" 200 ltre, configured for handling polluted material, protecting personnel from exposure; owned by Dredging Intl.***

when excavating polluted material. Dumping at sea and impounding ashore, were the accepted methods of disposal and thus continued the use of the CS dredge. However, environmental agencies began to demand that polluted materials once dredged, could not be re-deposited in the same area. ***(Gustafson, Appendix).***

Many studies have been made and continue to occupy Universities, laboratories and engineering firms who have gravitated to this source of funding and emphasis, attempting to determine better ways of coping with the problem. Dumping at sea, dispersing in fine layers in swamps as **wetland creation**; digging deep holes and covering with clean sand, pumping into **geotextile bags**; are some of the methods being used to dispose of polluted material.

The question was: What type of dredge can best remove polluted material with the least disturbance to the bottom material. By 1977, after considerable research and testing, the EPA removed its restriction on dredging due to turbidity. In the 1990s, however, turbidity returned as a supposed threat. A TH dredge still has difficulty because of overflow, but in some cases, eliminated by retaining the excess water until dumped or pumped ashore in a *safe* area. In one application, the Japanese use a box device around the suction head of a CS dredge (removing the cutter), to contain the polluted material.

Similarly, **IHI, Japan**, developed a system in 1976, called an overboard discharge system for TH dredges to remove the turbidity plume. The method de-energizes suspended solids and discharges through a vertical pipe to the bottom.

Bucket Ladder Dredge-BL

This paper centers upon the unique characteristics of the BL dredge, to offer a capability to remove pollution in the process of dredging. The call for healing of environmental wounds, may herald a renaissance for the BL dredge.

For mining such as with gold, tin, and diamond, it is particularly important not to leave valuable minerals (i.e., gold) behind (Richardson, 1992). A major benefit of the BL dredge is its manner of digging on the catenary of the bucket line, leaving a uniform cleaning of the bottom *(See Figure 1)*, as the ladder makes its lateral swings. This feature is applied to removal of polluted material, insofar as there is no need to resort to repeated dredging of a single course.

Features of Design

Buckets. The durable bucket design is one that is cast, using typically 19% manganese steel. Depending upon the nature of the material to be dug, the shape of the bucket can vary from vertical to flared sides with integral or attached digging lips. The usual system of interconnecting buckets is by using solid pins that interlock each of the buckets *(See Figures 1&2)*.

When using a **Caterpillar system**, lighter, fabricated buckets are used and bolted on to the chain. This results in a quieter operation but limits the ability to dig hard material, since they are unable to sustain severe forces, including swinging in an arc.

Bucket Ladder. The ladder carries the load of the catenary of the bucket chain underneath and the travel along the top, passing over ladder rollers. This is the heaviest component of the dredge when in operation with all buckets, pins, rollers and associated equipment installed.

The **Upper Tumbler** is mounted at the top end of the ladder, with inter-connected buckets. It is shaped like a six or seven-sided ratchet, providing the force to pull the buckets in the chain, and at the digging end. The upper tumbler is usually driven by either one, or, two bull gears with electric or hydraulic motors, driving through reduction gears. The **Lower Tumbler** is round, absorbing the digging forces and rotating the bucket line to travel upward along the top of the ladder.

General. The hull of a BL dredge is usually a barge, with one or two spuds on the stern; or, operates off of wires to permit positioning and swinging in an arc. The bow is swung on wires to anchors to port and starboard. An open well permits the ladder to be lowered vertically to the bottom. It is also not unusual for the BL dredge to be self-propelled.

Production Capabilities

The production variations and capabilities of BL dredges are illustrated in ***Figure 6.*** The largest BL dredge was built in Scotland and used in the digging of the **Panama Canal**; with **54 ft^3/1530 litre buckets**. The highest efficiency mining dredges (in Colombia, South America), have used standardized **14 ft^3** buckets that have achieved up to **550,000 yd^3 per month.**

If large boulders and other hard materials are encountered, larger buckets may be used. However, this increases the size of the dredge system, power and capital cost. The size of bucket and its speed in buckets per minute (bpm), determines the volume of production. While the speed of the bucket line regulates production, it also creates a corresponding increase in wear, maintenance and repair that must be balanced economically. Total installed power may be as high as 8,200kW for the largest dredge in a mining mode (BL/M), or as low as 130kW for digging soft material *(Table 1.)*.

TABLE 1.

LISTING OF SELECTED BL DREDGES

NAME/YR BUILT	BUCKET LITRES/FT^3	DEPTH m	SPEED BPM	HULL L/W/D(m)	POWER KW
ECODRAG/	195/7	8	35	29X5X2.1	130
BREMEN/'74	486/17	22	-	47X9X-	368
KUMOTORI/'86	486/17	18.5	39	73X15X5	400
FRIGG R	515/18	18	24	48X7X3	529
MINA NO.2/'76	535/19	23	49	58X12X4	1038
YUBA 21/'37*	622/20	40	28	68X21X3.5	2700
EMDEN/'56	632/22	22	-	55X11X4	294
AJAX/'90	681/24	36	27	52X12X4	380
SINGKET 1*	681/24	50	28	-	8200
B1/'72	681/24	22	24	60X11X3	1100
RHEA/'77	681/26	25	32	58X12X4	294
BAGERMEISTER GRUSHIN/'90**	759/27	18	16	80X14X5	1926
ADRIATICO/'67	875/31	32	28	77X13X4	516
MAAS/'60	973/34	21	32	55X10X4	404

NOTES:
1. The above Table is intended to show variations in bucket ladder dredges used today in various parts of the world. BL dredges usually discharge into hopper barges that shuttle alongside carrying the dredged material to a dumping or pumping station. *Mining applications discharge by stacker/conveyor belt. **Self-propelled, sea-going vessel.

2. This table was generated from WDMC(1997) and Dayton(1997).

Navigational Dredging Applications

BL dredges in navigational applications include ports, harbors, and waterways with easy access. Long recognized as a workhorse, it has had a faithful following in most countries outside the US. Applications where the BL dredge was effectively used for environmental cleanup are described below.

Project Scheveningen, The Netherlands

The Port of Scheveningen consists of three separate harbors, the innermost of which was heavily polluted **(Van de Graaf, 1995)**. The **Dutch Dredging Co.** executed the project and **chose a BL dredge**. **"-rather than develop something new, the dredge *Ecomolen* was used**

in order to retain the advantages of already known and proven techniques—".

Pilot Project *De Nieuwe Merwede '9;* Cleanup Dredging with Environmental Bucket Dredge and Grab Dredge.

The **Dutch Government (Rijkswaterstaat),** started a series of full scale pilot projects. One was situated on the river **Nieuwe Merwede** and consisted of clean-up dredging, re-use of dredged material, capping with clean sand over the remediated zone, and monitoring of the re-colonization of organic life on the cleaned bottom **(Van de Sluijs, 1995).**

Two different dredging techniques were employed, BL and BG dredges. The BL dredge is named the ***Ecodrag*** *(Figure 2)*, owned and operated by **Dredging International**, Belgium. The characteristics are summarized as follows:

Buckets: 200 litre/7 ft³ cleaned by water jet after dumping with air holes in the buckets. Ladder**:** Enclosed to avoid spilling on deck and at upper tumbler while dumping.

Chain: Closed with steel plates and a sail between deck and ladder; Caterpillar chain and fabricated buckets, hydraulically driven; reduces noise.

Reservoir: Collects material spilled from buckets, and pumps into the chute. Depth/digging: 1.5-8m; width of cut 120m.

Discharge: Dumps material dredged into hopper barges alongside.

Project Description: Nieuwe Merwede River I

A representative groyne section of the whole contaminated area was chosen. The objective was to create the conditions for ecological recovery. Therefore the zone had to be as clean as possible. The contaminated sediment had to be removed completely; to reduce the dispersion to ground and surface water. This was done by removing 50% of the contaminated layer, covering the remaining part with clean sand from the river.

In order to reduce the dumping costs and to dredge no more silt than was necessary, a very accurate means of dredging was required. A design level dredging profile with a slope from 115 to 1/200, a horizontal accuracy of 30cm and vertical of 10 cm were stipulated in the contract. Spillage, turbidity and the dispersion of pollutants in the water had to be prevented. The *EcoDrag* was used to dredge the main part of the groyne section. The BG dredge *EcoGrab* was used to dredge near the banks.

Before 1970 the daily ebb and tide washed out of the Dutch river **Nieuwe Merwede** and the national nature park ***De Biesbosch***. With the construction of the **Haringvliet Lock** downstream, the lower reaches of the Rhine and Meuse became a sink. Sediments contaminated with polycyclic aromatic hydrocarbons (PAH), PCB and heavy metals settled in areas with low flow velocities. The environment around the banks was more affected by contaminated sludge.

Design Tradeoffs

While most BL dredges are *stationary*, non-self propelled, some countries specify sea-going vessels, or self-propelled **(WDMC, 6/'79)**. For TH dredges and particularly as they are increasing in size, a significant benefit can be realized; i.e., large volume, open ocean where wave conditions make it impractical for barge hull dredges, long distances in exposed waters to discharge or dumping sites.

Because of its heavy ladder structure, the BL dredge is seldom operated in open waters; or, is towed to calm waters when storms approach. The recent application of the Caterpillar chain to replace close-connected, pinned buckets may be justified for two basic reasons:

1) Reduce the noise level;

2) Reduce the capital cost of the dredge.

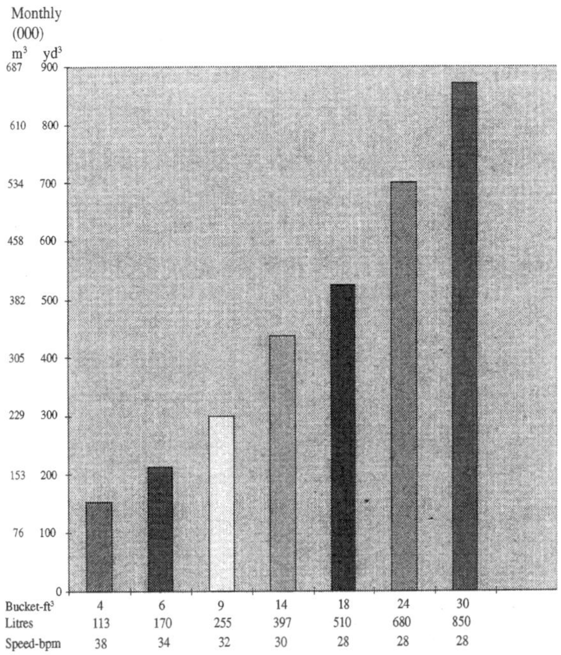

FIGURE 6

BUCKET LADDER DREDGE PRODUCTION

However, the loss of the ability to handle hard material, swing in an arc to clean the bottom and the scaling down of the overall structure, have produced some significant problems in dredge performance. The buyer may be led down a road of failure by his desire to meet budget constraints but should be aware of possible consequences.

In soft material without large rocks, boulders or trash to deal with, the Cat chain may satisfy a need. By restricting the dredge to small projects in terms of volume, it may offer an opportunity to be a cleanup dredge for some harbors, waterways and lakes ***(Burgett and Ouwerkerk, 1995).***

Some dredging projects of recent years could have used the BL dredge to advantage because of the mixture of materials; i.e., very hard and large fragments to fine silt and polluted material. In most instances, there would be a requirement for discharge to a barge system, towing to either pump-out or bottom dumping locations. This type of dredge performs best when not interrupted by stopping to move to the dumping area. Bucket dredges of BC/G or BB, also have the same need but cannot leave a smooth bottom without overlapping.

Conclusions

The use of BL dredges for environmental remediation has been demonstrated to be advantageous over other dredging systems, in certain applications. The projects where this has been recently proven were both in Holland. Other areas of the world including North America, are yet to try the BL dredge for that purpose.

In recent years the **Canadian government** funded a similar project as Holland, inviting proposals and tests of various types of dredges to clean up the environment. In Canada's case, they have serious toxic deposits that require specialized equipment for remediation. However, there is no evidence that the BL dredge was considered.

There are obstacles to choosing a BL dredge in North America where there is a lack of familiarity with their operation as well as construction. Add to that the competing dredge equipment such as the CS dredge, which is the main dredging system in the US. The capital cost of a BL dredge is also a problem unless the lighter weight, Cat chain is used which lowers the overall cost of the dredge considerably.

BL dredges in operation in the U.S. are for mineral mining of gold, or, sand and gravel. Many applications, however, would justify serious consideration, in contrast with other types of dredges. For instance, harbor and river areas, estuaries, reservoirs, and lakes, offer opportunities for the BL dredge. While a BL dredge is not usually considered to be portable, it can be dismantled and transported by truck and barge as it has in years past within the U.S. and other parts of the world for tin and gold mining.

Dredging Ooze

A systematic study of the environmental effects of dredging was conducted in Japan by three engineers. The purpose was to develop a suction device to minimize turbidity in dredging polluted sediment. Their results showed that a new type of suction mouth could be used in handling polluted mud, or, ***ooze***.

The suction device consists of agitating rotors and water jet nozzles, both contained in a case or box. The tests aided in developing the optimum shape of the agitating rotor and various main mechanisms forming the dredging characteristics.

The laboratory experimentation using scaled components, was followed by full scale dredging measurements to demonstrate its capability of dredging with less water contamination at a high concentration of *ooze*. The results confirmed the validity of the laboratory testing on scaled equipment.

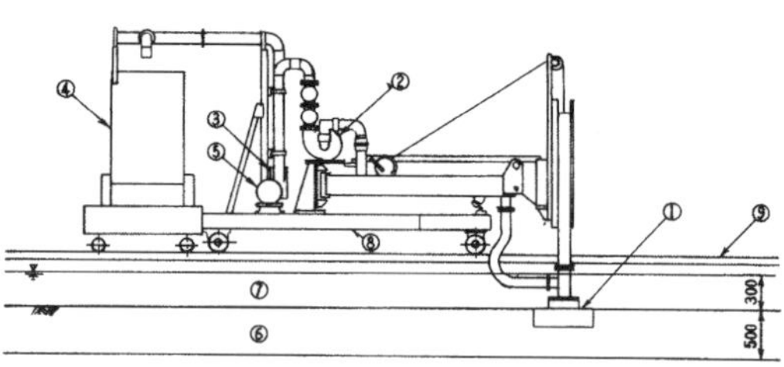

① suction mouth ② pump ③ magnetic flow-meter
④ ooze collecting tank ⑤ driving DC motor ⑥ test soil
⑦ clear water ⑧ carriage ⑨ rail

Fig.1 Outline of dredge test apparatus

The development of the ***Ooze Dredge*** was the result of many projects in Japan that were being planned to remove ooze by dredging, deposited on the bottom of rivers, lakes, harbors, and bays. Such ooze includes so-called polluted mud containing organic substances, causing water contamination and offensive odor. It was clear that some sort of new dredging device was necessary.

The principal criteria to be met for the new suction dredging device included removal of turbidity at the surface, to leave no ooze behind, and to increase the effective concentration of solids. Since the ooze is generally deposited in thin layers, the dredging work should be accurately planned for the swing width and spud stepping.

Laboratory Test

Testing Equipment

The outline of the dredge test apparatus is shown in ***Fig. 1.*** The apparatus, provided with a test mouth, consists of a slurry pump, flow meter, ooze collector tank, and water jet pump, installed on a DC-motor driven carriage. Its main particulars are:

Speed of carriage:	0-30 m/min (continuous variable)
Slurry pump:	Centrifugal pump, 40 m^3/hr capacity
Water jet pump:	Centrifugal pump
Hydraulic drive unit for	
Agitating rotors:	110 kgf/cm^2 max. pressure
Ooze collecting tank:	2 m^3 capacity

Test Suction Mouth.

Fig. 2 shows the suction mouth developed in this test. This feature is to have agitating rotors and water jet nozzles provided in the mouth.

Test Soil. Taken from the bottom of Hiroshima Bay. Its physical properties are:

Table 1. **Test Tank**. Provided by **Offshore Experimental Tank of Mitsubishi Heavy Industries, Ltd., Hiroshima Technical Institute.** The dimensions are: 40X4.5X2.5 m.

1. Relation between Dredging Speed and Dredging Characteristics.

The relation between Dredging Speed and Dredging Characteristics The relation between the dredging speed (swing speed), v, and the flow rate of mixture, Qm, is shown in *Fig. 3;* and the relation between the dredging speed, v, and the apparent solid concentration, Ca, in *Fig. 4;* in which A and Qo are the mouth opening area and the fresh water flow rate of the apparatus, respectively.

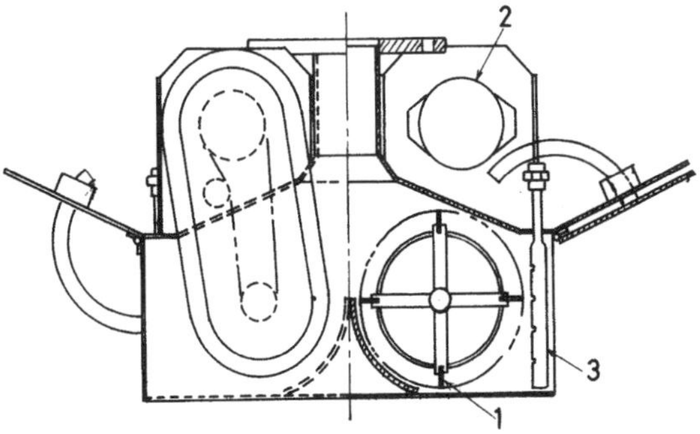

1. agitating rotor 2. hydraulic motor 3. water jet nozzle

opening area A = 990 cm^2

Fig.2 Suction month developed in laboratory test

Table 1. Soil used in laboratory test

Specific gravity	Liquid limit	Plastic limit	Symbol	Moisture content
2.600	89.5%	37.1%	CH	110%

Table 2. Main feature of KORYU

Length x breadth x depth	25.0 x 8.0 x 2.4m
Displacement tonnage	About 260t
Dredging capacity	100 m3/h (at w=200%)
Maximum dredging depth	7 m
Maximum discharge length	2,000 m
Generator	350KVA, 450V x 2

Table 3. Results of the trial of KORYU (1978)

Date	Moisture content of bed ooze (%)	Moisture content of mixture (%)	Apparent solid concentration (%)
March 25	150	161.3	94.3
March 25		174.0	88.6
March 26		158.3	95.8
March 26		163.6	93.2
March 26		168.8	90.9

Apparent solid concentration, Ca, is the percentage of in situ volume of ooze to the volume of mixture being dredged, as given by Eq. *(1)*

$$Ca = \frac{\gamma m - \gamma w}{\gamma b \quad \gamma w} \times 100(\%) \quad (1)$$

Where

γw: unit weight of water

γb: unit weight of bed materials

γm: unit weight of mixture being dredged.

When Ca is 100% it means theoretically that only the material bed is dredged without any inflow of space water.

As shown in *Figs. 3* and *4,* when the dredging speed is increased, the flow rate of mixture, Qm, is decreased and the apparent solid concentration, Ca, increased. The reason for this is that the amount of soil entering into mouth opening (Av) is increased at a higher dredging speed. However, there is a limit in concentration of mixture flowing through the pipeline (including the pump) of dredging apparatus.

If the moisture content of bed soil is lower than such limit (less fluidity than such limit) the pipeline is clogged by bed soil when the dredging speed exceeds a certain velocity.

2. Water Jet and Dredging Characteristics

The low moisture content of the bed soil is required to be adjusted within the mouth to an appropriate concentration so that it can pass through the pipeline. For this purpose, water must flow into the mouth by means of: 1) natural flow, and 2) positive flow. The natural flow-in method can be achieved by controlling the dredging speed.

The water jet method using the test mouth, as shown in *Fig. 2,* is one of the positive flow-in methods. The water jet attached to the test mouth jets water inwards to adjust the water content of soil. The test using the water jet of S.3 m3/h with the test apparatus showed a continuous suction even at Av/Qo = 1.5 or over, whereas without the water jet, Av/Qo = approx. 0.9 was the limit of continuous suction.

On the other hand, the use of water jet tends to give a thought that it lowers the solid concentration, but the solid concentration shows not much difference either with or without the water jet used below the limit suction speed, as shown in Fig. 4.

The reason for this is that the amount of water jet used below the clogging speed replaces the amount of natural flow-in water keeping nearly constant the entire amount of water to be sucked into the mouth.

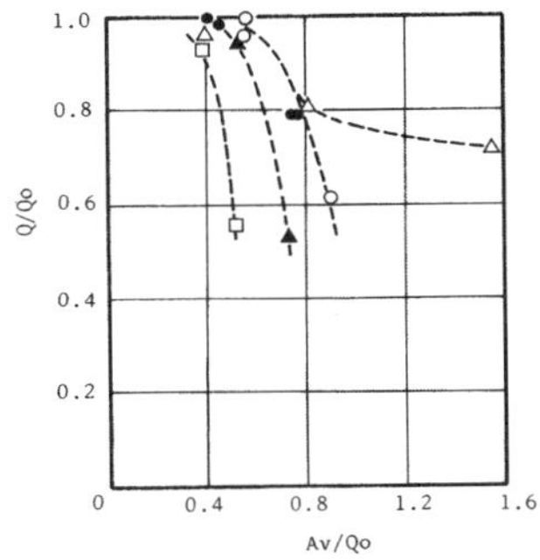

mark	water jet (m^3/h)	rotor (rpm)
○	0	515
●	3.2	515
△	5.3	515
▲	0	350
□	0	260

Fig.3 Relation between swing speed and suction flow rate

3. Agitating Rotors and Dredging Characteristics

For the bed soil with low moisture content, the dredging performance of the suction mouth is greatly affected by the presence of agitating rotors and its shape. That is, in the test, without using a rotor, the pipeline was clogged when Av/Qo> 0.4; while at a low speed (Av/Qo = 0.4 or smaller), it was not clogged but only water was sucked. This is because of the great difference in fluidity between water and bed soil. The type of agitating rotor must be capable of mixing the bed soil with water entering into the mouth to produce mixture of high fluidity. Thus, the agitating rotor finally employed was of basket type.

By means of this rotor, the dredging performance was tremendously improved. That is, without using agitating rotors, clogging was observed at Av/Qo = 0.4 or over. However, as shown in *Fig. 3,* each time the number of revolutions of the agitating rotors was increased to 260 rpm, 350 rpm, and 515 rpm, the clogging speed moved in its higher direction.

4. Generation of Turbidity in Dredging

Only turbidity caused by the developed suction mouth in the test was limited on the contact surface between the suction mouth case and the bed soil, which could be almost prevented by the use of normal dredging speed.

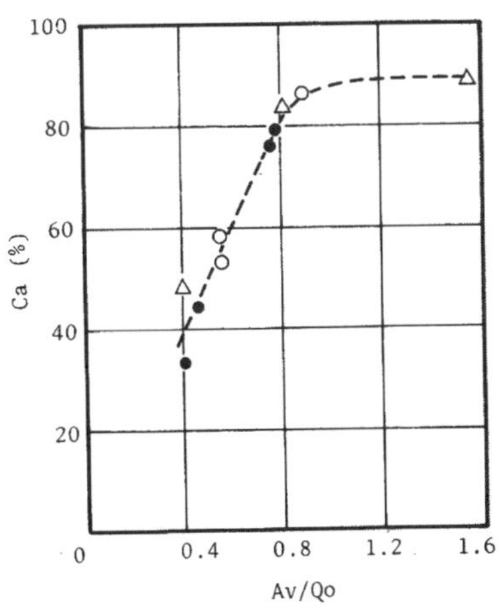

Fig.4 Relation between swing speed and apparent solid concentration

5. Summary of Laboratory Test

The suction mouth is capable of dredging at high solid concentration even without causing clogging. The moisture content of bed soil used in the test was lower than that of *ooze* deposited on the bed surface layer in the Japanese port and.harbor water. Therefore the test soil is harder than the usual ooze and is more difficult to dredge. However, dredging at the apparent solid concentration of 80 - 90% was even possible at an appropriate speed in the test. Therefore, in an actual dredging, it will be considered possible to perform dredging at a apparent solid concentration of 90% or over.

Ooze Dredge *KORYU*

An exclusive ooze dredge, *KORYU,* equipped with the suction mouth developed by this laboratory test and build by Mitsubishi Heavy industries, Ltd., Hiroshima Shipyard and Engine Works for the Ministry of Construction, is now operating at the Lake of Kasumigaura. *Fig. 5* shows *KORYU*..

From the results of the trial of *KORYU,* some of the measurements taken on the solid concentration are shown in *Table 3,* in which the values are given for the half swing. This test was con ducted for the purpose of making a comparison between the results of the laboratory test and those of the trial of an actual dredge. The apparent solid concentration of approximately 90% shown in both tests is in good agreement with each other. However, in the actual dredging, the average apparent solid concentration is approximately 25%, because of a great amount of water sucked at the end of a swing and at the point of re-driving of the spud, etc. This problem is discussed in the next Section.

Fig.5 Ooze dredge "Koryu"

Ooze Dredging Apparatus *DREX*

In case of actual dredging by the dredge, the average solid concentration is reduced due to suction of only water at the end of a swing and at the point of re-driving of the spud. Especially, when dredging ooze only deposited on the surface layer, the increased number of re-driving of the spud has greatly reduced the average solid concentration. An ooze dredging apparatus, *DREX,* as shown in *Fig. 6,* consists of a device installed on the end of the ladder which can move the high solid concentration suction mouth back and forth and can tremendously improve the average solid concentration.

Figs. 7 and *8* show the dredging efficiency in case of dredging only surface layer by means of the conventional method and *DREX,* respectively. The average dredging efficiency of approximately 40% by means of the conventional method is doubled to approximately 80% by means of *DREX.*

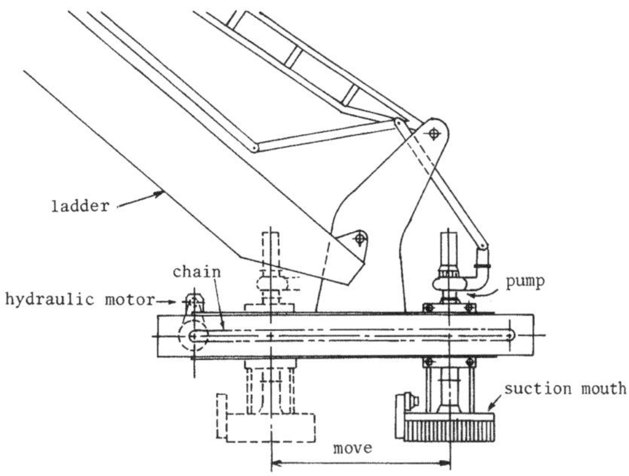

Fig.6 Schematic sketch of ooze dredging appatus "DREX"

By defining the average apparent solid concentration, Cam, by Eq. (2), let us consider Cam by means of the conventional method and *DREX,*

$$\text{Cam} = \alpha \bullet \text{Cao} \bullet \eta \qquad (2)$$

Where

α: ship. controlling coefficient (depending on the operator's skill and experience)

Cao: apparent solid concentration (when $\eta = 100\%$)

η: dredging efficiency (average value of those shown in *Figs. 7* and *8*)

First, Cam by means of the conventional method is obtained.

When

Cao = 0.9 from *Table 3,* $\eta = 0.4$

from *Fig. 7*, and making some allowance for the controlling coefficient, $\alpha = 0.8$; then

Cam = α.

Cao . $\eta = 0.8 \times 0.9 \times 0.4 = 0.29$

Cam is approximately 30%.

On the other hand, in case of *DREX,* when

$\eta = 0.8$ from Fig. 8.

Cam = 0.8 x 0.9 x 0.8 = 0.58

Cam is approximately 60%.

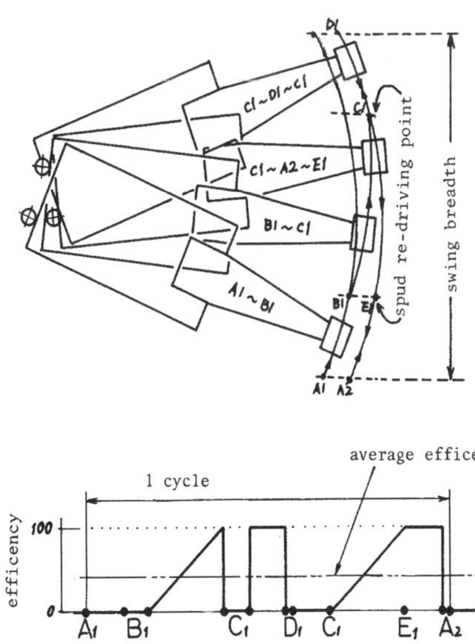

Fig.7 Dredging efficency by means of conventional method

Conclusions

The conclusion of this paper may be summarized, as follows:

1. The ooze dredge is required to be capable of:

   1) causing little turbidity during dredging,

   2) leaving no ooze behind, and

   3) dredging ooze at a high solid concentration

2. As a result of a number of labo-

ratory tests conducted, a new type of high solid concentration suction mouth has been developed. This suction mouth is provided with water jet nozzles and agitating rotors, capable of dredging ooze at the apparent solid concentration of 90% or higher if an appropriate dredging speed is used.

3. The high solid concentration suction mouth developed by the laboratory test was installed on an exclusive ooze dredge built by MHI and delivered to the Ministry of Construction. The apparent solid concentration of approximately 90% for a half swing obtained at the time of its trial showed a good agreement with that obtained in the laboratory test.

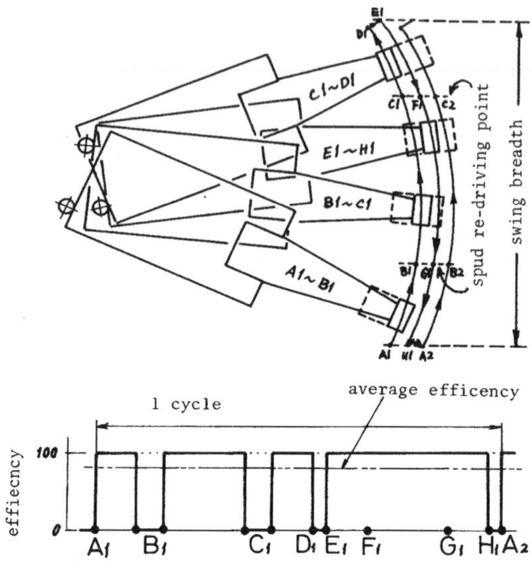

Fig.8 Dredging efficency by means of "DREX"

4. In case of actual dredging, even with the high solid concentration suction mouth, the average apparent solid concentration is reduced to approximately 30% due to suction of a great amount of water at the point of re-driving of the spud and at the end of swing. However, the use of the exclusive ooze dredging apparatus, *DREX,* consisting of a high solid concentration suction mouth and a device moving the mouth back and forth will improve the average apparent solid concentration to approximately 60%.

REFERENCES

1) Iwai, K. et al: *Development of a New Type Suction Mouth for Swing Type Mud Dredgers,* Mitsubishi Juko Giho, Vol. 15. No. 4, 1978 (in Japanese)

2) Kato, M.: *Ooze Dredge:,* Sagyo Kikai, June, 1979 (in Japanese)

3) MHI: *100 m^3/h Ooze Dredge "KORYU,"* Sagyosen, No. 118, 1978 (in Japanese)

4) Okayama, Y. et al: *A Study on the Drag Head Applied for Ooze,* Proc. of Third Int. Symp. on Dredging Technology, pp 421 - 436, 1980

5) Turner, T.M.: *The Bucket Wheel Hydraulic Dredge, The modern dredge mining tool,* World Dredging, February, 1977

Cleansing of Sludge & Dredged Material

After the founding of the DEME holding structure in 1991, the group went ahead with a further diversification in the field of pre- and post dredging activities - establishing and developing specialized companies, such as **Silt and Soils** for environmental treatment; **Hydro Soil Services (HSS)** which specialized in quay wall renovation and directional drilling; and the expert contractor for bituminous, environment-friendly embankments and capping of landfills, **Bitumar**.

In the course of time, these specialised contractors became respected players in their own right, both in Belgium and on foreign markets. Treatment plants for contaminated soil were built and technology for sludge handling, -recycling or -disposal was developed - attracting business from all over Europe. Bitumar sent its experts to **Bangladesh,** for the **World Bank** sponsored shore protection works at the **Jamuna Multipurpose Bridge. HSS** realized a *world-first* directional drilling between two offshore platforms at **Manslagt** in **North-Germany**. The same company carried out a *world-longest*, precarious directional drilling in the midst of the busy river **Scheldt** fairway. Its engineers received international admiration for in-house developed grouting techniques, which enabled port activity to continue during renovation of a quay-wall in the **Lithuanian port of Klaipeda**.

By the time the millennium had changed, the environmental activities of DEME had become so important that an umbrella structure was set up named ***DEME Environmental Contractors*** **(DEC),** which was the actual merger of the former DEME companies **SILT, SOILS, O.E.C.** and **Bitumar**. Specialized offshore contracts for the oil and gas industry became a significant activity, capitalizing on the core-business know-how in the field of pre- and post trenching, shore approach and landfall construction, pipe-pulling and installation, directional drilling, pipeline stabilization (stone dumping through **D.P. fall pipe vessels**).

This paper presents an overview of a program developed by **DEC Environmental Contractors,** Belgium. The company is designed to cope with the myriad of environmental problems faced by industry today. The emphasis is on applications of dredging technology to tackle these problems. "We focus the extensive environmental expertise and resources of our successful shareholding companies, DEME NV (Dredging, Environmental and Marine Engineering) and VMH NV (Flemish Environmental Holding)." *The paper is included in* ***APPENDIX B***.

Instrumentation of Environmental Dredgers

Introduction

Tighter controls and requirements have been introduced in recent years in environmental dredging projects, including: avoiding the re-suspension of fine grained material once it has settled; limiting the volume of material on disposal sites; storing excavated material within restricted boundaries and isolated from the surroundings to avoid leakage and diffusion of contaminants; continuing to operate in an economically viable way. These higher demands impact particularly on the accuracy, turbidity and spillage incurred in dredging. Innovations have been made to traditional dredgers to meet these tighter environmental parameters. These have included mechanical innovations and new instrumentation and automation to measure and precisely control a number of distinct variables within the dredging process.

The Development of the Sweep Dredger

Two cutter dredgers have been specially designed for environmental dredging, combining the characteristics of a classic draghead on a trailing suction dredger (high density and low turbidity) with the movement and pumping characteristics of a cutter suction dredger (accurate horizontal and vertical position, no rehandling and transport through a closed pipeline).

While hydraulic dredgers generally have the disadvantage of requiring 40-60 per cent water added (compared to 0-25 percent for mechanical dredgers) and a special disposal site facility, they do have certain advantages over mechanical dredgers that recommend them to different situations.For example, they offer greater efficiency since they are stationary and can dredge continuously. This in turn lowers the cost and also means that they require smaller crews. Trailer suction dredgers are particularly well suited to mud, silt, sand and generally to dredging loose materials. But since they have no cutting device they are less able to dredge gravel or well packed sand. Their pumping mechanisms also make them particularly sensitive to debris.

The Scoop Dredger

The Scoop dredger *BRABO* is an upgraded cutter suction dredger specially adapted for dredging contaminated materials. A two-sided functional draghead is mounted to allow dredging in two opposite swing directions. This is achieved in practice by using a turning blade that scrapes the material from the water bottom into the suction head of the dredger. At the end of the swing, the blade is turned in the opposite direction, the dredger walks forward (1.5 to 2.5m) and the draghead continues scraping in the opposite direction. It is because of this functional scraping that the draghead has been named the "scoophead."

The scoophead has no rotating cutting devices, so turbidity and the addition of transport water is minimised. At the same time no rotating parts means very low interference with debris in the dredging area and damage is almost impossible. The outer casing prevents dilution with water and the creation of turbidity in the surrounding water. On top of the scoophead, an adjustable water inlet system has been installed to prevent clogging and to ensure a regular feed of silt into the scoop dredger. Since this dredger can work at depths varying from 3 to 28 m, the angle of the scoophead to the ladder is adjustable with the help of a support frame.

The presence of gas bubbles in the dredged material provokes cavitation of the ladder pump and reduces the performance of the ladder and delivery pumps. Therefore, a specially designed degasification system was added to the basic Scoop dredger which draws off gas and some silt just before the ladder pump. This mixture is in a vacuum tank from where the silt is pumped back into the suction pipe. A vacuum is maintained in the vacuum tank by means of waterjets and a set of venturi nozzles.

Its first project (January to April 1993) involved the maintenance of the whole access channel to the **Kallolock, Antwerp**, removing 500,000 dry tonnes of silt. Further maintenance was carried out in the lock the following year removing 400,000 tonnes of dry silt and again in 1996/7 a further 900,000 tonnes.

From Scoop to Sweep Dredger

However the Scoop dredger has its limitations. Since the height of the dredge head opening is not adjustable, a great deal of water is pumped when dredging thin layers. In addition the Scoop dredger *Brabo* has a draught of 2.65 m and is thus more suited for applications in deeper water.

To improve the performance of the scoophead for clean-up dredging, attention was paid to:

- improving the removal of thin layers without excessive water addition during dredging;
- controlling the addition of transport water,
- reducing the minimal dredging depths;
- tuning in the dredged and pumped volumes;
- optimising the process steering (even computerising it);
- visualising and storing of all process parameters;
- implementing the latest technology developed in recent years with trailing suction hopper dredgers (*Antigoon, Pearl River*) in silt to cutter suction dredgers.

For the development of the new Scoop dredger, now referred to as the Sweep dredger, the following conditions were set out:

- the concept of the dredge head is based on the known current characteristics

of a classic trailerhead;
• the dredge head can be positioned with high accuracy, horizontally as well as vertically;
• turbidity generation is minimal;
• the thickness of the dredged layer is adjustable between 20 and 60 cm;
• a degasification system must be installed;
• addition of transport water is adjustable;
• the sweep-head is cutting mechanically avoiding (hydraulic) erosion;
• the suction (dredging) process can be separated from the reclaiming process.

Based on the above-mentioned requirements a classic stationary dredger equipped with a spud carrier and dredging by swinging was again chosen. In this way a controllable and steerable positioning of the dredge head was made possible. Furthermore it is now possible to control and steer the dredged volumes (layer thickness and swing speed).

The Sweep Dredger

The existing cutter dredger *Vlaanderen XV* has been rebuilt to become a Sweep dredger. The most important change was the removal of the cutter and the installation of the sweephead. Further, the dredger was equipped with a purpose-built silt degasification system.

The *Vlaanderen XV* has a total installed power of 2,550 kW and overall dimensions of 52 m x 11 m with a draught of 1.75 m. The maximum cutwidth

Figure 1: The sweep dredger ideal for the removal of thin polluted silt layers.

equals 80 m and the maximum swing speed equals 50 m/min. The principal characteristics of the sweephead can be deduced from Figure 3.

A visor rather than a turning blade (Scoop dredger) visor make it possible to operate in the two opposite swing directions. The same visor makes it possible to adjust the cut height. The transport of the dredged material is hydraulic.

In November 1995, the new sweep dredger *Vlaanderen XV* started its first trials on the Brussels Sea at Hingene (below). Maintenance dredging works are being carried out for **N V Zeekanaal** (Flemish Community).

The test programme involved the evaluation of swing-speed, production, and automatisation, and the specific environmental aspect concerning clean-up dredging, such as turbidity generation, spillage, accuracy and selectivity. The results were very good. From March 1996 through May 1996 the Sweep dredger performed the maintenance dredging works in the **Port of Nieuwpoort.** In June the Sweep dredger moved to the **Ketelmeer** in The Netherlands, where very elaborate tests were executed for the Dutch Government. The quality of the Sweep dredgers performance meant the maintenance works in the Port of Nieuwpoort, were resumed with the same dredging equipment. In Australia (Brisbane) another cutter suction dredger *The Bilba* was rebuilt and mounted with the Sweephead for a clean-up dredging operation of 650,000 cu m.

Accuracy of Dredging Operations

High accuracy dredging is now considered essential for successful execution of clean-up projects. It leads to less volume being transported, sorted and treated and more effective removal of contaminated sediments. Furthermore, it generates less turbidity and spillage and consequently produces better end results. Finally, it brings down overall project costs.

In order to obtain extreme accuracy in dredging, the major system used by Dredging International is the **DGP Kart** (Kinetic Differential Global Positioning System). This on-board system uses satellites and a ground station at a known location. At the ground station, corrections are calculated by phase measurements and sent to the dredger via a
Radio link. This is used to give the dredging point on the x, y and z axes to an accuracy of 5 cm.

Instrumentation on the Sweep Dredger

New instrumentation has been introduced on the Sweep dredger to optimise the use of the head. These instruments measure four parameters of particular importance in environmental dredging - turbidity, layer thickness, dredging depth and position:

1. **Two turbidity sensors** - mounted on each side of the Sweephead. They can

measure the generated turbidity in the dredging area before and after removal of the soil. These two sensors have to be calibrated for the specific material being dredged. The result of this measurement is the amount of solids in suspension expressed in mg/1.

2. Three altimeters - two mounted on the Sweephead, one mounted on the fixed cutting edge, the second on the moveable cutting edge. This provides an on-line verification of the dredged layer thickness of each swing. For extra control, the third is placed behind the head to measure the depth of the former swing. This generates a delivered profile to compare with the design profile.

3. Three dredging depth measurement systems - ensuring accuracy through comparison against one another:

i. **Measurement of** - **the ladder angle**
- **the trunion draught**
- **tide gauge**

The ladder angle gives the inclination of the ladder. The length of the ladder is fixed so the exact depth of the excavation point can be calculated. The trunion draught gives the position of the ladder rotation point relative to the water surface. The tide gauge gives the level of the water surface referring to the fixed reference level (e.g. NAP, TAW).

ii. **Pressure transducers and tide gauge**. A pressure transducer gives the exact depth of the Sweephead under the water surface. The tide gauge gives the level of the water surface to the reference level (NAP, TAW).

iii. **DGPS-Kart**. This system calculates the vertical position of the antenna to an accuracy within 3.0 cm.

4. The Digital Terrain Model databases - two databases are installed. One contains the depth of the top of the contaminated layer which was received by means of an echo sound survey. The second contains the depth of the bottom of the contaminated layer which was received by means of drilling different samples. Using these databases ensures only the contaminated soil layer is removed. During operations the position of the head is measured accurately using the DGPS-Kart system. Once the position is known, the system calculates from the database the set-points for dredging depth (depth of the cutting edge which follows the bottom of the contaminated layer) and bottom depth (depth of the bottom following edge which follows the top of the contaminated layer).

The automated Control System on Sweep Dredgers

The early deployment of the Scoop dredger revealed that two crucial elements are required in the complete environmental dredging system. One consists of the mechanistic developments required to dredge contaminated silt layers. The second, however, is automated control. The number of measurements and quantity of data required to control the dredging process sufficiently to achieve environmental targets, can exceed the operational capabilities of manual control. This is especially true of the Sweephead which includes additional variables, such an adjustable cut height.

All information from the Sweep dredger which can include in-situ silt levels, water content and silt density, generated turbidity, desired concentration, production and accuracy, is now managed in a highly automated steering system and displayed on a computer screen. This enables the dredge master to see at a glance, exactly where and how to dredge, and the works manager also has a clear view at all times of all the parameters of the dredging process. This is the in-house developed **Dredging International Positioning System**, linked with the dredger's in-house specified, automated steering system.

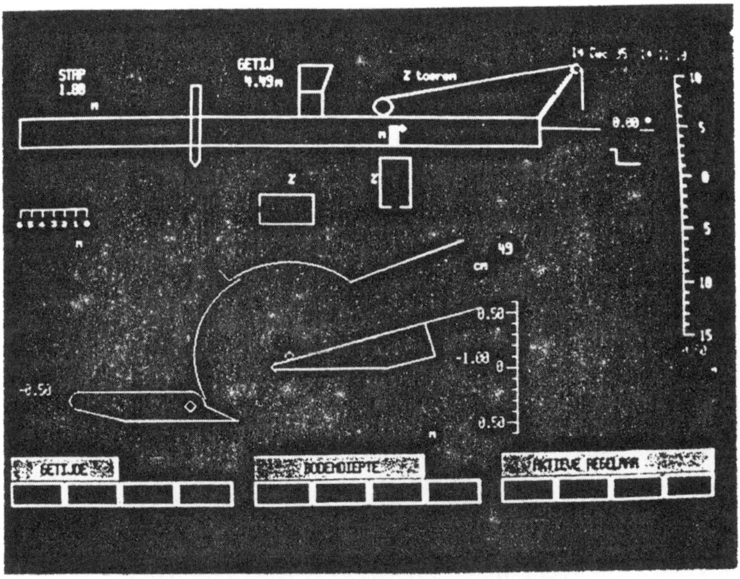

Figure 2: A single computer screen displays the automated control system.

The automated system consists of five computerised control mechanisms, all of which appear on a single VDU screen:

1. Sweephead cylinder speed control - controls the Sweep head angle by means of a hydraulic cylinder mounted on the ladder. The Sweep head must be kept horizontal at all times.

2. Design depth and cutting depth controller - ensures the layer thickness being cut corresponds with the design layer thickness required.

i. set points: set points for this controller are generated by accurate position measurement of the sweep head in the x and y plane and by using the two digital terrain models which generate set points for the depth of the top and the bottom of the contaminated layer.

ii. feed back: depth of the fixed cutting edge is measured accurately using three different methods *(see 4.3)*, depth of the movable edge is measured by offsetting the fixed cutting edge depth with the depth of cut measured by means of the visor cylinder position.

The controller:

i. ensures the cutting edge follows the bottom of the contaminated silt layer. The cutting edge can be either the
fixed edge (during PS swing) or the movable edge (during SB swing).

ii. ensures the bottom following edge follows the top of the contaminated silt layer. The bottom following edge can be either the fixed edge (during SB swing) or the movable edge (during PS swing). The depth of the fixed edge is controlled by means of the ladder winch, the depth of the movable edge is controlled by means of a combination of ladder winch control and visor cylinder position control.

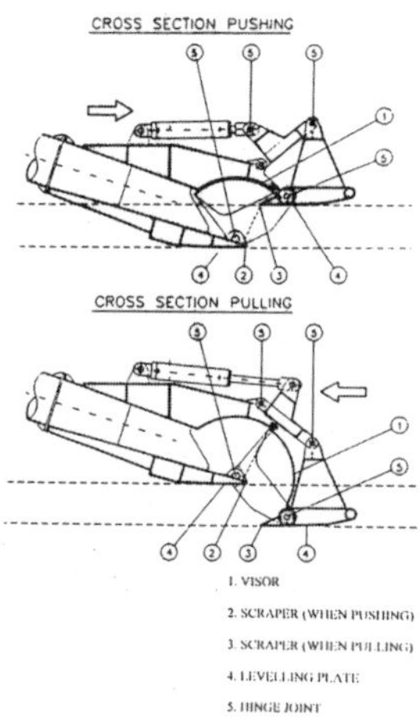

Figure 3: ***Controlling the fixed and moveable edges determines the cut height.***

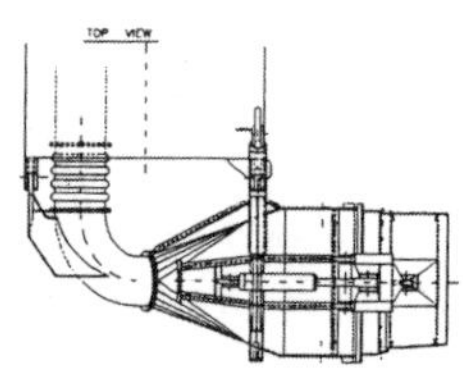

The sweep head (general lay-out)

3. Production in-situ controller - controls the amount of material being removed in cu m/hour. To achieve this it takes in measurements on:

i. Visor cylinder position giving the depth of cut (thickness);

ii. Spud carriage position to calculate the step (usually each spud carriage consists of three steps). The spud carriage is 5.6 m and the step 1.8 m;

iii. Heading, using a gyrocompass to calculate the swing speed. From changes in the heading the swing speed is calculated;

iv. Turbidity. A maximum value of the turbidity is set. If this set point is reached, the swing speed is adjusted to lower the turbidity .

The hauling winch speed control and veering winch brake control are operated through this controller to optimise the swing speed and therefore the production rate without exceeding required turbidity levels.

4. Mixture density controller. Since the actual production is known by measuring the depth of cut (layer thickness), the step and the swing speed, it is possible to control the amount of water added to the silt and therefore to control the mixture density by adjusting the mixture flow. This can be achieved by means of pump speed control, and in this specific case, since the pump is driven by a diesel engine directly, by a discharge valve position control in case a lower mixture density is required. Also when the swing speed is reduced i.e. at the end of cut, the mixture flow is decreased proportionately. This control ensures that the precise density is achieved and the amount of added water is minimised for storage purposes.

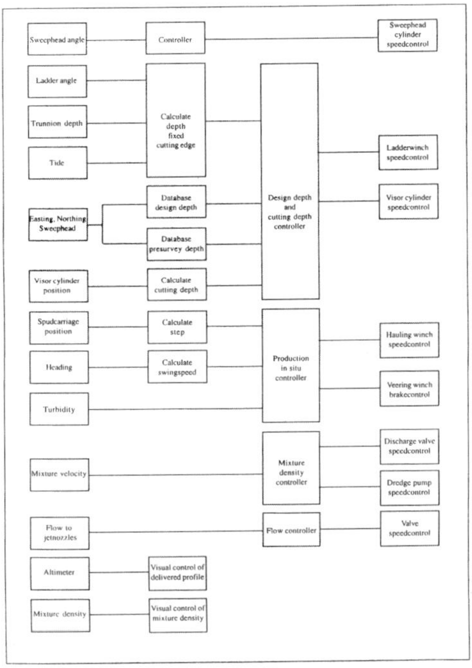

Figure 4: ***Five computerised controllers manage the dredging operation.***

5. Flow controller - valve speed control ensures that jet water (six Bar) nozzles in the Sweep head do not add too much water. The jets ensure a good mixture inside the head. If too much water is added the system takes measures to reduce the water added.

The VDU screen also includes visuals from the altimeter measurements giving the delivered profile and visual control of the mixture density.

Conclusion

New stringent criteria and requirements for environmental dredging projects have led to the development of a number of specially adapted dredgers, capable of removing polluted materials from harbour areas and waterways. The new dredging systems require the precise control of a number of variables. This has led to the introduction of a range of new instrumentation on the Sweep dredger to measure these variables. This has resulted in a dramatic increase in the amount of data being managed during the dredging process. Automation has been added to process this data and maintain precise control of the dredging operation at all times.

S. Vandycke, Dredging International NV'
J. Lefever, Dredging International NV'

CHAPTER 10

DREDGE MINING

Precious Minerals

Alluvial Gold is the primary precious mineral for which dredging is practiced today. However, with the depression of gold prices, alluvial gold dredging has diminished greatly. It only survives where the equipment is fully amortized and the deposit well proven. The BL/M dredge has proven to be the only effective type of dredge for mining placer/alluvial gold due to the need for uniform cleaning of bedrock and recovery of fine particles of gold. (Note: Dredging for gold is necessarily practiced in un-consolidated material, such as alluvial or eluvial deposits, referred to as ***placers***.)

Should the price of gold return to a reasonable level there could be a revival of gold dredging. There are many areas in the world where *placer* gold could still be mined should the price rise to $350/Troy oz. However, prior to mobilization of a dredge, every prospective deposit should be:

1) Evaluated by competent drilling and sampling methods, combined with appropriate calculations of the proven reserves; fundamental economics must apply;

2) Average values for a deposit should be determined by calculating gold content in all the material to be mined, from bedrock to surface;

3) The cost of excavating and processing the mine material must be less than the values recovered.

Artisan mining of alluvial gold by hand has and will continue through time. It is usually conducted by illegal miners in under-developed countries; estimated at three million persons. They may employ simple suction dredges but usually conduct their mining by hand with shovel and sluice. *(See Photo Plate 12)*

Dredge Mining for Diamond

Diamond mining with high volume (BL/M) dredges, was first realized in the 1960s with the invention of the ***Cleaveland Circular Jig*** (**Norman Cleaveland**). His mineral jig was installed on **five (5) BL/M dredges** mining alluvial diamonds in **Minas Gerais, Brazil**. The function of mineral jigs in a mining dredge after classification of the ore, is to perform the separation and concentration to the extent that the majority of the target mineral is recovered before being lost overboard into tailings.

In the 1950s, crude methods to mine diamond in rough seas offshore **SW Africa,** were conducted from barges and some equipment was lost as a result of those conditions. The primary organizer of these efforts were by a **S.V. *Sammy* Collins** and his company, The Marine Diamond Corp. In 1961, a grab samle of 4.5 tons was recovered by airlift pump containing 9.0 carats of diamonds. This was taken from a concession offshore near Luderitz, between the mouth of the Orange River to Diaz Pt. within the 3.0 mile limit.

Dredging experts were called upon to evaluate and develop methods to recover diamond in those severe conditions. The result, was the development of specialized self-propelled vessels, equipped with airlift pumps and elaborate processing and screening systems for final recovery of the diamonds. Experiments with bottom-crawling tractor vehicles have been made and reported to be successful. The capital cost for this equipment is high but the quality and quantity of diamond has apparently justified the investment.

Offshore **Namibia** has become one of the leading gem-quality, diamond producing areas contributing some 30% of world output. About 98% of the diamonds that are mined are of gem quality, according to company reports. In 1996, diamond production was **1,486M carats**. The leading diamond mining company with an output of around **1.3M carats** is **Namdeb Diamond Corporation,** created in **1994** by **De Beers**. The Namibian government has a 50% shareholding. In 1996 total production increased by 1.2% over the previous year.

In 1998, Namdeb commissioned a **Bucket Wheel Suction (BWS)** dredge and floating treatment plant, to operate on the inland area of the **Orange River** at a cost of **US$40M**. Their plan was to increase its full time production fleet (offshore) to five ships. A new vessel, ***Voyager,*** will operate in the deep waters off Namibia's coast. It will make use of a 120-t remote controlled, seabed crawler.

In 1995, ***Debmarine*** recovered **457,397 carats**, 34% of total output. **De Beers Marine** commissioned the ***Debmar Pacific*** in 1998 as a drill ship. They also converted the ***Coral Sea*** from a production vessel to a dual purpose, evaluation sampling and production vessel the same year. The vessel ***Zealous*** carried out geo-survey work during 1997 and deployed a two-person submersible in Namibian waters.

Namibian Minerals Corporation (NAMCO), was established in **1991** and is the second largest holder of marine diamond concessions in Namibia. NAMCO holds exclusive prospecting rights over five marine concessions totaling **5600 km^2** off the west coasts of Namibia, at **Luderitz, Hottentots Bay** and **South Africa**.

NAMCO has an output of about **750,000 carats** per year. Most of their discoveries are marine diamond and more than **95%** are gem quality. Their **120-t seabed crawling machine,** called ***NamSSol*** has gone into production and is expected to recover **1.0M m^3** of seabed material, mining an area of **0.3-0.4 km^2** each year. Its companion processing vessel, ***SS Kovambo*** was fitted with a **50-t/hr** processing plant with an annual full production target of **150,000 carats**. *(See Photo Plate 11).*

Industrial Minerals

The minerals of **tin, titanium, sand & gravel, phosphates**, and other peripheral mining applications such as concentrates of copper, coal, and iron, are in the family of Industrial Minerals that are mined with various types of dredges. About every type of dredge is used in various industrial mineral mining.

Production of sand and gravel constitutes about 40% of the construction aggregates produced in the US. The remainder is produced from crushed stone. The top states that produce S&G, in descending order are: California, Texas, Michigan, Arizona,

Ohio, Colorado, Washington, Utah, Minnesota and Wisconsin. About 4,000 companies with about 6,200 operations produced S&G in 2000.

Large volumes of Sand and Gravel have been dredge-mined in the **English Channel** for years. Most dredges are of the trailing suction (TH/M) type. The production serves both sides of the Channel, including UK and Europe. There have been as many as 60 such dredges operating in the channel for S&G in the past. The procedure is to mine and fill the hoppers then proceed to off-loading stations ashore. This system has been a model for the massive reclamation dredging projects in SE Asia; in Hong Kong and Singapore.

Other dredges used effectively for industrial materials such as oyster shells in the Gulf of Mexico have included cutter suction (CS), plain suction (S), and bucket clamshell (BC) dredges. The BG/C dredge has been popular for S&G dredge mining in the US and Europe.

The BL/M dredge was introduced into tin mining in Thailand and Malaysia in the early 1900s, after its development with placer gold mining. The alluvial gold processing plant with mineral jigs, is adaptable to tin except for the last stage of recovery. It continues to be the primary type of dredge mining tin in SE Asia. The largest fleet of BL/M dredges mining tin is owned by **PT Timah**, operating offshore from **Bangka Island, Indonesia**. There are **30** mining dredges varying in individual bucket capacity from **7.0 to 24.0 ft.**[3]

In the 1970s, the bucket wheel suction (BWS/M) dredge was introduced into newly discovered cassiterite (tin) fields in Brazil. Because of the large volume and rich values of the tin reserves, production from Brazil became the largest in the world for a period of time. The BWS/M dredging system has also been used mining mineral sands in Australia and South Africa and other areas including the east coast of the U.S. (i.e., New Jersey, South Carolina, and Florida).

India

The first minerals sand production in India was **c.1911**, at **Manavalakurichi** in **Tamil Nadu,** southern India**.** These are based upon large coastal deposits of **monozite. Ilmenite** was first mined and shipped in 1922 with the development of the **Chavara** deposits near **Quilon,** near the state of **Kerala**. The mineral sands industry is dominated by **Indian Rare Earths (IRE),** which accounts for 95% of output. Dredging of industrial minerals is mined with cutter suction or plain suction dredges.

CHAPTER 11

TECHNOLOGY DEVELOPMENT

Dredge Types

Accomplishments of the dredging industry during the 20th Century, would not have been possible without advancements in a variety of technologies. As with many older industries, technology transfer has contributed greatly to improvements in the design and operational capabilities of modern dredges. Civil engineering has likewise made advancements that have contributed to the design, evaluation and project management of dredging projects. Some of the significant equipment and technology improvements are listed as follows.

Suction/Hydraulic Dredges

The suction/hydraulic dredge system gained popularity from its combined function of excavation and transportation of material. Various innovations have amplified its advantages which are apparent from the following configurations. Each type of suction dredge has evolved into new productive systems through the adaptation of features including: cutting, agitating, submerged pumps and jet systems, to different types and sizes of pumps.

Cutter Suction Hydraulic Dredge (CS)
Dredge *Triton*
42" CS, 30,000 hp Steam turbine
Digging depth: 100 ft

CS Dredge "Triton."

One of the largest CS dredge, was the ***Triton,*** built in the 1960s to dredge a new channel in Florida. The unique aspects of the dredge included a 46 in. suction, 12 ft diameter cutter head, digging depth of 100 ft. The pump discharge diameter was 42 in. and built by **Mobile Pulley & Machine Works**, with production in hard material of **8,000 yd³/hr**.

Total installed power was a **30,000 hp steam turbine**, a WWII US Navy destroyer power plant that was purchased from Navy surplus. The dredge barge was built around the section of hull from the destroyer containing the steam plant.

The ***Triton*** performed a major dredging project building the new **port of Tuburao, Vitoria, Brazil** in the early 1970s. The depth in the berthing area was specified at 26.3 m through rock and hard clay along with alluvial materials. One of the problems included pinnacles of granite that were present but not detected in a pre-dredging survey. The cutter teeth were often changed every 15 minutes as they attempted to dig the granite.

Other problems included periodic swells, coming into the "sheltered" harbor up to 10 ft in height (in a few cases measured at 14 ft). As a result of the swells, running at 7 to 8 second frequency, the spuds were removed from the dredge and an A-frame/*Christmas Tree* system was used with 1-5/8" cables spread out and attached to 30,000 lb. anchors.

The total volume of dredging in the harbor was about 8.0M m^3, with approximately 25 % in rock. As the dredge dug into the soil, it found that in a single sweep the material could change from sand to hard clay and into boulders of 30" or more. A number of cutter designs were tested, each one failing in its own way to dig the harder material. It was later decided to change reduction gears on the cutter drive in order to apply higher power at a lower rpm and to apply greater cutting forces without glazing the hard material. Digging depths were 80 ft on some cuts.

In a post analysis of the project and questioning of many dredging contractors worldwide, no one had ever dredged with a CS dredge under such extreme circumstances; and able to complete the project. It was said that "the ***Triton*** may go down in history, as having pioneered the way for new design concepts to apply to dredging under extreme weather conditions." The project constructed an iron ore export port with the ore being transported from a mine some 200 miles inland to a pelletizing mill. From there it was loaded onto bulk carriers, shipping the iron to Japan.

The owner of the ***Triton*** was **US Industries, Inc.,** working in a joint venture with **Bauer Dredging Co./Texas Gas Transmission, Inc**. The project also involved dredging of an outer channel using the TH dredge ***Delta Bay,*** owned by **COBLA (Costain-Blankevoort Intl Dredging Co. Ltd.),** The Netherlands.

Dredge *Florida*
36" CS, 16,115 hp
Digging depth: 100 ft

The CS dredge ***Florida,*** owned by **Great Lakes Dredge & Dock Co**., includes a 36" discharge, total power of 16,115 hp, and is one of the most powerful CS dredges in the US. It has operated on the West Coast for many years, upgraded several times (i.e., submerged pump, gears and other equipment for modernization) and performed the primary dredging and reclamation of the Port of Los Angeles in the 1990s. It uses shore electric power to meet local air emission restrictions. Equipped with a submerged pump on the ladder, its production allows dredging at depths of 100 ft, discharging long distances for reclamation.

The dredge *Florida* was originally built in 1954 by Ellicott Machine Corp. At the same time, Ellicott built a *sister* 36" CS dredge, for the **Steep Rock Iron Ore Mine, Canada**. It was shipped to California several years later and has continued to be operated under different companies, dredge names, conversions and locations. This illustrates the versatility of the CS dredge to extend its life and change its configuration.

Dredge *Mashhour*
36" CS, 22,795 hp
Digging depth: 115 ft

A Cutter Suction (CS) Dredge supplied to the **Suez Canal Authority (SCA)** in 1996, by **IHC Holland**. The total installed power is **22,795 hp, with a 900 mm (36 in.)** pump and discharge. Digging depth is 35 m (115 ft).

This was the **32nd dredge built by IHC** and former companies for SCA since c.1901. It was purchased to work on the widening and deepening of the Suez Canal, to accommodate 250,000 TDWT vessels. The *Mashhour* can work in all sorts of soil, including rock, abrasive sand and sticky clay. Combining all types of material dredge production has averaged over **538,000 yd^3 per month**.

Some of the equipment is also impressive:

- Ladder pump motor: 2-1,200 kW electric motors on the upper end of the 1,150 t ladder;
- 4-six bladed cutter heads;
- 2-1,200 kW electric motors driving the 2,400 kW submerged single walled pump;
- 4-Diesel engines 2,650 kW at 600 rpm.
- 2-double-walled, 5,400 kW dredge pumps in series, powered by 2-turbocharged after-cooled 4-stroke diesels, with 5,400 kW at 500 rpm boosting delivery to 6 km with 900 mm pipeline.

Dredge *Leonardo Da Vinci*
35" CS, 27,524 hp
Dredge *Marco Polo*
35" CS, 21,913 hp

Both are owned by **Jan De Nul Group**, Belgium and used overseas for pipeline trenching and port construction projects.

Dredge *Vlaanderen XIX*
36" CS, 15,950 hp
Dredge *Vlaanderen XI*
32" CS, 13,400 hp

Trailing Suction Hopper Dredgers Designed for Improved DP/DT Capabilities A concept for next generation hopper dredgers.

Results of trials with knowledge based DP/DT system for dredgers.

In this introduction a short definition of DP/DT and a historical review of its evolution with regard to equipment and operation are given. DP/DT, short for Dynamic Positioning and Dynamic Tracking, are operational modes by which the ship, although affected by constantly changing currents, wind and waves, can be accurately kept in or moved to a fixed position (DP), or kept on an accurate track (DT). For this special control equipment is needed, called DP/DT-systems.

In the late 1960's, offshore drilling vessels were designed and built for operations in even deeper waters, even where they could no longer be held in position by mooring systems. These drilling vessels therefore got sets of dynamically controlled omni-directional thrusters. Automatic pilot systems had to be adapted for keeping the vessel exactly in its drilling position. The DP system was born.

Deeper exploration also needed larger offshore structures, such as semi-submersible crane and drilling vessels and a varied auxiliary fleet, such as suppliers and shuttle tankers. All these vessels and smaller craft needed dynamic positioning systems.

Other offshore operations, like stone disposal and placement, pipe and cable laying, required an accurate forward movement of the ship along a track. Dynamic tracking was born and the DP control equipment was adapted to provide automatic control in DT-mode.

In the early 1990's trailing suction hopper dredgers needed DP/DT systems to qualify for offshore dredging oftrenches and back filling works. Three prime reasons for demanding DP/DT equipment: safety, accuracy for meeting the exact position specifications of offshore structures to be built and, last but not least, accurate dredging is efficient dredging, important for both contractor and client.

Up to now most DP/DT systems used on board dredgers are basically the same as those used for other offshore purposes. They are additional to the ship control equipment and only slightly adapted for dredging application, which has its disadvantages:

- commissioning and testing of the system during sea trials takes unduly long;
- if not used continuously, general purpose DP/DT systems are too complicated, which frustrates their operational use;
- accuracy is often doubtful.

Inadequate accuracy may have several causes:

- insufficient propulsive and transverse (bow) thruster power available, in other words:
- is the ship capable for accurate DP/DT operation, or:
- are the accuracy specifications possible with the ship?
- poor mathematical modelling of the ship (a mathematical model is part of the control algorithm of the DP/DT system);
- insufficient measuring of the ship's position and the fluctuating environmental conditions (wind, current), dredging forces, etc;

• badly tuned DP/DT control parameters.

This paper deals with above items. It demonstrates how a vessel can be specially designed for optimal DP/DT behaviour, and it demonstrates what can be achieved by a dedicated design for a DP/DT system for the dredging industry.

DP/DT Developments in the IHC Group

Before going into more detail, we have to explain why our group of companies has special interest in this subject.

From the early beginnings, **IHC Gusto Engineering** has been involved in developing DP/DT. First as the shipyard where *Pelican*, the world's first **DP-drilling vessel**, was designed and built, followed by many similar offshore vessels. Later, after Gusto had stopped actually building ships, it evolved to become a major engineering company for the offshore industry, highly skilled in design of floating structures such as drilling ships, pipe-laying vessels, crane barges, etc. All these vessels are 'all-electric' ships with one or two central power plants. The power plants feed the operational equipment (like cranes, drilling tools or pipe-laying equipment), the propulsion units (used for dynamic positioning) and the hotel and ship-services.

In 1994/'95 IHC Holland designed and built the trailing suction hopper dredger '*Pearl River*', the first of the so called *Jumbo* hopper dredgers, for DEME of Belgium. This vessel was equipped with a LIPS/IHC Systems DP/DT system. While developing this, the need surfaced for a specially adapted DP/DT system that would meet the high accuracy requirements for DT-dredging. New algorithms had to be developed in order to predict current, which is complicated by the restricted water depth at which dredging usually takes place, and to cope with trailing forces, even without direct measuring. Also the DP/DT capability of the dredger was studied in order to determine the balance of design specifications, environmental conditions and accuracy expectations.

In the wake of *Pearl River's* DP/DT system, priority was given to further development of a dedicated dredging DP/DT system, DP/DT capability prediction and DP/DT ship design tools. As a result of this, IHC Systems designed a new DP/DT system, highly integrated into the ship's control systems for navigation, power management and dredging. 'Dedicated for dredging' implies that only functionality related to dredging operation is applied. In particular much effort has been devoted to designing a user-friendly man-machine interface. At the moment of writing this paper, the new DP/DT System has been commissioned and tested on board the TSHDs *Lang Wapper* and *Queen of Penta-Ocean.* The first experiences and test results will be presented and discussed in this paper, together with an explanation of system philosophy and special features.

Contrary to common practise in offshore, DP/DT capability has up to now rarely been specified in hopper dredger design. Current developments in dredging may indicate that in the near future more emphasis will be put on accurate positioning and dredging than before, which will be translated into design specifications. For this, a two steps study will be presented showing how the DP/DT capability can be upgraded by adapted ship design.

Dredger Design Study
General

In each ship design a balance must be found between desired and attainable properties (specifications). The design study, presented in this paper, is demonstrating step by step the effect of adapting parameters, aimed at increasing position and tracking capabilities.

Before starting this study, a reference ship had to be selected, best suited for being a starting point for the study. The desired reference ship should be a modern and powerful dredger, with excellent, not easily improved, manoeuvring capabilities.

The recently built medium large trailing suction hopper dredger *Lange Wapper* of Dredging International (DEME) was the obvious choice. Designed and built within our group, all necessary data were easily available. Moreover, the ship is equipped with a *Knowledge based DP/DT system*, for which extensive capability studies and sea trials have been carried out already.

Basic design, trailing suction hopper dredger *Lange Wapper*

This dredger is of modern design, with high loading capacity at relative shallow draught. As can be seen from the main particulars, the length to beam ratio L/B is only 4,55 and the beam to draught ratio BIT at design draught of 8 [m] is as high as 3.35.

Main particulars:

Length o.a.	129.8	[m]
Length pp	122.1	[m]
Beam	26.8	[m]
Depth	10.9	[m]
Draught int. freeb.	7.88	[m]
Draught dredging	8.0/9.0	[m]
Hopper volume	13700	[m^3]
Deadweight	20000	[ton]
Light ship weight	7350	[ton]
Sailing speed	16.0	[kn]
1 suction tube	1200	[mm]

Fig 1. DI's *'Lange Wapper'*

With a pump room aft, dredge pumps directly driven by the main engines and a single dredge pipe, the overall layout is very cost effective, with a high Deadweight to Displacement ratio (DWT/D). (For a General Arrangement Plan see Annex 1)

Lange Wapper has two main engines of 5,760 [kW] each, both driving a controllable pitch propeller (CPP). Two flap rudders, two bow thrusters of each 525 [kW] and an additional bow jet system of 1,500 [kW] ensure excellent manoeuvrability. Combined with a high sailing speed thanks to the above mentioned hull dimensions and the described propulsion system, very good DP/DT behaviour is experienced.

First alternative design, *Lange Wapper* with azimuth thrusters

In the first alternative design was researched if further improvements of manoeuvrability could be achieved by replacing the conventional CPP propellers and rudders by azimuthing thrusters, while otherwise maintaining the same overall layout as in *Lange Wapper*. Azimuthing thrusters are propeller units that can rotate in a horizontal plane, and can deliver thrust to any direction. Those thrusters are used in the offshore for propulsion (and replace a rudder) and for DP/DT purposes. The existing engine room, however, lacked space for directly driving azimuthing thrusters, so an electric alternative was chosen *(see Annex 1).*

Second alternative design, DP/DT dredger design

The second alternative design was a completely new concept based on an all=electric ship philosophy, to reach the optimum DP/DT performance. An internal development study resulted in a DP/DT dredger which hull is derived from a well-intervention/well-servicing offshore vessel. The dredging arrangement See (see Annex 2) is based on the reference dredger *Lange Wapper*.

A propulsion plant, thruster and thruster drive configuration with associated lay-outs was designed. This resulted into a relatively short engine room, positioned as far aft as possible, with three diesel-generator sets for optimum power distribution. The power of the bow tunnel thruster has been increased compared to the *Lange Wapper* design, and for reasons of better DT accuracy, the suction tube is placed in a forward position. Having the freedom in design, also the hopper and dredge-pump room layouts were changed. For this design, the dredgepump room is located forward of the hopper, shifting the centre of buoyancy of the hull aft, so reducing the differences in trim between fully loaded and empty vessel. It also gives a finer entry of the hull, which reduces wave resistance and pitching, with an increase of ship speed and reduced ship wash as a result.

Further advantage obtained from the short engine room and the dredge pump room further forward is that a longer hopper can be made. This is favourable because of its large surface area, which increases efficiency of the loading (settling) and unloading processes. Besides, the longer hopper gives a larger hopper volume. As the loading capacity will remain about the same, the specific mass of the hopper will be decreased, which generally is favourable for maintenance dredging. A longer hopper space also gives a more equal load distribution, with decreased light ship weight and an increased deadweight as results. This layout furthermore makes the dredge pump room more accessible for repair and maintenance work.

Besides a more flexible arrangement of the main components, advantage is tak-en from the forward location of the dredge pump room, which allows for a more flexible suction tube arrangement. As shown in a next section, also the DT accuracy is better for a suction tube arranged forward. Also with the suction tube located forward, relatively longer suction tubes can be accommodated, giving the possibility to dredge in deeper water. The all-electric design is provided with an inboard dredge pump and a booster pump, but can easily be equipped with a submerged dredge pump.

It will be clear that the investment costs for this *all-electric* design are substantially higher than for the basic version. This may be returned rapidly, however, through a more flexible arrangement and operation and higher accuracy, which results in more efficient dredging.

Power plant

The power plant consists of three medium-speed diesel-generator sets, each delivering approx. 4,000 [kW], at 1,000 [min $^{-1}$]. The choice for three diesel-generator sets increases redundancy. These faster diesels are smaller and lighter than the reference dredger's diesels, allowing a more compact engine room. Also the power of these diesels is relatively larger than that of the reference dredger, because those diesels were specially designed for driving the dredge pumps, resulting in relatively less power.

The medium-speed diesels can run on either HFO or MDO. MDO is used for starting up and for very accurate DP/DT operations, when the diesels must react more quickly on varying loads. HFO may be used when lower accuracy is required, as in transit conditions, normal dredging, or pumping ashore etc.

The generators feed the electric power to the high-voltage (6.6 [kV]) switchboard, which is located at main deck level, directly aft of the engine room. This switchboard feeds the high-voltage frequency converters, which control the speed of the electric motors of the thrusters and the dredge pumps.

The converters for the aft thrusters are located in a room directly aft of the engine room and adjacent to the thruster rooms. The converters for the fore thrusters and the dredge-pumps are located in a room directly after the forepeak, at tween-deck level.

Thrusters

The vessel has two 5,500 kW fixed pitch azimuthing thrusters aft for propulsion and steering, as well as for DP/DT operations.

Because these thrusters can be rotated freely in any direction, the DP performance and manoeuvrability is significantly better than the conventional arrangement with rudders.

The advantage of fixed pitch propellers (FPP) over controllable pitch propellers (CPP) is that FPP's are cheaper (however, this profit is largely compensated by the additional price of the frequency controllers), FPP's are less vulnerable and require less maintenance. Also, FPP's need less power at the lower thrust ranges, resulting in lower fuel consumption during operations in which thrusters are not fully loaded. (Note that FPP's cannot be used or have severe disadvantages when used in a diesel-direct propulsion system, as the rpm of the propeller than directly have to be controlled by rpm variations with the diesels).

The thrusters and motors are placed in one vertical arrangement, which makes it possible to easily remove or install (without dry-docking) the electric motor and thruster for maintenance and / or repair.

For DP/DT operations and for manoeuvring in ports, two fixed-pitch tunnel thrusters of 750 [kW] each are installed in the fore ship.

Pump room arrangement

The pump room, with one 3200 [kW] suction pump and one 5760 kW booster pump in series for pumping ashore, is located forward of the hopper. In case a submerged pump will be used, the frequency converter of the booster pump can be used to control this submerged pump.

The dredge pumps are each driven by an electric motor through a gearbox. The booster pump is located to port on the tween-deck. The forward arrangement gives a relatively short delivery pipeline to the shore connection, while allowing for easier extension of the suction tube. Because it remains well within the boundaries of the coaming deck, the view from the bridge on the suction tube is better.

The pump drives are from the constant power type, which leads to very flexible operation in different working modes and a higher pump out production in relation to conventional drives.

DP/DT Behaviour Comparison

General

In order to compare the DP/DT behaviour of *Lange Wapper* (LW), the alternative of *Lange Wapper* (LW) equipped with azimuthing thrusters and the new all-electric DP/DT dredger (DPD), a static DP analysis is carried out, resulting in capability plots for the three vessels. Also the dynamic behaviour of the vessels is studied by means of time-domain simulations. In the comparison, *Lange Wapper* and the alternative *Lange Wapper* have the suction tube located aft, whilst the DP/DT dredger has the suction tube located forward. For comparison of the effect of suction tube forward versus suction tube aft, the positioning accuracies are also given for the DP/DT dredger with suction tube located aft.

From the capability plots the maximum environmental conditions, which the vessel can withstand, can be seen for various directions of those forces.

The capability plots indicate whether a vessel can withstand a certain environment, but give no information on positioning accuracy. For this, the following time-domain simulations were carried out for all three vessel alternatives:

- DP simulations (shore pumping) in which the position and heading of the vessel were controlled;
- DT simulations (dredging) with free heading (ground speed 2.5 [knots]), in which the transverse track offset was controlled by controlling the heading of the vessel. (Steering is done both by rudders or azimuthing thrusters and by bow thrusters);

• DT simulations (disposal) with fixed heading (ground speed 1.0 [knot]) in which the forward speed was controlled together with the transverse offset and heading.

All simulations were done for an environment with wind speed fluctuating up to 10 [knots], current speed up to 1.5 [knots] and a wave height of 0.5 [m], and for a range of environmental directions.

DP capability and accuracy

The DP capability for the three vessels is expressed in capability plots given in *fig. 2*. In those plots, the wind and wave forces are kept equal, while the current velocity is varying. In this way, the figure shows the maximum current by which the vessel is capable of performing its DP tasks.

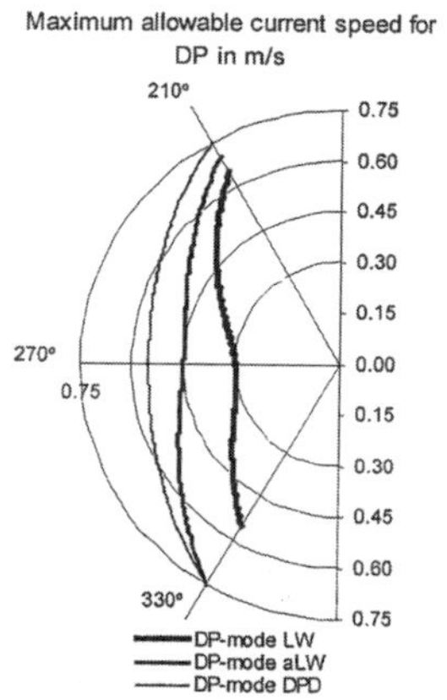

Fig 2: ***DP Capability***

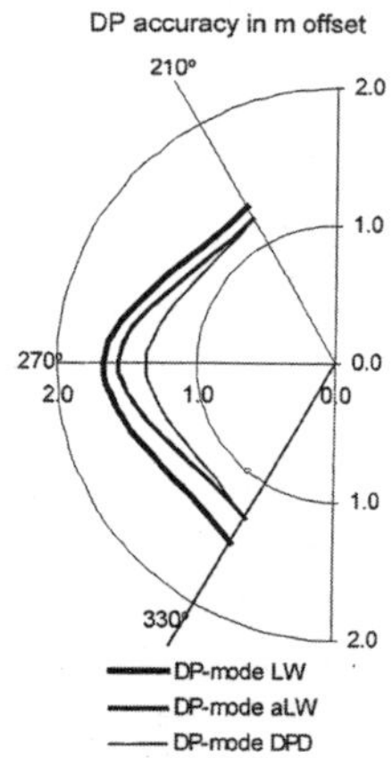

Fig 3: ***DP Accuracy***

As can be observed from this figure, *Lange Wapper* copes well with environmental forces on bow or stern, but, due to the conventional propeller/rudder arrangement's limited capacity of generating transverse thrust, has limited DP capability with forces on the quarter or the beam. The alternative *Lange Wapper* (LW, equipped with azimuthing units) has a larger capability for those directions, whilst the purpose designed DP/DT vessel has an even larger capability. This improvement is mainly due to the increased bow propulsion power and a more flexible power management; i.e. power is shifted from the azimuthing thrusters, which are more effective in transverse direction than conventional thrusters with rudders, to the bow propellers. Also the forward located suction tube is contributing positively to a more balanced transverse force distribution.

For the most demanding environmental conditions the positioning accuracy is determined and expressed inaccuracy plots, as given in *fig 3*. This figure shows that the positioning accuracy for vessels equipped with azimuthing thrusters is better than of vessels with conventional propulsion/rudder (accuracies given are for the maximum environmental conditions, which differ for all vessels).

For above maximum environmental conditions, the associated thrust and power levels are given in *table 1*. From this table, it catches the eye that for the off-stern environment (330[°]), *Lange Wapper* (LW) has one propellerrunning forward and one running astern, in order to generate transverse forces with the rudders. The vessel requires 2800 [kW] in an environment with 0.55 [m/s] current, while the DP/DT design (DPD) requires 1700 [kW] for compensating 0.75 [m/s] current. Further can be seen that the DP/DT design generally needs less thrust, while larger environmental forces are compensated.

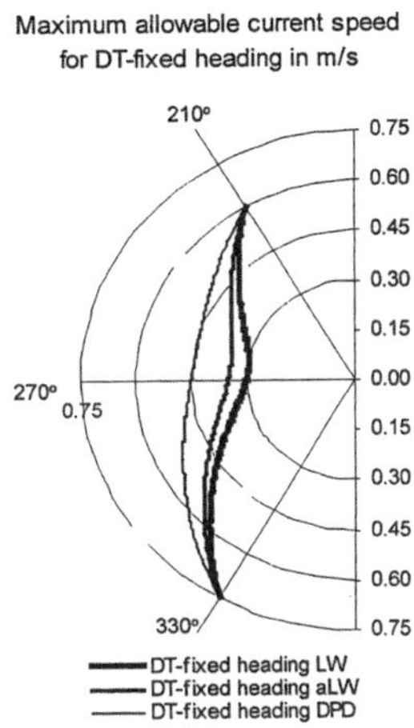

Fig 4: ***DT Capability***

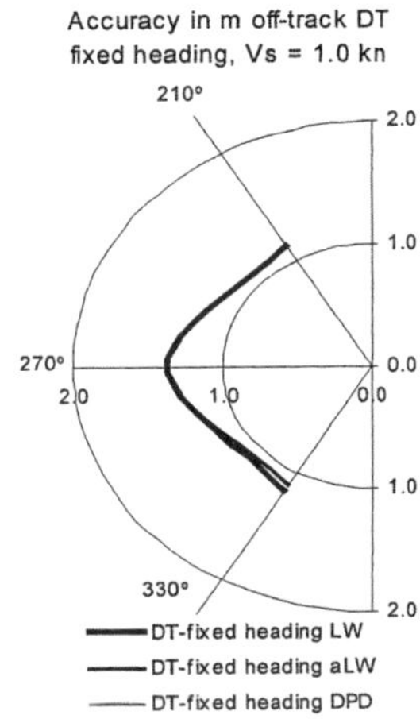

Fig 5: ***DT Accuracy***

DT capability and accuracy

Two DT modes were investigated, i.e. free heading (dredging) and fixed heading (disposal and dredging within defined tracks or at limited depth).

In case of free or optimal heading (depending on current, wind and dredge pipe forces, while dredging along desired track), all three vessel alternatives can in all cases compensate the maximum design current velocity of 0.75 [m/s]. Also the obtained accuracies differ slightly for vessels with the suction tube aft. In case of a design with a suction tube forward (as the DPD) the accuracy is improved (the steering moment is increased with the positive moment contribution of the suction tube in that position).

However, the delivered thrust values and associated required power are in all cases lower for the DPD than for the aLW, which in turn is lower than for LW.

Differences between the vessels are up to 1000 [kW], on a mean power consumption for the DPD of 3800 [kW], so resulting in a power saving of 20 %. This power saving can be used to increase both the dredging force at the draghead and the available power of the dredgepump, which allows more efficient dredging.

In case of fixed heading, the capability plot for the three vessels is given in *fig.4* and the associated accuracy in *fig 5*. Results are comparable to the DP capability, only less pronounced. A comparison of the required thrust levels is given in *table 2* and *3*.

Although capabilities and accuracies obtained here are similar for the three vessels, also here the DPD needs less power than the aLW, which in turn needs less power than LW. Differences are up to 50% less power for the condition when LW must thrust astern with one of the main propellers, as explained in above chapter on DP.

Table 1: ***Dynamic positioning***

	Unit	210 [°]			270 [°]			330 [°]		
		LW	aLW	DPD	LW	aLW	DPD	LW	aLW	DPD
Current speed	[m/s]	0.65	0.70	0.75	0.30	0.45	0.55	0.55	0.75	0.75
Longitudinal-offset	[m]	1.32	1.21	1.21	1.66	1.05	1.06	1.50	1.29	1.29
Transverse-offset	[m]	1.14	1.14	1.14	1.59	1.56	1.36	1.23	1.12	1.12
Thrust portside max.	[kN]	44	37	35	165	46	56	119	41	35
Thrust starboard max.	[kN]	46	39	38	-125	46	57	-121	41	35
Bow thrust max.	[kN]	44	50	58	50	57	80	34	32	30
Peak power (total)	[MW]	2.0	2.1	2.2	3.2	2.2	2.5	2.8	1.9	1.7

Table 2: ***Dynamic tracking, dredging with free heading (Vs = 2.5 [knl)***

	Unit	210 [°]				270 [°]				330 [°]			
		LW	aLW	DPD		LW	aLW	DPD		LW	aLW	DPD	
Pipe position		aft	aft	aft	fore	aft	aft	aft	fore	aft	aft	aft	fore
Current speed	[m/s]	0.75	0.75	0.75	0.75	0.75	0.75	0.75	0.75	0.75	0.75	0.75	0.75
Track offset	[m]	1.96	1.87	1.74	1.51	1.83	1.73	1.52	1.36	1.91	1.82	1.64	1.43
Heading max.	[deg]	6.3	6.3	6.8	5.4	19.5	19.5	20.2	15.3	6.4	6.2	6.7	3.7
Thrust portside max.	[kN]	541	484	432	436	436	397	365	372	367	328	322	323
Thrust starboard max.	[kN]	502	482	429	431	436	397	367	374	363	324	319	319
Bow thrust max.	[kN]	26	26	26	23	22	22	21	20	26	26	28	26
Peak power (total)	[MW]	5.8	5.3	4.8	4.8	4.5	4.0	3.7	3.7	3.6	3.1	3.0	3.0

Table 3: ***Dynamic tracking, disposal with fixed heading (Vs =1.0 [kn])***

	Unit	210 [°]			270 [°]			330 [°]		
		LW	aLW	DPD	LW	aLW	DPD	LW	aLW	DPD
Current speed	[m/s]	0.60	0.60	0.60	0.30	0.35	0.45	0.75	0.75	0.75
Track-offset	[m]	1.22	1.17	1.14	1.40	1.37	1.36	1.17	1.13	1.12
Thrust portside max.	[kN]	79	61	47	149	36	44	212	31	27
Thrust starboard max.	[kN]	81	64	51	-79	37	45	-178	31	26
Bow thrust max.	[kN]	52	51	55	51	53	75	48	35	33
Peak power (total)	[MW]	2.2	2.2	2.2	2.8	2.1	2.2	3.6	1.9	1.8

210[°] = environment 30[°] off bow
270[°] = beam on environment
330[°] = environment 30[°] off stern

Knowledge Based DP/DT-System for Dredgers

General

Most DP/DT-systems in use on board of dredgers have been designed for the offshore industry, with slight modifications for coping with the peculiar forces involved in dredging. This part of the paper reports the shortcomings of these systems, listed through interviews with almost the majority of the European dredging companies. One of the outcomes is that such DP/DT-systems, adapted but not fully optimised for dredging, are quite complex to use and need uneconomically extended sea-trials for proper tuning.

From this starting point, IHC Systems - in co-operation with half-sister company Imtech Marine & Industry - undertook to develop a *Knowledge based DP/DT-system for hopper suction dredgers*, which would have to be easy to use and tune, in order to increase efficiency and drastically cut back on trial time. The requirement *easy to use* could be met by studying what exactly a dredger needs to do and to equip the DP/DT-system with a Man-Machine Interface that concentrates on directing it to do just that. So, no unnecessary information is presented. Special DP/DT-control algorithms have also been designed to meet the peculiar requirements of dredging operations.

Far-reaching integration with dredging, navigation, survey and propulsion control equipment results in an integrated bridge, in which the DP/DT-system takes a central place, so that all actions related to a ship's movements can be performed through the DP/DT-system. To cut back tuning time on board, as well as to improve the DP/DT-control algorithms, advanced extensive computer ship models have been developed. IHC Gusto Engineering calculated for each ship the specific parameters that define the behaviour of these models, such as hydrodynamic coefficients including the peculiarities of actuators as rudders, propellers and thrusters, and the effects of suction tube(s) and windage.

On the basis of these dynamic models with parameters specific for each ship, the DP/DT-system is tuned in the laboratory, which enormously saves time in tuning on board. In fact, performance under specific sea conditions can be accurately predicted, so no further time needs be wasted after installation on board.

By May 1999 the first specially designed DP/DT-systems have been installed, tuned and tested on board two dredgers. Experiences relating to performance, user friendliness and short-tuning-time are presented.

Study of existing DP/DT-systems

Recently a study was carried out by **TU Delft/IHC Systems** into existing DP/DT-systems on board of hopper dredgers *(Koolhaas, 1998)*. These systems all suffer from being universal, based on a common denominator that is suitable for a variety of users.

A hopper dredger, however, differs significantly from the average vessel working in the offshore industry. Differences are, for instance:

- the very large and varying pulling forces of the suction tube(s), which may easily

require some 50% of the propulsion power, this means that these varying pulling forces greatly dominates the behaviour of the vessels movements;
• the lack of thrusters; up to now hopper dredgers are expected to come by with rudders, propellers and bow thrusters only;
• the system is used only for special dredging jobs where accuracy is required, so that the operator will enjoy too few occasions for gaining experience with it.

Within the above mentioned study the major Dutch and Belgian dredging companies were interviewed, from which the researchers learnt that:
• operating the off-the-shelf DP/DT-systems on board hopper dredgers is thought difficult, partly because the system is so little used, but also because the systems often can do much more than is needed. Lack of experience inspires to not using the system at all;
• the DP/DT system mostly stands alone, quite apart from the navigation console. As a result switching from hand to DP/DT operation is even less inviting;
• the DP/DT system does, in the eyes of quite a few users, not sufficiently compensate the forces of the suction pipe;
• calibrating the DP/DT-system for its operational environment usually takes unduly long.

Above points were considered as great disadvantages of the off-the-shelf DP/DT-systems.

New dedicated hopper dredger DP/DT-system

The results of above study motivated the development of a new DP/DT-system, with as main requirements:
• simple to use;
• integrated concept for smoothly switching from hand to automatic operation;
• better and more dependable calculation of suction pipe forces;
• less time lost calibrating when making the system operational.

These requirements are met as follows:

Simple operation

Restraint in the number of operating options, concentrating on a hopper dredger's functions, did a lot to make operation easier, as did organising these functions properly:

Auto Dredge dredging with desired heading and speed.
DT-Dredging dredging with desired speed along desired track.
DT-Dump disposal with desired heading and speed along desired track.
DP dynamic position keeping at desired point with desired heading.
Joystick manoeuvring with joystick.
Transit manual steering with wheel and throttles.
Auto Sail sailing with desired heading and speed.
DT-Sail sailing with desired heading and speed along desired track.

The operator can select these 8 functions, each with its own screen presentation, with one single button on a specially developed DP/DT-desk, where further combinations of manual and automatic operation can also be set. Desired heading and sailing speed can easily be set with two buttons; rarer used options, easily monitored, usually on the screen's left part, can be set and changed by trackball, for instance for:

• values for desired pivot points (each to be set separately for DT-Sail, DT-Dredging, DT-Dump and DP/Joystick);
• values for DT-operations;
• values for DP-operations;
• values for suction pipe forces.

In other words, the number of pages that will normally be used is purposely kept as few as possible. Extra pages can be called in for maintenance purposes.

Integrated concept with simple switch from manual to automatic operation

As mentioned before, manual operation can also be selected at the DP/DT-panel, and switching between manual and automatic requires the pressing of only one button.

During manual operation (navigational) information is displayed on the monitor. A theoretical study into a possible solution for optimising the integrated concept for use on board trailing-suction hopper-dredgers, was presented during the 1997 CEDA dredging days *(Klugt et al, 1997)*. Such integration may be achieved by placing the DP/DT-panel on a central position in the navigation console. All controls and systems for manual navigation can be placed on the left, survey and dredging systems on the right. The chosen lay out, with only the most needed tools for manualoperation -rudder, propellers and thrusters- minimises the number of levers, buttons and keys. All controls for emergency operation can be brought together on the left wing of the navigation console.

Fig. 6: ***Navigation console on "Lange Wapper"***

Right in the heart of the navigation console, all relevant information for navigation can be presented on the screen in front of the DP/DT-panel, where virtually all DP/DT handling takes place.

The DP/DT-system gets its desired track from ECDIS (Electronic Chart Display System), placed to its left, for instance for shuttling between dredging and locations, or for navigating towards an entirely new dredging location. ECDIS's charts make planning of the desired track quite easy.

For optimising actual dredging and disposal tracks, data from the survey system, right of centre, are fed into the DP/DT-system. A display for dredging information on the far right and a radar system on the far left, complete the integrated console. DEME's hopper dredger *Lange Wapper*, commissioned in May 1999, has an integrated DP/DT-system as described above. An equal system will also be placed on board another new DEME-trailer, the *Nile River,* due for commissioning by end 1999.

Better and more reliable calculation of suction pipe forces

Experience on board a variety of dredgers has taught that reliably measuring the pulling forces due to the suction pipe is absolutely indispensable for accurate trench dredging. Experience further has proven that pull measurement in the trunnion's pivots is vulnerable while also its accuracy deteriorates in the wake of wear and the large shocks encountered here. Therefore we assumed, when designing the DT-Dredge calibration, that automatic calibration would be necessary *(New sensing system ..., 1997).*

This patented automatic (and while dredging continuous) calibration has proven to run so well that far less accurate measurement signals can be linked also. Each dredger master knows there is a certain (although unclear) relation between suction power on the draghead and the encountered pulling force, which depends on the soil, whether or not jet water is being used, condition of the chisels, etc. The automatic calibration compensates such variations automatically.

Trials in 1997 *(New sensing system..., 1997)* and in May 1999 on board *Lange Wapper* and on board *Queen of Penta-Ocean* gave no measurable differences in precision between measuring pulling force in the trunnion's pivots and those derived from the differential pressure over the suction head. By using both methods of measuring, a reliable assessment can be obtained, in which drop out of one does not influence progress or accuracy.

Knowledge based DP/DT-system for 'plug and play' installation

In order to achieve the desired DP/DT-system for hopper dredgers the DP/DT and survey systems had to be developed in-house, assisted by IHC Systems' half-sister **Imtech Marine & Industry** (who also conceived the ECDIS) and DEME (co-operating in the development of the survey system).

Developing the DP/DT-system started from a Knowledge based system, for which Gusto provided the necessary hydrodynamic ships' parameters and models for rudders and propellers. This information provides the basis for designing the control algorithms and fine-tunes these before installation. As a result the DP/DT-system can be installed rapidly, with no time lost in fine-tuning once it is fitted on board. This *plug and play* ideal has indeed been achieved during the installation of the integrated DP/DT-system on board of *Lange Wapper (First fully integrated navigation bridge...,1999)* and as a stand-alone system on board of *Queen of Penta-Ocean.* Within the timeframe of the planned sea-trials the DP/DT-systems were put into operation successfully.

Results of Newly Developed DP/DT-System

Before actual installation, a study was carried out for each vessel for which a new DP/DT system was to be installed. This study was envisaged to establish the specific ship's parameters and predict the accuracy (for various DP- and DT-modes). During commissioning and trials, in May 1999, on board *Lange Wapper* and *Queen of Penta-Ocean*, it turned out that:

- goals set regarding user friendliness were achieved as concluded from crew's experiences;
- the predictions concerning accuracy were indeed achieved;
- the specific ship's parameters were so accurate that the *laboratory* values barely needed fine-tuning on board, resulting reduced commissioning time.

Four examples of the trail results prove the point:

Fig. 7: Rotation around bow in DP-mode.

In this figure the bow was selected as rotation-point. During a rotation from 340[°] to 250[°] the bow of the *Queen of Penta-Ocean* remained within 15 [m] of the desired point.

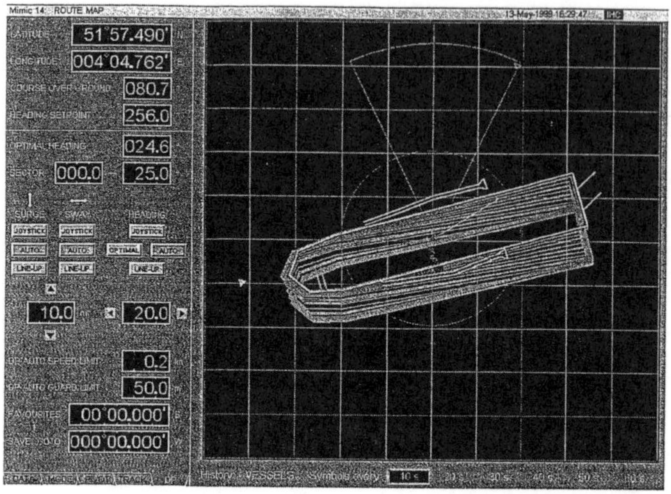

Fig. 8: ***Making a step of 20 meter to starboard in DP-mode***

Queen of Penta-Ocean reached the new desired DP-point via a straight line and stops within a few meters.

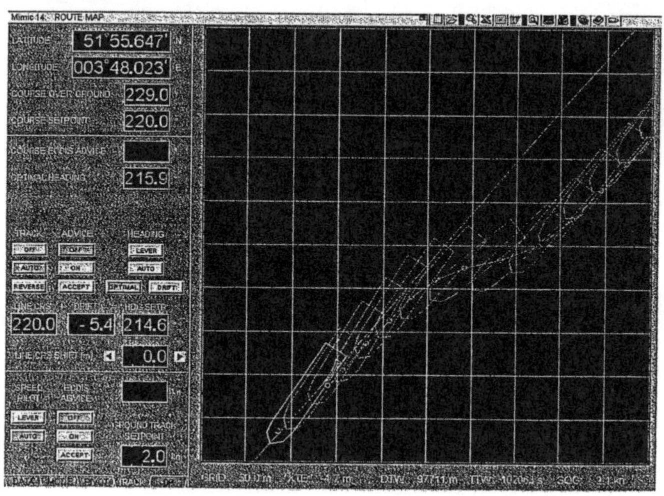

Fig. 9: ***Reaching a new track during dredging.***

How fast a new track could be reached was demonstrated clearly by shifting the desired track 100 [m] to starboard. *Lange Wapper* reached the new track within about two ship lengths.

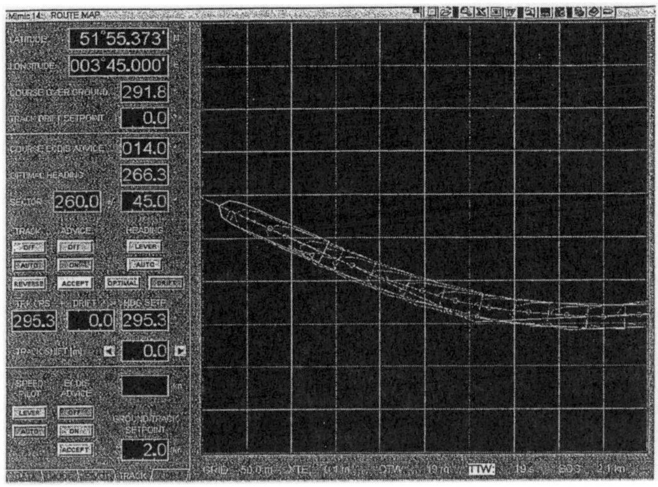

Fig. 10: ***Sailing parallel track over an arc from 270[11 to 295[°1 with a speed of 2 [kN].***

By means of the DP/DT system, *Lange Wapper* appeared to be able to follow all kind of tracks with pre-defined turning circles. At last some accuracy figures are given for Dynamic Track Mode (without pipe forces (with heading parallel to track)). The accuracy values are compared to the theoretical values calculated in the same way as done in the design study presented earlier in this paper, but now for other headings and wind forces. The practical accuracy can be read from the following figure. Three successive track directions were sailed; to 270, 295 and 080 [°]. The main disturbances were from the wind (180 10-17 [kn]) while the current was estimated as 160 [°] and 0.5 [kn].

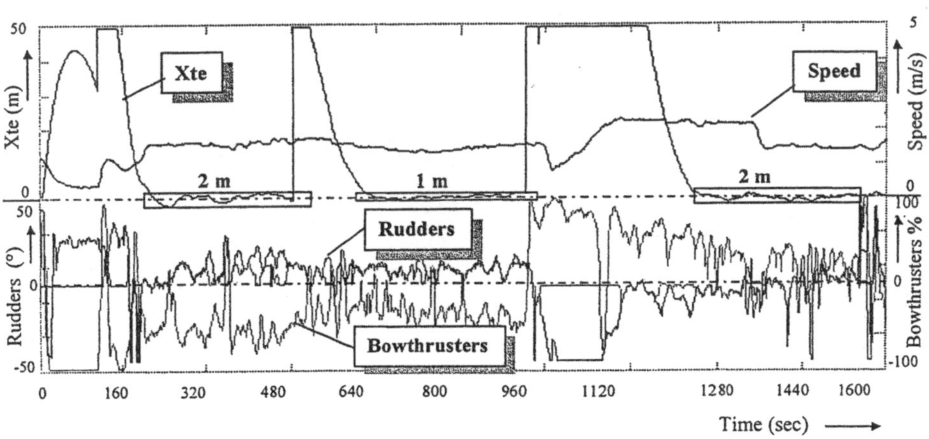

Fig. 11: ***Cross track error (Xte) as function of wind angle of attack***

As can been seen from this figure, the accuracy seems to be quite depending on the angle of attack from the wind. In the table below, these values are listed together with the theoretical calculated values. The differences might be explained by the fact that the real measured values are less accurately read from the graphics over a relatively short period (to short to draw accurate statistical conclusions), while the calculated values are taken over a statistical correct period of one hour.

Table 5. ***Measured and calculated accuracy as function of wind-angle of attack***

angle of attack [°]	measured accuracy [m]	calculated accuracy [m]
90 (from portside)	2	1.7
75 off stern (from portside)	1	1.5
80 off stern (from starboard)	2	1.6

From the above can be concluded that the real measured accuracy meets the calculated accuracy quite well.

Conclusions with respect to dredger design study

• The basic design *(Lange Wapper)* is very capable for DP/DT operations within the observed environmental conditions (as also observed during the sea trails).

• The positioning accuracy for a hopper dredger equipped with azimuthing thrusters is better than that of a dredger with conventional propulsion/rudder arrangement.

• The DP/DT design generally needs less thrust, resulting in larger environmental forces that can be compensated, and/or power savings of up to 50% that are possible. This power saving can be used to increase both the dredging force at the draghead and the available power of the dredge pump, which allows more efficient dredging.

• The DP/DT Dredger allows for a more flexible and efficient arrangement with a larger hopper space resulting in more efficient loading and more deadweight (load capacity).

• The suction tube placed forward results in an improved track keeping accuracy, while also a relatively larger suction tube can be accommodated, giving the possibility to dredge at deeper water.

• The DP/DT Dredger allows for a more flexible power allocation, relative more power is available at each desired location.

• Investment costs for this *all-electric* design are considered substantially higher than for the basic version; it is assumed that this additional investment may be returned rapidly through more flexible operations and higher accuracy, which yields more efficient dredging. This should be subject to further investigations.

Conclusions with respect to *Knowledge based* DP/DT system

• The goals set when developing the new DP/DT-system have all been met according to the crew experience during trials and after delivery.

• The following advantages have materialised compared with conventional systems:

• simpler to use;

• integrated in navigation desk, which facilitates shifting form manual to automatic operation;

• equipped with extra pull-measurement of the dredgepipe, based on differential pressure measurement over the draghead. This economical and reliable way of measuring can be used for DT-control quite as good as any normally used measurement system.

• Use of previously established ship's parameters makes pre-tuning of the system possible and allows 'plug and play' installation on board resulting in reduced commissioning time.

• DP/DT operation's accuracy by using the system can be predicted beforehand.

References

Klugt, Peter van der, and **Keizer Cees de** (1997). *The Intelligent Fully Integrated Bridge for Hopper Dredgers.* CEDA Dredging Days 1997, Amsterdam.

New sensing system for measuring draghead pull. Offshore Visie - May/June 1997.

First fully integrated navigation bridge on DEME hopper suction dredger. Offshore Visie - Aug/Sept 1999.

Koolhaas, C.P., *Design of Man-Machine Interface for Dynamic Tracking / Dynamic Positioning Systems of a Trailing Suction Hopper Dredger.* M.Sc. thesis TU Delft (Dec. 1998).

C.H.M. Kramers, J.P. Koert, C. de Keizer, S.A.W. Janse

IHC Holland NV Dredgers

Owned by **Dredging International NV,** Belgium and used internationally for all types of work including construction and maintenance.

Jumbo TH Dredge

The Jumbo Hopper Dredges are offering the greatest increase in dredging and transport capability seen in the history of dredging. Jan de Nul

European dredging companies improved upon the TH dredge design and began a construction program, building fleets that gave them supremacy in dredging worldwide. The capacity of hopper dredges was limited to about 10,000 m^3 until the early 1990s when a race to build large, *Jumbo* hopper dredges began.

The *Jumbo* TH dredge has emerged as the premier configuration for major reclamation. The advent of the **first *Jumbo* TH dredge *Pearl River*** in **1994** has been coupled with intense competition in Europe to build more and larger Jumbo's. The prime market for this expensive dredge has been in Southeast Asia. What was previously a debatable economic question, the tradeoff between sizes and their effect on projects, is now dominated by the advantages of large TH dredges.

The advantages and applications of new technology has been demonstrated as being effective on the efficiency of all sizes of TH dredges. Being a self-propelled sea-going vessel there are more opportunities for improvements to be realized with technological innovations. The driving force for its application to major reclamation development such as in Singapore and Hong Kong, is its ability to *mine* sand at long distances (i.e., <100km) from the reclamation site at increasing depths, transporting it *(see **Reclamation**)* to the site.

The *Pearl River,* with a capacity of 17,000 m^3 was built by IHC Holland and commissioned in 1994, by **DEME/Dredging International** of Belgium. Until that time the largest TH dredge was less than 12,000 m^3.

The essential driving force for Jumbo TH dredges, are the massive reclamation projects being conducted in SE Asia. The economic advantage of a larger volume in a single hull was a compelling factor. Projects such as the new Hong Kong Airport and Singapore's industrial and housing expansion, required new sources of sand and extended distance from the reclamation sites.

A summary of Jumbo TH dredges is shown in the following table for an illustration of this new trend. Prices for these mammoth dredges have varied from US$60M to $120M. Needless to say, the projects must be substantial in order to amortize these investments. (*See Photo Plate's 4,5,8* and *9).*

Jumbo Trailing Suction Hopper Dredges (TH)

Name	Volume-m^3	Owner
Vasco da Gama	33,000	JAN DE NUL NV
HAM 318	23,700	HAM, Dredging & Marine Contractors

Queen of the Netherlands	**23,350**	**Royal Boskalis Westminster Dredging**
W.D.Fairway	**23,347**	**" ' "**
Rotterdam	**21,500**	**Ballast Nedam Dredging NV**
Volvox Terranova	**20,015**	**Van Oord Dredging ACZ BV**
Queen of Penta-Ocean	**20,000**	**Penta-Ocean Construction Co. Ltd.**
Amsterdam	**18,000**	**Ballast Nedam Dredging**
Gerardus Mercator	**18,000**	**Jan De Nul NV**
Nile River	**17,000**	**DEME Group**
Pearl River	**17,000**	**" "**

Applications for this new class of dredge appear to be dominated by major reclamation projects such as Singapore, Hong Kong, and other parts of SE Asia. The other is for deep trenching for offshore pipelines, to depths of over 100 m.

The leading builder of dredges, IHC HOLLAND has classified hopper dredges into four categories:

Class	Range-m^3	Category-Dredge	World Quantity
A	**750-4000**	**Small sized maintenance**	**120**
B	**4000-9000**	**Medium sized general purpose**	**58**
C	**9000-17000**	**Large sized general purpose**	**11**
D	**17000-35000**	**Very large reclamation ("Jumbo")**	**11**

The Jumbo's are at this time owned by companies in The Netherlands, Belgium, and Japan who are currently dominating international dredging programs. With their major investment in such high cost capital equipment, there appears to be little chance of being overtaken in the mega dredging projects of the near future. However, other national factors come into play where private enterprise is not a controlling element such as in China where there are many government-owned dredges.

One interesting factor in the Jumbo dredges was dramatized after delivery of the *Pearl River* in 1994. It began work on small projects, locally in Belgium and then progressing around the globe ending up in Asia. It was able to dispense rapidly with projects of 1.0M m^3 and up and proceed to the next; proving its operational efficiency and preparing it for the major dredging programs that lay ahead. This is not to say, however, that the Jumbo can compete equally where vessel draft and nearby location of dumping areas, can be better performed by the smaller TH dredges.

The following table by IHC shows a comparison of characteristics of four different TH dredges that illustrate the advantages of each. Dredges A through D belong to one family; i.e., based on the same economics oriented design philosophy. Dredge A is a typical shallow water dredge. Dredge C design has been based on the B design; A, B and C have a single drag arm; the Jumbo D has two drag arms. The dredging depth of C-1 is restricted to the maximum of B. In that case C is competing with B at the same limiting condition.

All of these dredge factors are fictional, however their designs are realistic; their building prices and operating costs have been derived using ***VGBouw*** (4) (Dutch standardized operating cost for construction equipment).

Comparison of Dredges

Particulars	A	B	C	C-1*	D
Volume-m^3	3,500	8,500	13,500	22,500	
Load-tons	5,000	12,000	18,300	15,300	31,200
Length-m	85	107	121	151	
Beam-m	16.8	22.4	26.6	30.4	
Depth-m	6.30	9.80	10.50	13.0	
Dredging-m	5.72	8.30	9.45	8.30	11.22
Speed-knots	11.9	13.5	14.3	14.4	16.1
Power-kW	5,400	9,700	13,400	22,500	
Loading time-min.	45	53	68	58	69
Pump shore-min.	80	101	111	94	113

Technical Developments

Shipbuilding facilities have been lacking in handling the larger hull size of the Jumbo TH dredges and have had to be enlarged. The leader in dredge building, IHC Caland Group has as a result of the need for larger facilities, enlarged through acquisitions and mergers with other shipyards in The Netherlands.

Some of the improvements that have made possible more rapid engineering and construction of larger dredges, include computer-controlled systems like Cad-Cam, Finite element method (FEM). The introduction of PCs, on a large scale in all stages of dredge design and building process, have contributed to increased efficiency.

In the fabrication and construction of dredges, cutting and welding technology have improved also applying CAM. Using high tensile steel the hull weight has been reduced substantially. Quality has improved by building in covered slipways.

Improved efficiency has been realized through some of the following:

- dredge pump improvements;
- hull and propulsion efficiency;
- optimized Dead-weight to Displacement ratio (Dw/D) in relation to operating cost and overall performance;
- mean dredged mixture density increased from about 1.25 to 1.35 (t/m^3), which means an efficiency increase of about 40%;
- deep dredging, using under water pumps;
- mean pump-out mixture density increased (up to 1.75 t/m^3 have

been reported).

To a certain extent, larger hoppers can compete with smaller ones, even when operating under the same constraints such as restricted draught. It will be obvious that the shorter period a dredge works with restrictions, the more efficient it becomes.

Dustpan Dredge (SD)

The **US Army COE** invented the Dustpan Dredge and produced the first SD dredge, named the *Alpha* in 1895. This was motivated mainly as a solution to emergency, river channel maintenance on the Mississippi River. The *Alpha* had a wooden hull and a draft of 59 in. The dredges were not self-propelled but rather used bow winches to work against river current, gaining control for removing sand.

SD dredge "Beachbuilder."

" I did not fully comprehend the extent that the Economic Development of America had been in the past and was still, to a considerable degree, the result of the use of the Mississippi River and its Tributaries for transportation. Actually, the use of these rivers for transportation was almost entirely dependent on the work of the Dustpan Dredges, which kept the rivers navigable throughout each entire year."

Subsequent dustpan dredges followed with five new dredges being ordered in 1898. After considerable experimentation the system design was standardized on using jet agitation along the leading edge of the dustpan suction head. The suction head was increased to 39 ft in width with production capability of 1600 yd^3/hr. Further development built to COE specifications continued into the 1900s.

"Other countries either ignored or were not aware of the Dustpan Dredge until 1955, when a Combination Dustpan and Cutterhead dredge was constructed by the **Ellicott Machine Corporation**-. This was a 20" dredge used successfully in South America. –In 1965, the Ministry of Public Works in Argentina entered into a contract with Ellicott—to design a 28" Dustpan Dredge for use on the Rivers Parana, Uruguay, and Paraguay. The dredge has a speed of 6.25 knots."

The Dustpan Dredge principle was later adapted to the design of the ***Side Caster*** dredge. It has been useful in small harbors and river inlets and the COE ordered three of them. These are usually large dredges that are self-propelled similar to TH dredges, with a long discharge pipe rigged at the center of the superstructure and pivoted to spray the dredged material as much as 90 ft to the side.

The Side Caster dredge was built and operated in Venezuela on the Orinoco River and Lake Maracaibo, where annual maintenance dredging exceeded 69M m^3. The latter area has been the source of 70% of the country's oil production requiring access from the sea to the lake for tankers.

The first Side Caster dredge was furnished by **National Bulk Carriers, Inc**., under contract to the government to perform the necessary channel dredging; the ***Zulia,*** a combination TH dredge with a side casting discharge. The basic dimensions of the dredge included the hull: 549'x95'; pump suction- 35", discharge 32" with side caster pipe of 57" x 376 ft and steam turbine propulsion; 6,600 m^3 hoppers; displacement loaded of 29,181 t.

A second side caster dredge, a duplicate to the ***Zulia*** was built by **Kure Shipyard**, Japan in 1961 and named the ***Icoa.*** Its function was to maintain the channel on the Orinoco River for iron ore carriers from the **Orinoco Mining Company**. In 1971, the ***Icoa*** dredged over **40,588,000 m³** from the Orinoco River channel.

The Netherlands application

An adaptation of the SD dredge was made in Holland in the 1970s. This was on a project for the construction of the closed dam in the **Oosterschelde**. There were some complexities about the project since it involved the cutting of previously laid filter cloth and other materials.

The need was to dredge a trench after removal of the cloth and related materials. The trench was dredged in depths varying from 16 to 38 m below MSL. Various types of dredges were considered and it was decided to take a cutter suction dredge and install the dustpan suction head in place of the cutter head.

Some of the problems involved dealing with the tidal action, reversing direction twice daily and thus installing anchors both forward and aft. The dustpan head was 10 m wide and 24 jet nozzles on top of the head. Model testing showed that in sand when dredging thick layers up to 4 m, jets on top of the head would be more efficient than on the bottom of the head like those used in the US. Also, the US version uses four parts of the dustpan head

The dustpan system was mounted on the CS dredge ***Sliedrecht 27.*** The hull dimensions were: 79 x 13 x 2.95 m. There are six winches of 600 kN pulling force and 1000 kN holding force; a suction pipe of 1.05 m dia., a discharge pipe of 80 cm; a 1300 kW pump on the ladder which is necessary for dredging in those tidal waters.

The refilling of the trenches was accomplished with a TH dredge dumping sand in the open trenches that were made by the Dustpan system. This was either by direct dumping from the hoppers, or, discharging back through the suction pipes into the trench. Results under the conditions imposed by the project were not particularly encouraging due to lower average production than experienced by the Mississippi-type SD dredges.

Bucketwheel Suction Dredge (BWS)

The BWS dredge was first patented **c.1926**, showing two parallel helical cutting wheels. An adaptation of the giant surface mining systems used for coal and iron ore mining, dredge builders including LMG, Lubeck, Germany; Ellicott Int'l, USA; and IHC Holland have been building the BWS (or, UCW-Underwater cutting wheels, as they are also called) dredges.

The first designs used a single row of buckets as in the surface systems. After a number of single wheel dredges were built and placed into operation in a number of locations, it soon became evident that there were limitations to that configuration. The next evolution produced the double wheel thus making the left and right swings of digging more effective.

BWS dredge, mining mineral sands in Australia.

Some of the most successful applications of the BWS dredge have been in mineral mining; i.e., mineral sands, cassiterite (tin), stripping of overburden for placer gold and diamond mining. They are used for navigational dredging projects in some areas.

Portable Suction/Auger Dredge (CS/A)

The "Mudcat" dredge was invented by a mechanic in Florida in 1971, a small suction dredge similar to the concept of the dustpan dredge. It was first developed to remove the imported *South American* hyacinth plant that was choking small canals in Florida.

In its first stage, the National Car Rental Company purchased the rights to the *Mudcat* dredge design. After building several hundred dredges with the idea of renting them like cars, they sold their rights to Ellicott International. Over 500 *Mudcats* were subsequently built. Other manufacturers of similar small dredges have emerged. They are widely used throughout the world for canals, ponds and lakes for maintenance and environmental cleanup.

Mechanical Dredges

Design development in BG/C dredges with new technology has been a combination of bucket size and hoisting equipment. The Japanese have demonstrated leadership in both R&D and construction of this type of dredge. The award-winning technical paper at WODCON III, 1970 in Singapore illustrates the work that has been done: *Research on Dredging Grab-Buckets,* by Hisao Iwata, Port & Harbor Research Institute, Ministry of Transport, Japan.

Research on Dredging Grab-Buckets

Preface

This research was conducted for the improvement of the dredging efficiency of grab-dredges. Dredging grab buckets are usually operated under the following conditions which are not given to general grab-buckets used on land; first they are used for the excavation of various kinds of soils; second they are operated under water into which we can't look, third they receive water resistances as they are hoisted up or down.

So for the planning and designing of dredging grab dredges, it is necessary for us to determine proper types of grab-buckets, considering not only general problems on the sizes, materials strength and forms of shells and teeth, but also such problems on various problems in the process of re-windings, diggings and hoisting up of the grabs. The decision of which type of grab-buckets have direct influence upon their excavation operations. Today, there are many unsolved technical problems of deciding the type of grab-buckets for each digging operation.

This report presents the methods of the decision of the excavated facilities of the soil which are dug by grab-buckets, on the determination of well-adopted type of grab-buckets for each soil, and also on the digging characteristics of grab-buckets.

The author cleared some theories on grab-buckets and studied the applications of these theories for actual dredging operations of grab-dredges.

Some studies on the methods of decision of dredging materials on the investigation and test methods of dredging materials.

In applying general soil investigation methods and soil test methods for the decision of kinds of soils at the sea bottoms which are dug by dredges, it is necessary for us, first of all, to get a clear relation between the records of soil tests and digging characteristic of the dredging equipments of dredges.

There are two ways expressing their relations. One is the direct expression of combining test records with excavation characteristics as shown on *Table-1.* The other is the expression of relation using a standard scale which is common to test methods and digging characteristics as shown on ***Table-2. Table-1*** means, for example, that the sandy soil-whose N value of standard penetration tests is 30-are difficult to dig by grab-dredges, but are easy to dig by dipper-dredges and shows each dredging ability. The expression shown as *Table-2* is convenient to combine excavation phenomena which were observed in actual dredging operations with the operations carried out in the

Table-1 Expression for characteristic relation between dredgers and soils. (a)

Tests for original grounds.	
Dynamic Pnetration tests.	Characteristic of grab-dredger
Static Penetration tests.	
- - - - - - - - - - - - -	Characteristic of dipper-dredger
Tests for picked soils.	
	Characteristic of pump-dredger
Tests for grain size analyses.	
Unconfined compression tests.	- - - - - - - - - - - - - - -
- - - - - - - - - - - - -	
Physical investigation tests.	- - - - - - - - - - - - - - -
- - - - - - - - - - - - -	

laboratories.

By this expression, we can get the correct decision of soils. So first of all, the author provides some results on the decision of dredging materials.

For the standard scale used for the expression of *Table-1,* the author used horizontal components of excavation resistance acting on the testing excavation blades which move horizontally. This horizontal component Fh shows the same tendency as shown by the records on the tangential components of excavation power of the inversion blades, and following relation was established between components Fh and Ft.

Fh = 1.82 Ft

Table-2 Expression for characteristic relation between dredgers and soils. (b)

Tests for original grounds.		
Dynamic penetration tests.		
Static penetration tests.		Characteristic of grab-dredgers.
Vane tests at original grounds.	Basic digging characteristic	
Jet boring tests.		Characteristic of Dipper-dredgers.
- - - - - - - - - - - - - - -		
- - - - - - - - - - - - - - -		Characteristic of Pump-dredgers.
Tests for picked soils.		- - - - - - - - - - - - - - -
		- - - - - - - - - - - - - - -
Tests for grain size analyses.		- - - - - - - - - - - - - - -
Unconfined compression tests.		
Shear tests.		
Consistency tests.		
- - - - - - - - - - - - - - -		
- - - - - - - - - - - - - - -		
Physical investigation tests.		

Table-3 shows that the relation between Fh - the horizontal components of excavation resistance - (Hereinafter this horizontal power is referred to as Fh simply, or excavation resistance) and N' blow values of the blow testing machine shown on *Figure-2*, by which we get about six times N values gained by the general standard blow testing machine for the same sandy soils. ***Figure-5*** shows the relation between Fh and thrust resistance values R gained by thrust resistance measuring machine with flat ends shown on *Figure-4.*

Figure-7 shows relations between Fh and T torque values or shear resistance by the small type vane testing machines.

These experiments were carried out using testing materials shown on *Table-3.* The measured values of sandy soils and clay soils for the relation between Fh and N' are different as shown on *Figure-3.*

For sandy soils, the relation N' = 0. 51 Fh is observed at the loose conditions, but as the soils are tightened, this relation and measured values are not on the line which shows the relation N' = 0. 51 Fh and they are plotted irregularly as dotted lines shown on *Figure-5.* From *Figure-5* we know that the measured values of sandy soils and clay soils, do not satisfy above relation between Fh and N'.

Figure-7 means that the relation T = 2.05 Fh is applicable in both clay sands and sandy soils.

There are some other test methods of the decision of the dredging materials except shown above; however, in view of the results of the experiments, they are the same in getting the relations between standard excavation resistance and N values of the soil or static penetration resistance values or shear resistance values.

So the fundamental tendencies gained from the experimental results may become important factors which decide the kind of the dredging materials.

For example, in the case of deciding the kind of dredging materials by N values of the standard penetration tests, it is necessary for us to distinguish whether they are sandy soils or clay sand, and to know the ranges of the variety of measured values for sandy soils. So to prevent the estimation of excavation resistance excessively, we have to carry out other test methods together. Next the author shows the relation between standard digging resistance and digging characteristics of grab-buckets.

Table 3.

Materials	Specific Gravity	Mark In the air			In the water
Sand A	2.62	○ Dry Sands	⊗ Wet Sands	⊕ Water Contained Sands	●
〃 B	2.71				
〃 C	2.86				
〃 D	2.71				
〃 E	2.76				
〃 F	2.68				
〃 G	2.73				
〃 H	2.72				
Glass-ball 4φ	2.48	△			▲
〃 8φ	2.51				
Steel-ball 3.2φ	7.70	▽			▼
〃 6.4φ	7.71				
Vinyl-grain 4φ×7	1.23	□			■
〃 8φ×13	1.23				
Gravel A	2.83	◇			◆
〃 B	2.82				
Clay	2.71	×			×

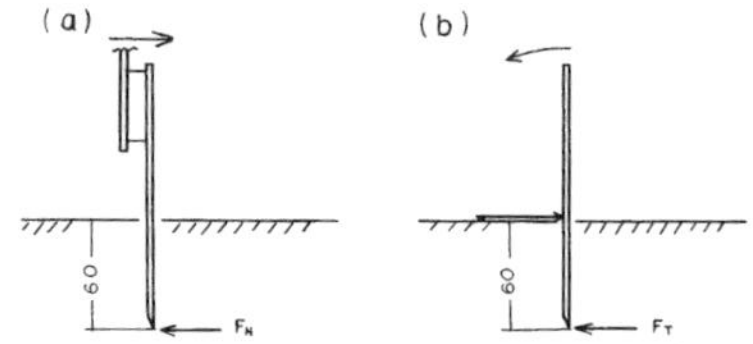

Figure 1. Standard digging resistance testing machine

The relation between digging resistance and grabbed weight of soil by grabs

Figure-8 shows the relation between grabbed weights of soils by model-grabs, and the basic digging resistance Fh gained from the penetration test and vane tests.

In the *Figure-8*, Wg is the weight of grab-bucket, Ws is the weight of grabbed materials.

From the *Figure-8*, we know that the measured points are not on the same line, but plotted separately

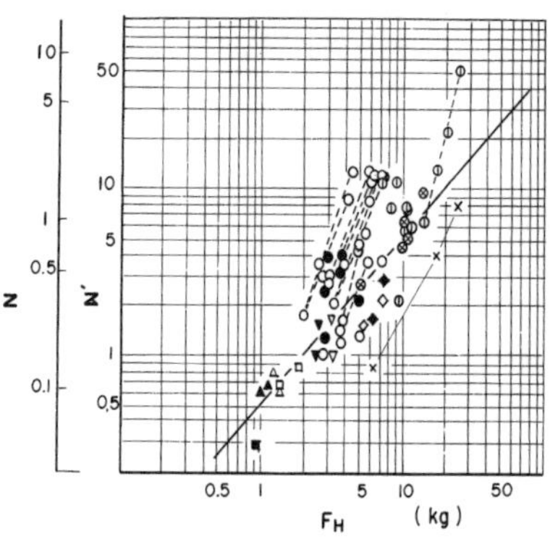

Figure 3. The relation between digging resistance values and blow values

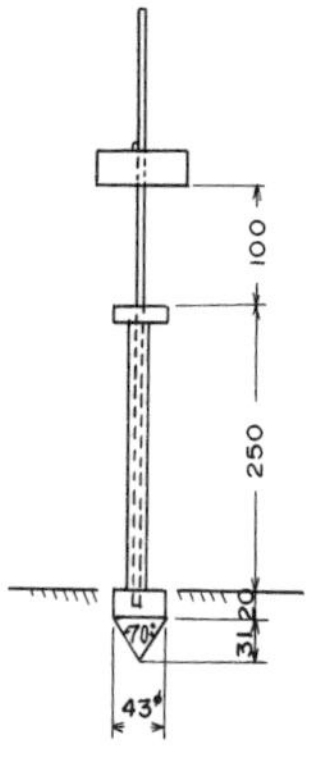

Figure 2. Blow testing machine

according to the test materials. But as for sandy soils and clays, the measured points are nearly on the same line and they have a same tendency.

Within the small values Fh, the plotted points of the values Ws/Wg show only small changes; this means that the buckets are full of grabbed materials.

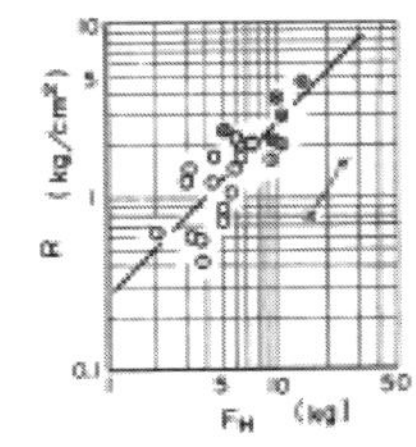

Figure 5. The relation between digging resistance values and thrust resistance values

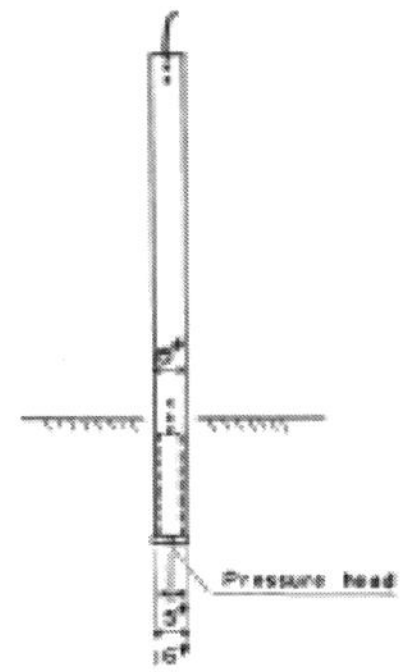

Figure 4. Thrust resistance measuring machine

Some plotted points in the *Figure-8* are included the measured values of the weight of grabbed materials which were grabbed in the water hoisted up and measured in the air. So these values of measured points in the air must be corrected by subtracting the weight of water contained in the grabbed materials from the measured values and also by adding the weight of materials which were washed away from the buckets as hoisted up in the water.

Except for these corrections, we have to correct the measured data by adding the weight which were subtracted by buoyancy, if we express the measured weights of the materials as those of the materials in the water.

The measured points in *Figure-8* are in the case of grabbing by a type of bucket. It is quite natural that the values Ws/Wg will be different in the case of other types of grab-buckets which have different volume V, grab-weight Wg, and reev-ings.

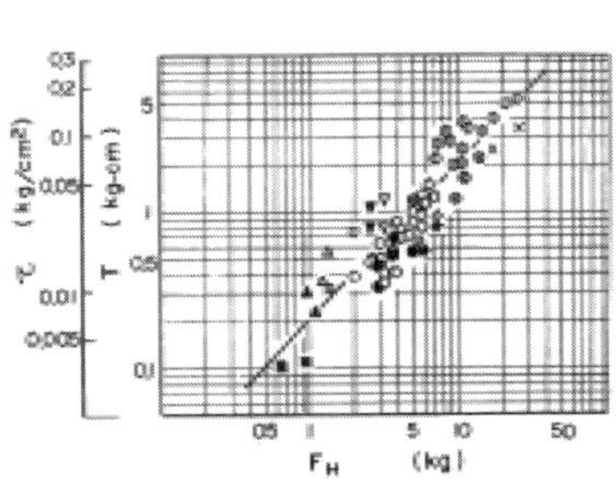

Figure 7. The relation between digging resistance values and torque values.

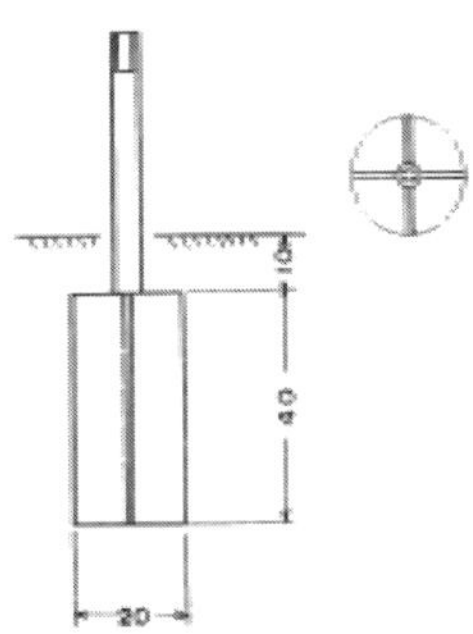

Figure 6. Vane testing machine

So in choosing the types of grab-bucket, we have to use another experimental results.

Figure-9 shows the change of the values Wg/Ws corresponding to each change of open length L, and grab weight Wg with general form.

From the figure, we know that the values $L\sqrt{B/2Wg}$ of the plotted points for the grab-bucket in *Figure-8*, being 0.053, they are on the vertical dotted line. The crossing points of this vertical dotted line with the characteristic curve of each dredging material mean the values Ws/Wg corresponding to each value L, B, Wg of the grab-bucket. From the *Figure-9,* we see that the weight of the dredging materials grabbed by lighter grab-buckets with bigger capacities are heavier than grabbed by heavier grab buckets with full of grabbed materials. It must be noted that this fact is one of the important factors in considering dredging operation efficiency of grab-bucket.

The relations shown in *Figure-9* are applicable only for the comparably easy excavated soils except hard strata and soft rocks.

So, as for the clay soil, the curves shown on *Figure-9* must be corrected sharply according to the capacity of the grab-buckets.

Figure-10 shows the change of the ratio Ws/Wg with the change of reeving in the case of grabbing gravels.

As for other kind of grabbed materials, we gained the relations with the same tendency. So the tendency of the values of Ws/Wg, corresponding to each reeving is considered to be determined by the tendency of the grabbed weight ratio Ws/Wg.

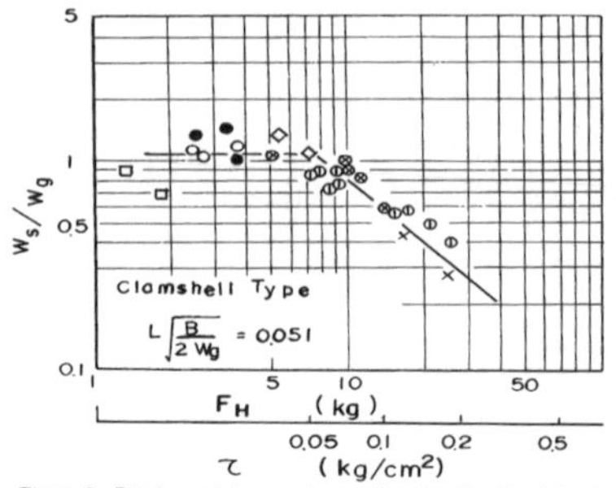

Figure 8. Digging resistance values and grab-ratio of grab-bucket (Capacity V = 0.005m³)

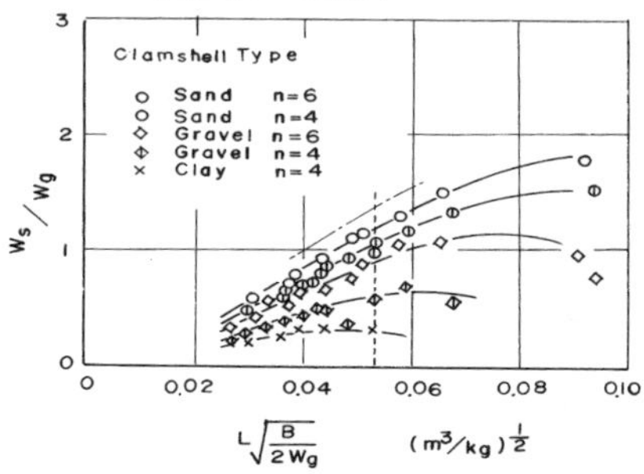

Figure 9. Characteristic between $L\sqrt{B/2Wg}$ and Ws/Wg

In general, as reevings are determined by the values B/2Wg and Ws/Wg, the relations shown above are used only in the first planning and designing of grab-buckets. Grab-buckets are often used for the investigations and tests of deciding the excavated facilities of hard strata and soft-rocks.

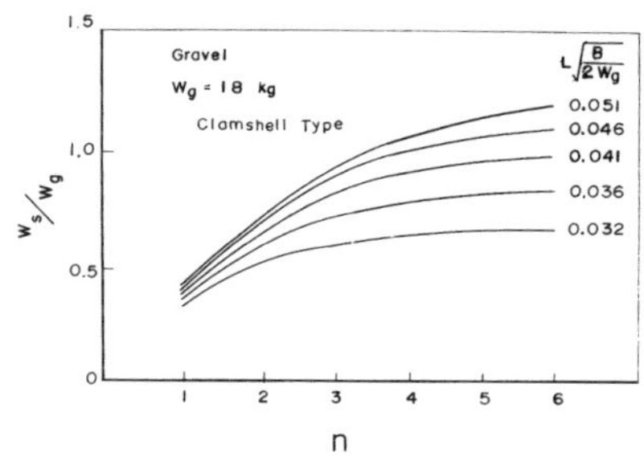

Figure 10. The relation between reeving and grab-weight ratio

Figure-11 shows that from the changes of hoisting were rope tensions, we know that whether the grab digging operations are suited or not for the dug materials. Through the above explanations, the author showed one method of relating the grabbed weight ratio Ws/Wg with the N values of standard penetration tests and the recordings of shear resistance for the original soils and grounds.

The next problem is how to carry out real investigations and tests for them. There may be various ways of carrying out their investigations, and the author shows some examples of them, in the following chapters.

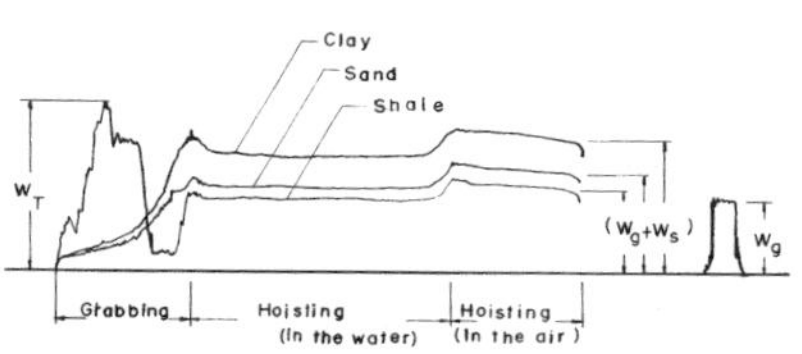

Figure 11. The change of wire rope tension at the dredging operation

Some examples of tests on dredging materials

It is not proper for the investigation of dredging materials to build such towers on the sea as are often constructed for the foundations, in view of the dredging periods and their building costs.

The reason is that for the investigation of dredging materials, it is necessary for us to get as many records as possible from the tests at the shallow places of the sea bottom.

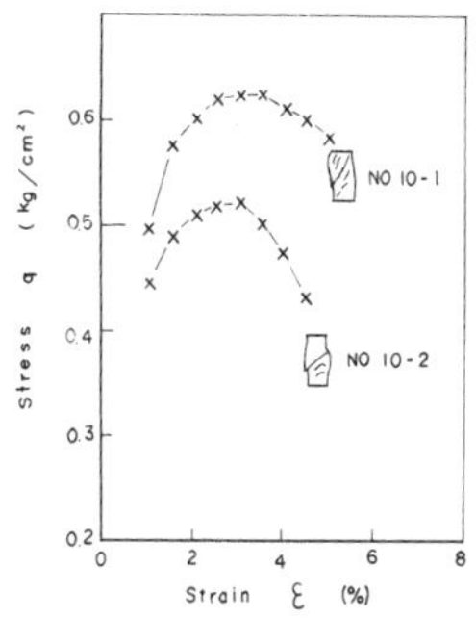

Figure 13. An example of unconfined compression strength of picked clay

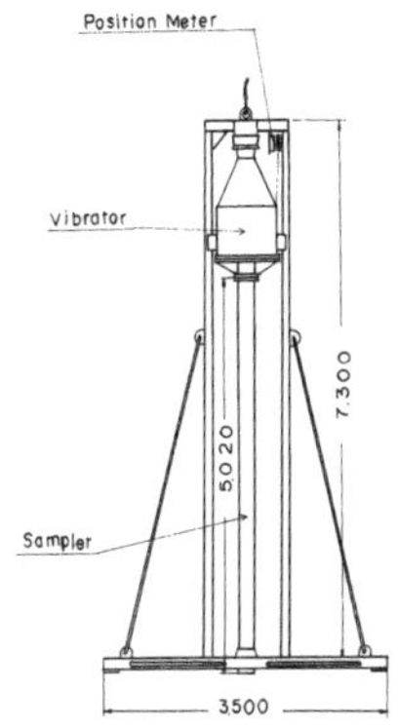

Figure 12. Vibro-core-sampler

The author used a **vibro-core- sampler** for the investigations of dredging materials and gained good results. This vibro-sampler is made up of a watertight vibro type pile driver under which double pipes - one is fixed and other is used core sampling and the length and diameter of them are 500 mm and 180 mm respectively - are connected. This core sampler pipes moves up and down along the guide shown on ***Figure-12***.

It is brought down to the bottom of sea by the winch on the surveying ship and the pipe section is penetrated into the sea ground with the help of the Vibrator. This pipe section is possible to cut into soft clay ground without the help of the vibrator and it is very important problem whether the sands which are picked by the penetrated pipe with the help of vibrator are disturbed, or not.

Considering the examples of the tests so for operated for clayey soil, we find that there are maximum values of unconfined compression strengths for each sampled material.

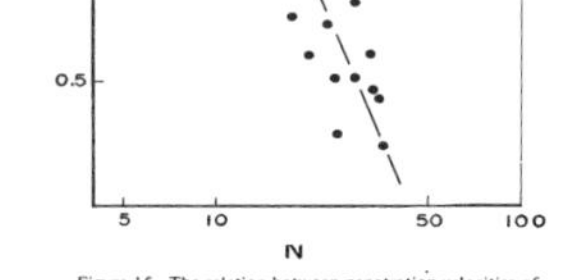

Figure 15. The relation between penetration velocities of the sampler and N values

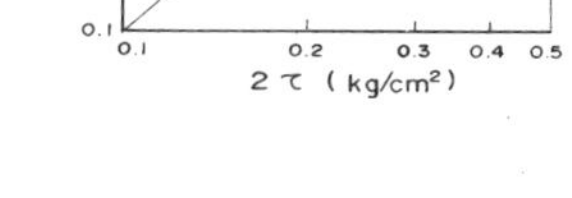

Figure 14. The relation between the results of unconfined compressions and shear stress

Judging from the general tendency of the curves shown on the figure, we can distinguish the excavated facilities of the materials roughly. ***Figure-14*** shows the relation between the shear resistance by the vane tests and unconfined compression strength q of the sampled clayey soil.

From *Figure-14,* we find that the values 2 η of the plotted points are greater than the values q.

For the clayey soil, the core-sampler was penetrated into the ground about 5 m, and the length of the gained soil in the sampler was more than 4.5 m.

Sampled sandy soil. being usually disturbed when pushed into a pipe by vibrator, they don't keep the same pressed state as the original grounds.

For the sandy soils, this core-sampler is used only for picking them and we analyze the soil grains and soil profits of them. For the judgement of excavated facilities, we have to presume them by the results of grain analyses, observations of experiments, and the penetration velocity.

Figure-15 shows examples of the relation between N values of the standard penetration test and V penetration velocity of the sampler for the sandy soils. We carried out some investigation tests by using the soil-sampler equipments in which vibro-driver are built and are operated.

But the results of them are imperfect for the data by which we try to get relation with the dredging materials.

Experiments for the digging characteristic of dredging grab-buckets.
On the grabbed weights of dredging materials by grab-buckets.

In designing and constructing grab-dredges, the most important problem is the decision of planning hoisting power W (In this case, W is not a maximum value), that is the sum of Ws weight of grab-bucket itself and Wg weight of grabbed materials. Expressing the above relations as follows:

$$W = Wg + Ws \quad \text{.................................... (1)}$$

we know that the efficient dredging operation is expected with the increase of the value Ws, including in value W, assuming that the time used for each grab-operation - dredging cycle time - is constant.

The weight of the soil which is washed away in the water, and the weight of the remaining water in the bucket when it is hoisted up in the air, should be added to the *equation (1)*; but they don't need to be considered at the designing and planning of grab-buckets.

From the *equation (1),* the value Ws included in the values W would be expressed as follows:

$$\frac{Ws}{W} = \frac{1}{1 + \frac{Wg}{Ws}} \quad \text{........................ (2)}$$

So with the decrease of values Wg/Ws, the weight of grabbed materials increases when the value W is constant. *Equation (2)* could be expressed as follows:

$$Wg/W = \frac{1}{1 + \frac{Ws}{Wg}} \quad \text{........................... (3)}$$

From the *equation (3),* with the increase of the value Ws/Wg the efficiency of grab-dredging will be expected. These values Ws/Wg are used in *Figure-8, Figure-9*

Table-4

		Grab-A	Grab-B
Planning hoisting power	W(Kg)	10000	10000
From characteristic curves (Reeving = 6)	K = Ws/Wg	1.0	1.5
	$L\sqrt{\frac{B}{2W_g}}$	0.04	0.06
Calculation values	Wg (Kg)	5000	4000
	Ws (Kg)	5000	6000
	$L\sqrt{B}$	4.00	5.35

and *Figure-10* so far explained.

So if we determine the value W for any soil which is dug by grab-buckets, the values Wg, L, and B which give maximum weight of grabbed weights would be determined. *Table-4* shows an example of the calculation explained above for the clam-shell type grab buckets.

These equations shown above are formed from the experiments on loose sands; but they are applicable to general sandy soil and clayey soils widely. Here the values L and B should be determined by equalizing the values of the product L and B, to the values on *Table-4,* considering the capacity of a grab-bucket and the central position of it.

When we finish the determination of the values Wg, L, B and reeving N, we start the design of the forms of shells, rods, sheaves and pins of the bucket.

These designs will be accomplished by deciding the design conditions and studying each shape and strength of the materials of buckets. These calculations are not applicable to such materials as the hard strata and soft rocks which are difficult to dig by grab-bucket.

They are not applicable to also the case when digging soils by dropping grab-buckets at high speeds on the soil and also the case when the shape of the teeth changes, or when similar grab-buckets are used.

By dropping grab-buckets at high speeds on such materials as hard strata, soft rocks, boulders, cobbles, which are difficult to dig, the weight of the grabbed materials will increase with the increase of the excavation depth of the bucket, but as for general soils and sands, which are easy to dig, the weight of dredged materials do not show any increase with high speed dropping of the buckets.

As for the problem on the shape of lippers, it must be considered especially for such materials which are difficult to dig; and if the decision of shapes of lippers is right, the weight of grabbed materials will be increased and we can get good results for them. So the conclusions of the problems on the dropping speeds and shapes of lippers is that these two factors have only small influence upon the soils to which the characteristic Ws/Wg- $L\sqrt{B/2Wg}$ are applicable.

Next, as for the relation between the size and grab weights especially for the clay, it was confirmed that the bigger the size of buckets become, the heavier the weight of the grabbed materials; so the use of bigger sizes of grab-buckets are advantageous for the digging of clayey soil.

These conclusions may be explained as follow. Expressing the width of digging blades as B and digging depth as D, the digging resistance Fh will be formed in the following expression

$$Fh \backsim BD^2$$

for sandy soils and for clayey soils

$$Fh \backsim BD$$

This difference of expression of Fh gives above different digging states for sandy soils and clayey soils.

Digging experiments by grab-buckets.

The dredging grab-buckets may be divided broadly into two categories, one is the **clam-shell** type and the other is **orange-peel** type: further more, clam-shell types are divided into **plate-type, half-tine** ones, and **hole-tine** ones. Some orange-peel types are classified by the numbers of lippers attached to them.

The grab-buckets presented in this paper are all half tine type.

In the early days of the author's researches on grab buckets, the author carried out the experiment on the comparison of hole-tine types ones with orange-peels ones. Judging from the results of the experiments there is no clear difference of the grabbed weight of the material between in the case by the hole-tine types and by orange-peel ones, and it was found that half-tine type grab buckets are possible to carry out sufficient dredging operations.

As for the digging experiments, the author usually took the measurements as shown on *Table-5.*

By these measurements records, we can calculate the digging curves of the grab, lippers, the changes of excavation resistance on the bucket in its progress of the digging, grabbed weights, and the stress of the materials of the

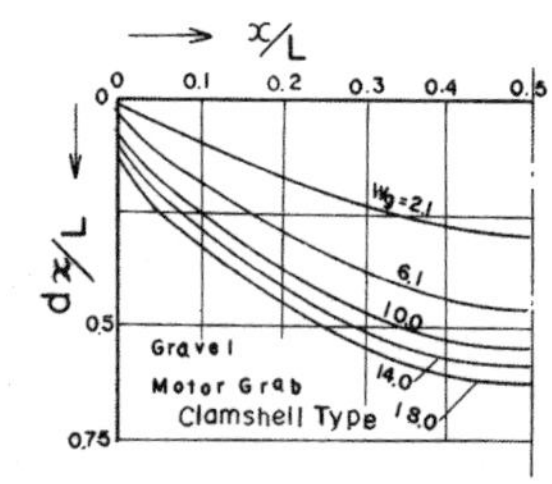

Figure 16. An example of the tendencies of digging locuses with the change of grab-weight.

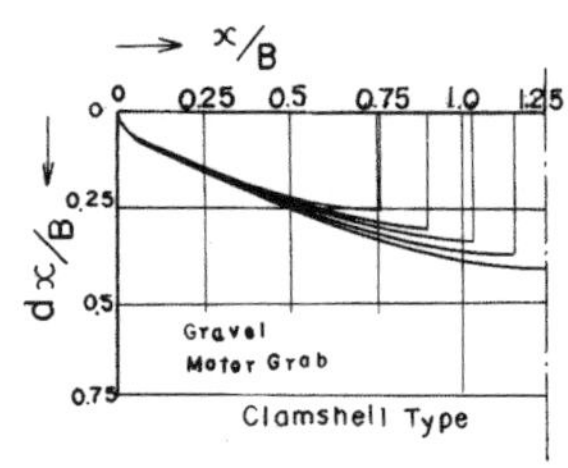

Figure 17. An example of the tendencies of digging locuses with the change of the open length

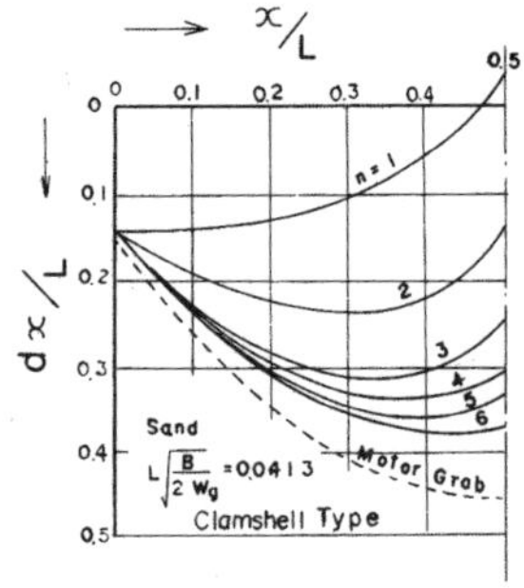

Figure 18. An example of the tendencies of digging locuses with the change of the reeving

rods and shells.

Figure-16, 17, 18 show some examples of digging curves of the lippers in the experiment in the air, X in the Figure means a travel distance of the lippers of buckets when they move from the perfect open to perfect close, and dx means the digging depth at the position of points of lippers. *Figure-19* shows the changes of wire rope tension in the process of the diggings.

Table-5 Content of measurement at the digging experiments of grab-buckets.

List of measurement	Measurement method
Tension of hoisting wire rope.	Load cell — Oscillograph
Stresses at rods.	Strain gauge — Oscillograph
Stresses at lips.	Strain gauge — Oscillograph
Distance between the upper sheave box and the lower one.	Position meter — Oscillograph
Heights from the surface of the ground to the upper sheave box	Position meter — Oscillograph

Figure-20 shows the changes of Ws (weights of grabbed materials) with the change of Wg (grab weight itself).

By arranging many experimental results shown above, the characteristic curves of the digging operations for the half-tine grab-bucket will be expressed by the relation Ws/Wg- $L\sqrt{B/2Wg}$ as shown on *Figure-9.*

Figure-22 shows the tendency of the direction of the digging resistance on grab-bucket in its process of the digging.

From the *Figure-22,* we know how the horizontal components Fh and vertical components Fv varies, and also find that the tips of the blade are cut into the soil with the operation of components Fh in the first process of digging.

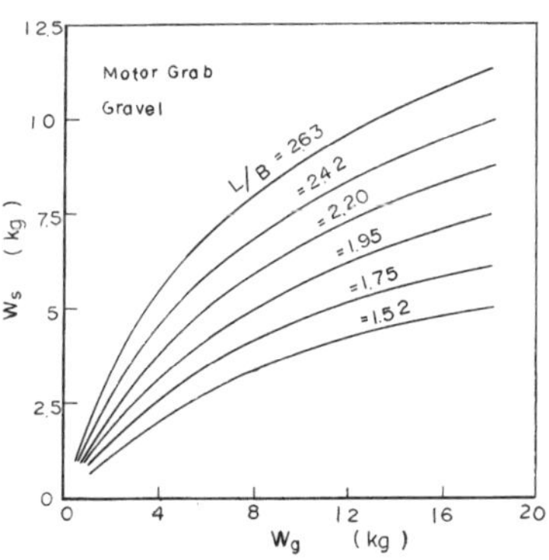

Figure 20. An example of the relation between grab-weight Wg and grabbed weight Ws

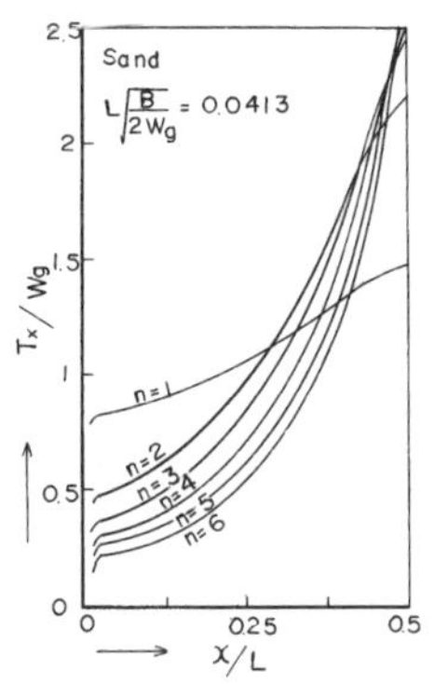

Figure 19. An example of change of hoisting wire rope tension Tx

If the directions and magnitudes of the digging resistance on the blade of grab are gained through the calculation for each test materials, the weights of grabbed materials will be estimated easily, and any complicated methods of decision of grab-buckets, so for explained, is not needed. But the fundamental theory on digging of soils is not still established in the author's research.

Figure-21 shows the differences of the experimental results between the case of digging in the air and that of the digging in the water, and also between the case of the digging by clam-shell type grabs and the case of the digging by orange peel ones. As for generally easy excavated soils, it was found that there is only a little differences between the measured records of digging in the air and in the water; and a little heavier weight is observed in the digging in the water.

Comparing the clam-shell type with orange peel ones on the weight of grabbed materials, there is only a little differences between them when each weight of buckets and reevings are equal.

At first it was anticipated that there would be a big differences between them by the influences of the buoyancies of grab-bucket itself and sandy soils in the water; but it may be considered that the various kinds of influence factors made the difference of each results small.

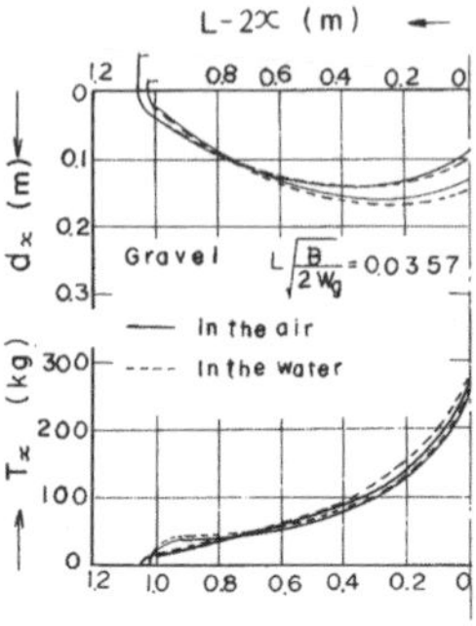

Figure 21. Comparison of the experimental results carried out in the water with those carried out in the air

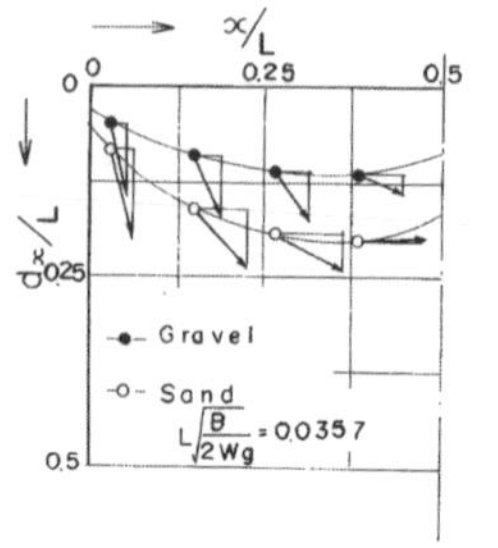

Figure 22. The direction of digging resistance exerted on the blade points of the bucket

Some problems on the digging of hard strata by dredging grab-buckets

Grab-dredgers are usually fitted for the dredging operations for easily excavated soils so far explained; however, in recent years, grab-dredgers have been needed for the dredging operations at hard strata and soft rocks which are difficult to dig.

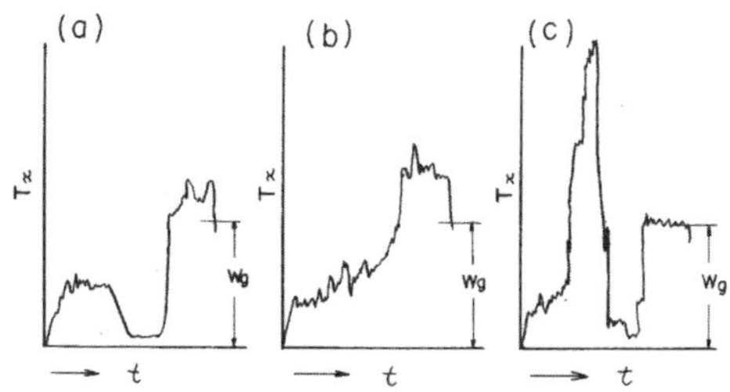

Figure 23. The change of hoisting wire rope tension at the digging of soft rocks

For digging of hard strata and soft rocks, it may be needed that the grab-weight should be increased, shape of lippers should be shaped, and the dredging efficiency should be improved by dropping the buckets at high speeds, producing the increase of the destruction effect of these materials.

Figure-23 shows the changes of hoisting wire rope tension as digging shale, unconfined compression strength of them are about 100Kg/cm^2, as an example of experimental results on this problem.

Following are explanations for each digging states of the case (a), (b), (c) on *Figure-23.*

(a) The parts of the rocks on which grab buckets are dropped were destructed by the shock of the lippers; but they were not excavated at the digging process, because lippers are not thrusted in, but just slide over the rocks.

(b) The materials were fairly digged.

The tension Tx becomes small for a moment just before the close of the bucket; and it was because that the rocks were cut both by right and left lippers, so the grab-closed instantly.

(c) This case means that the shapes of lippers are too sharp to dig these rocks. Lippers were thrusted in, causing excessive force on each member of the grab-buckets; so it is feared that the lippers and shells may be destructed.

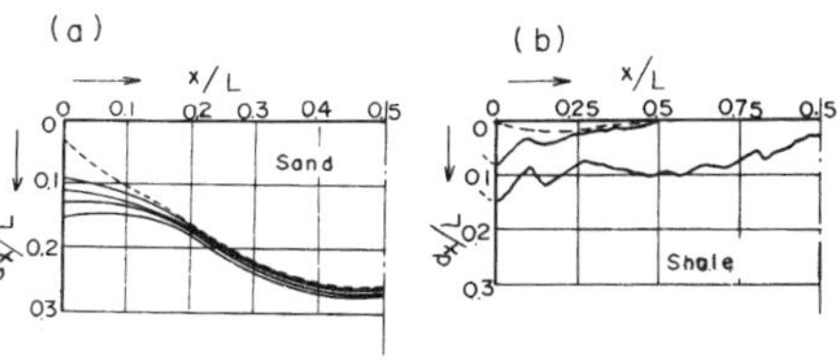

Figure 24. The change of digging locuses with increased dropping speeds of the grabs

The phenomena shown on *Figure-23* (c) are the most critical state of the grab-bucket at its digging process in considering material strength of shells and lippers.

But the determination of the members' strength of the grab-buckets, estimating excessive hoisting power at its planning stage, makes primary factors of producing such grab-buckets whose weights are larger than usual ones.

So, in general, real hoisting powers are limited to no less than two times as planned ones.

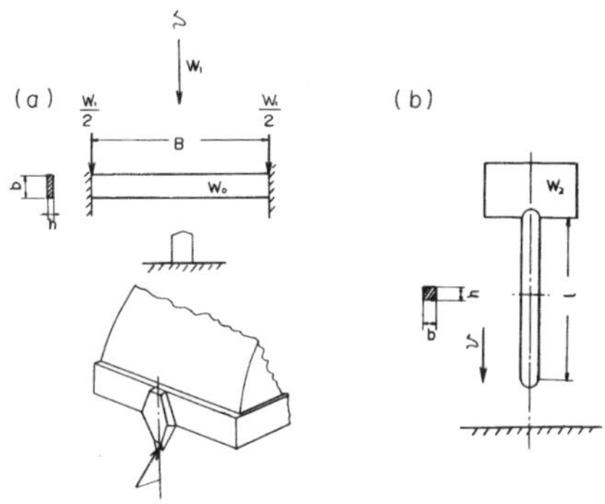

Figure 25. Marks standing for the sizes of lips and rods

Next problem is how to increase the dropping speed of the grab-buckets. For this problem, the author carried out some experiments on the digging curves, material strength of the buckets, and drags.

As for the digging curves, *Figure-24* shows each tendency of the curve, the case on easy dug soil and the case on shale. Of course, in each case, grab weights are determined from the soils which are most fitted to dig for them.

On *Figure-24*, the dotted lines shows the case that grabs were placed softly on the surface of the soil (or rocks), sea bottom and were started digging operations, while the solid lines mean the case that the primary penetration depth of the grabs into the soils were increased by the shock at the collision with the ground increasing dropping speeds of them.

Judging from the figure, the curves which express the digging states of ordinary soils are rising with the advances of the digging, and they are approaching the dotted lines, so the weights of soils expressed on the enclosed area by the dotted lines and the solid lines are small in comparison with the gross grabbed-weight of the soil.

On the other hand, in digging shale, the surfaces of the rocks were greatly destructed by the dropping of the grab-bucket; so deeper digging depth will be gained in the whole process of digging, increasing the grabbed weight of the materials.

Next it is necessary for the calculation of the material strength of the grab-bucket to know how much shock stress is produced in each material of grab-buckets when they collide with the rocks at high speeds dropping.

As for this problem, the author investigated the shock stresses produced in the materials of lips and rods, by dropping grab-bucket on the surfaces of concrete. Expressing the size of lips and rods as shown on *Figure 25*, (a), (b), the shock stresses produced in the center part of lips and rod are roughly within the limits shown as (a), (b), on *Figure-26*.

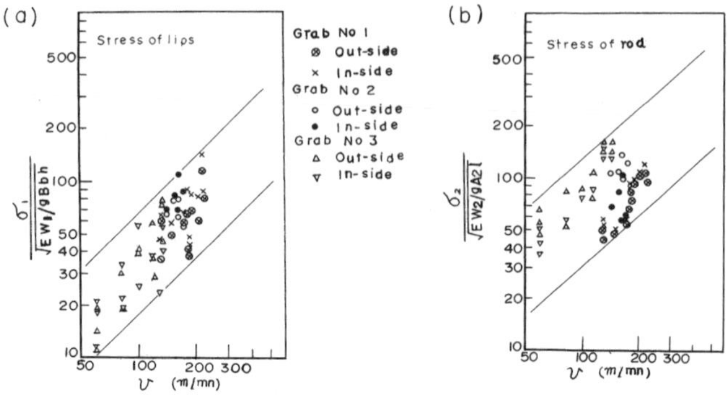

Figure 26. Results of measurements of shock stresses.

Here in this *figure*, E=elasticity of the materials, g = gravity acceleration, v = dropping speeds of grab-buckets. The problems on the drag of grab-buckets in water is important not only for learning how much dropping speeds is possible, but also for considering the added weight of the bucket at the process of hoisting and for the observation of the digging state of the bucket in the water from the measured records.

Table-6 Drag coefficients of grab-buckets at the rewinding and hoisting in the water

	At the rewinding		At the hoisting	
Type of grab-buckets	half- tine	Orange-peel	half-tine	Orange-peel
Speeds of the rewinding and hoisting at the experiments.	57.2	57.2	57.2	57.2
Calculation data of pressured area.	L = 0.55m B = 0.18m	D = 0.50m	L' = 0.385m B = 0.18m	D' = 0.34m
Pressured area (m^2)	(L B) 0.099	($\pi D^2/4$) 0.196	(L' B) 0.069	($\pi D^2/4$) 0.091
Drag coefficients (Cx)	0.42	0.60	2.38	2.67

The drag of the water P exerted on the grabs is expressed as follow,

$$P = Cx\ Ao \cdot \gamma_w \cdot \frac{v^2}{2g}$$

where the values of drag coefficient Cx are shown on *Table 6* gained from the results of model experiments.

Here Ao = projected area of grab-bucket, γ_w = weight of unit volume of water, g = gravity acceleration, v = dropping speeds of grab-bucket.

The results of the calculation for dropping speeds of grabs in the water, the capacities of which are about $4m^3$, using the values of Cx as shown on *Table-1* are shown on *Figure 27.*

The character m in *Figure-27,* is the value gained by subtracting rewinding resistance of the dredging equipments from the weight of the grab-bucket in water, and the character S is a dropping distance.

From the results of this calculation, we know that getting last dropping speeds 300 m/min is quite easy except for the extreme light buckets.

Through the experimental results, we know that the high speed dropping of the grab-bucket on hard strata and soft rocks is an efficient method increasing digging effects. For the practical points of views, however, we still have to investigate such improvements and developments of the mechanisms of the dredging equipments which stop the rotation of the drum simultaneously with the touch of the buckets with the grounds.

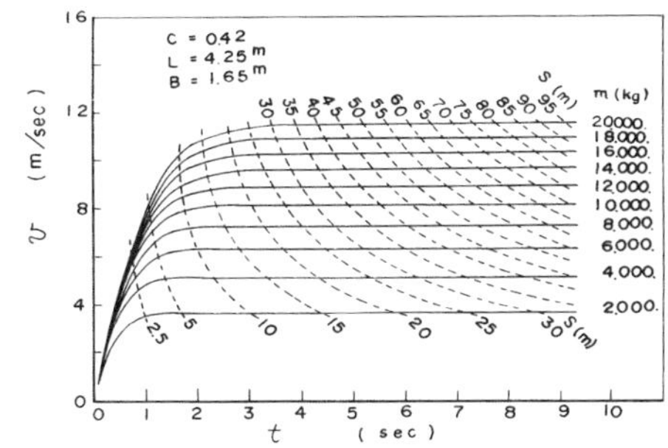

Figure 27. Results of calculation for dropping speeds of grab in water

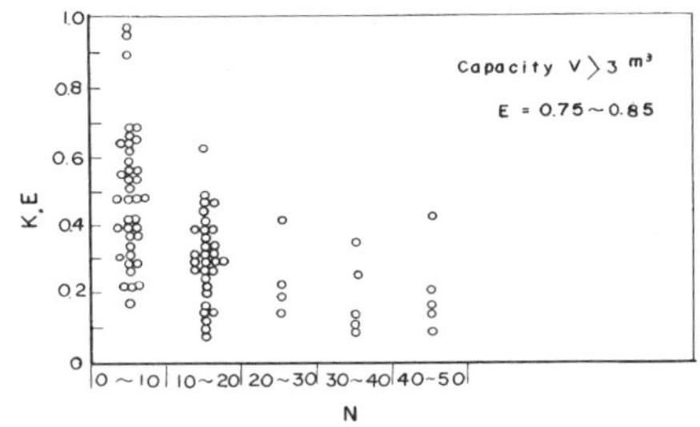

Figure 28. An example of dredging operation achievement of grab-dredgers

Some examples of the decisions of dredging grab-buckets

Under the existing circumstances, the expressions of the dredging materials are usually expressed by the N values of standard blow test.

In the experimental results so for explained, it was found that the excavated facilities for the sandy soils and clayey soils are different as shown on *Figure-3,* and points which the relation between the blow values and the digging resistance in sandy soils are plotted roughly.

A similar tendency in the case of experiment is also observed for the practical use. But at present, ordinary grab-dredgers have usually only two or three kinds of grab-buckets on them; so there will be no mistake in deciding and using grab-buckets.

Technical studies on grab-buckets explained so far in this paper are necessary at the planning and construction of grabs, and also at the dredging operation planning of them. They are used for the estimation of operation efficiencies corresponding to each type of grab-bucket.

Next the author explains what type of grab-buckets are proper to use, when equipped grab-buckets are limited in number.

Investigating the operation achievements of grab-dredgers in actual dredging area, as is known by the examples shown on *Figure-28,* we know that the grab-dredge was used for dredging the soil, N values of which are almost less than 20.

The character KE on *Figure-28* shows grab ratio (the ratio 37 of grabbed weight of soil to grab-capacity) in the actual dredging operations at the rugged areas where waves and tides act.

When we use same buckets, we found that the grab weight ratio for the soils, N values of which are 10-20, is about 65% of the ratio for the soil, N values of which are less then 10.

Table-7

Capacity V(m3)	Grab-weight Wg(ton)	Grabbed weights Ws(ton)	Weight-ratio Wg/V	$L\sqrt{\frac{B}{2Wg}}$	Kinds of grabbed soils
4.0	6.0	7.6	1.5	0.053	(Basic grab-ratio 100%) N < 10
3.0	7.5	5.8	2.5	0.043	(Basic grab-ratio 80%) N = 10~20
2.5	9.0	4.7	3.6	0.036	(Basic grab-ratio 60%) N > 20 broken rocks

So it is necessary for each grab-dredge to have at least two grab-buckets, one is for the use of digging the soil, N values of which are less than 10, and other is for the soil, N values of which are 10-20.

Another heavy grab-bucket may also be necessary for grab-dredges at the digging of the soils, N values of which are more than 20, and another heavy grab-bucket may also be necessary for grab-dredges at the digging of the soils whose N values of which are more than 20. So judging from all factors so far explained collectively, the types of grab-bucket which should be equipped on grab-dredges whose nominal grab-capacity is 4 m^3, may be decided, for example, as following three kinds of types of grab-bucket shown on Table-7.

For the ordinary types of grab-buckets, the relation between the values $L\sqrt{B/2Wg}$ and weight-capacity ratio is established as shown on *Figure-29;* while, as for grab capacity V, we can't find any clear relation, because it varies distinctly with the small changes of heights between the bottom and upper sides of shell at the closing. In considering the excavated facilities of soils, we can presume more clear relationships with the measured records for them, if we express excavated facilities by the values Wg/V, rather than by values $L\sqrt{B/2Wg}$. But, in general, excavated facility is usually expressed by the values Wg/V. When a grab-dredger is planned to have three kinds of grab-bucket, the values of weight-capacity ratio, Wg/V, or the values of $L\sqrt{B/2Wg}$ may proper to be determined as shown on *Figure-30.*

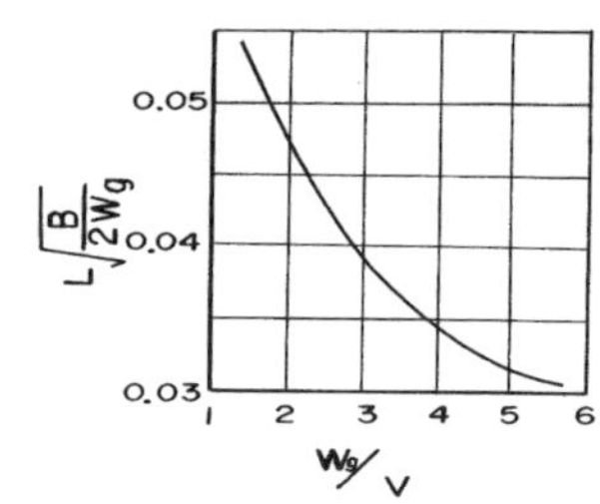

Figure 29. Relation between Wg/V and $L\sqrt{B/2Wg}$

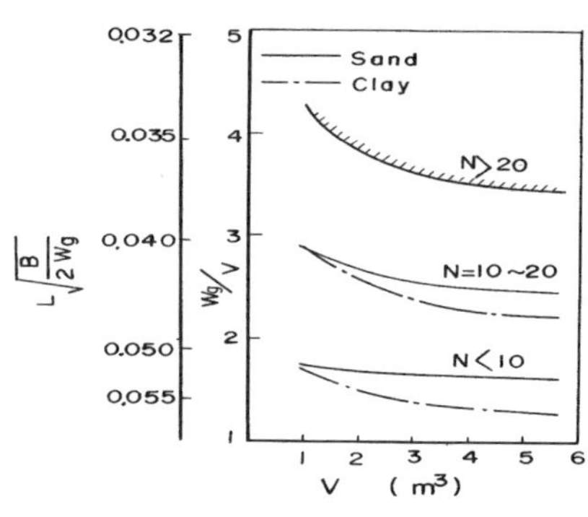

Figure 30.

Figure- 30 shows the results of laboratory experiments and actual dredging operations. From this figure, we know that the solid lines which express the digging states for sandy soils do not coincide with the dotted lines which express the digging states for clayey soils. So the weight-capacity ratio of grabs which are used for dredging both sandy soils and clays, should be determined from the average values of Wg/V, or rather those of heavy ones.

For the digging of wide ranges of the dredging materials, whose N values are usually more than 20, the types of grabs should be determined as the examples shown on *Table-7* by deciding the object dredging materials to be dug.

The above mentioned way of determination of grab buckets is one example of deciding the limit numbers of grab-buckets necessary for the digging of wide ranges of soils.

When object materials to be dug are determined, the most fitted grab-bucket would be chosen by the methods shown in this paper. The author will continue the investigations and researches on the relations between dredging soils and dredging efficiency of the grab-dredgers in practical dredging operations.

The postscript by the author

The contents of the research on grab-buckets above mentioned were written by contracting the contents of references shown at the end of this paper. Here in this paper, the author explained only some examples of experiments and their conclusions, so there may be found some imperfect points which the readers perhaps can't understand.

The author omitted the full explanation for material strength of the grab-buckets. So the author has to apologize for their imperfect explanations.

Biography

H. Iwata graduated from Aichi technical college in 1951; and worked in Port and Harbour Bureau Ministry of Transport. His main work there was the buildings and repairs of working crafts used for the port and harbour constructions. In 1962, he was transferred to Port and Harbour Research Institute and since then he has been working for the researches on dredges. His main research themes are *Research on water-jets apparatus for dredging* and *Research on Dredging grab-buckets.*

References

1. *Research on the dredging grab-buckets (first report)*, H. Iw ata
Report of Port and Harbour Research Institute, Vol. 4, No. 7
2. *Research on the dredging grab-buckets (second report)*, H. Iwata
Report of Port and Harbour Research Institute, Vol. 5, No. 7
3. *Research on the dredging grab-buckets (third report)*, H. Iwata
Report of Port and Harbour Research Institute, Vol, 6, No. 7
4. *The elucidation on excavation processes*, S. Hatake, The Journal of Construction Machines, No. 135

5. *Tests on dropping shock of grab-buckets and on grab characteristic (first report),*
H. Iwata, T. Takasago, Technical data of Port and Harbour Research Institute, No.38
6. *Tests on dropping shock of grab-buckets and on grab-buckets and on grab-characteristic (second report)*; H, Iwata, T. Takasago, Technical data of Port and Harbour Research Institute, No. 68
7. *Some experiments on the decision of dredging materials,* H. Iwata, T. Takasago, Technical data of Port and Harbour Research Institute, No, 94
8. *Investigation on the soils at the sea bottom of Malacan Strait,* Technical data for the 16th Research Society of Port and Harbour Machinery Techniques

Nomenclature

Ao = Projected areas of grab-buckets for the calculation of drag in water.
A = Cross sectional area used for the calculation of the material strength.
B = Width of shell lip.
b = Length of shell lip. Cx = Drag coefficient.
D = Diameter of a circumscribed circle determined by each tip of the blade.
dx = Digging length of a grab in the digging operations.
E = Elasticity of materials.
FH = Horizontal component of standard digging resistance.
FT = Tangential component standard digging resistance.
Fv = Vertical component of standard digging resistance.
Fh = Horizontal component of digging resistance at the digging process of a grab.
F = Vertical component of digging resistance at the digging process of a grab.
g = Gravity acceleration.
h = Thickness of lips of a grab.
KE = Grab-grab-ratio of a grab-bucket including actual dredging operations conditions.
L = Open width of a grab-bucket.
i = Length of a rod.
O = Weight of a grab-bucket in the water.
n = Reeving of a grab.
N = N value of a standard penetration test,
N' = Blow value of a blow penetration testing machine (=N/6)
P = Drag exerted on a grab-bucket,
q = Unconfined compression strengthes of soils.
R = Thrust penetration resistance of a thrust penetration testing machine.
S = Dropping distance
T = Torque value of a vane test machine.
Tx = Hoisting wire rope tension at the digging process of grab-bucket.
t = Time
V = Capacity of grab-bucket.
= Speed of a grab in the water.
W = Hoisting weight of dredging equipment,
W1 = Load exerted on lips shown on Figure-25 (a).
W2 = Load exerted at the center part of rods Figure25(b).
WT = Rewinding load at the process of digging.
Wg = Weight of grab.
Ws = Grabbed weight of the dredged materials.
x = Travel distance of the point of the blade at the process of digging.
= Weight of water per unit volume.

Giant Grab Dredge-*Tosho*

The World's largest grab dredge, ***Tosho*** carries a bucket that has a capacity of **200 m³** with a production capability of **6,000 m³/hr**. This dredge was designed and built in Japan by **Sumitomo Heavy Industries, Ltd**. to meet the demand of major dredging projects.

In **Nagoya Bay, Aichi Prefecture**, the ***Tosho*** dredged over 5M m³, deepening channels and moved on to **Nakayama Channel** in **Ise Bay.** The dredge's large capacity and energy savings are the result of the self-contained electro-hydraulic grab bucket and counter weight-balanced lifting/lowering system. It has a traveling spud system that facilitates the dredging operation without any assistance from auxiliary vessels or mooring lines, with attendant cost savings.

Photo 2 Grab Bucket

Some of the features of the *Tosho* dredge are listed as follows:

1. Counterweight-balance and hydraulic-operated mechanism;
2. Simplified dredging gear reduces consumable parts and maintenance cost;
3. Strong hydraulic seizing power when the grab bucket dredges hard soil;
4. Mooring system without

Photo 1 "TOSHO"

an anchor-chain but a walking or traveling spud system eliminates anchor work;

5. Operation system of dredge including winches and spud winches, are controlled from a remote operation room in a unified way.

The dredge Features are listed in the following table.

Type: Counterweight-balance and hydraulic operated system

Main hoisting capacity 690 t
Max. dredging capacity 6,000 m^3/hr
Max. dredging depth 30 m
Working radius (fixed) 31 m

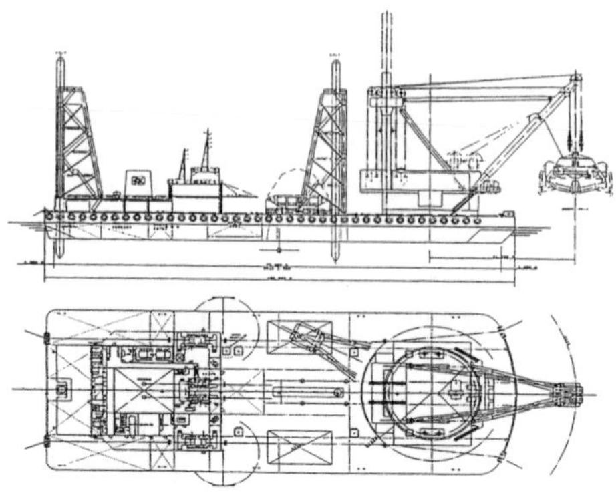
Fig.1 General Arrangement

Hoisting speed	46-66 m/min.
Lowering speed	66 m/min.
Slew speed	0.6 rpm
Grab bucket	self-contained electro-hydraulic
Capacity	200 m^3
Self-weight	370 t
Wt of mud	320 t (design density of mud 1.6 t/m^3)
Closing time	26 sec
Opening time	18 sec
Hoisting rope	60 mm
Electric motor for hydraulic pump	8 sets-132 kW, 2 sets-7.5 kW
Electric power	600 v. 60 Hz, 3 phase

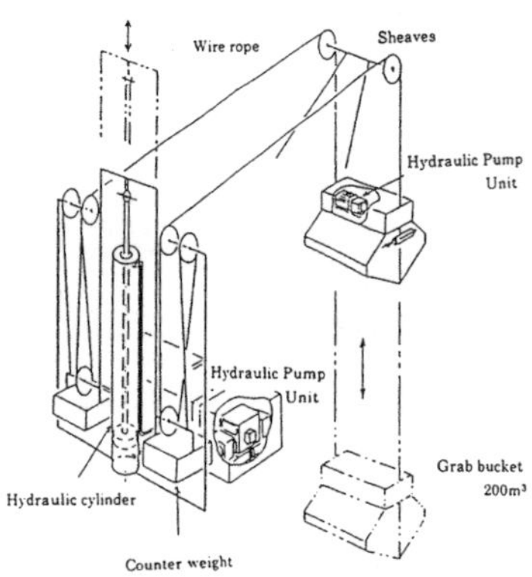

Fig.2 Hoisting Mechanism

Hull dimensions (moulded)

Length	100.0 m
Breadth	36.0 m
Depth	6.0 m
Draft	3.60 m

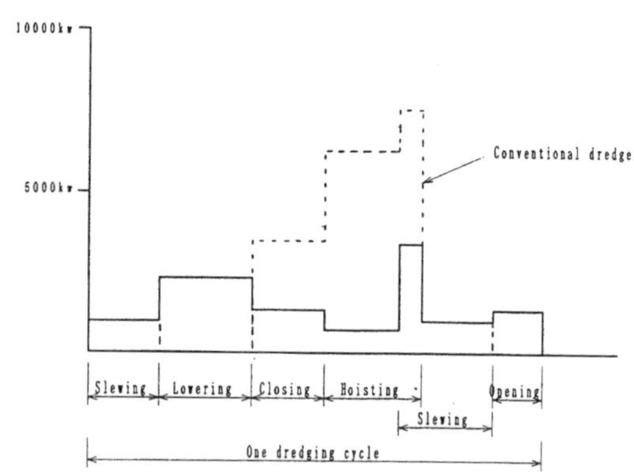

Fig.3 Estimated Load

Machinery

Main generator engine: Three sets 3,800 PS, Alternator 2,500 kW (one stand-by). The counterweight is driven by a hydraulic cylinder with a stroke of 13.5 m and pushing load over 1,400 t.

The traveling spud system consists of two spuds; one on the stern and the other at the mid-ship well. The dredge can move by the spud and hauling winches and thus does not require an anchor, mooring ropes or any assistance from auxiliary boats.

The traveling distance of the dredge is about 15 m, which corresponds to the dredging breadth of the grab bucket. The cycle time for one traveling motion is about 3 min. to minimize loss of time at the dredging operations.

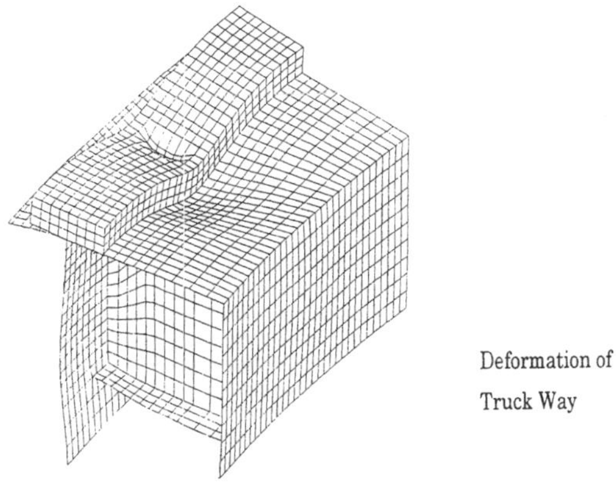

Fig. 4 Fatigue Strength Analysis

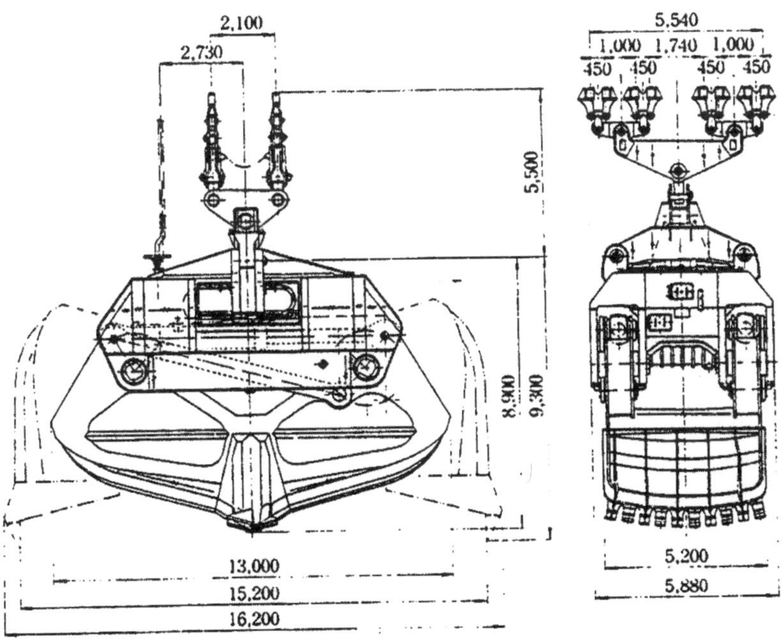

Figure 5 Grab Bucket

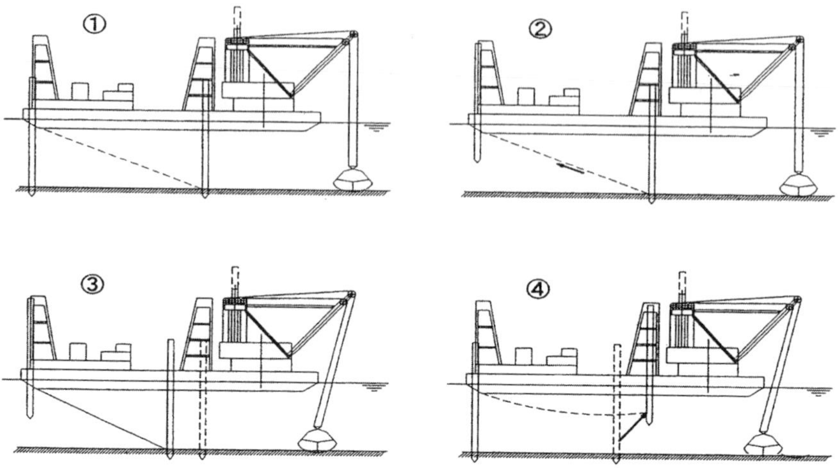

Fig. 7 Traveling Spud

Table

Dredging Results-Average

No.	Depth-m	Volume-m³	Vol-m³	Cycle-sec.	Rate-m³/hr
1.	13 (mud)	500,000	230	135	6,100
2.	10 (mud)	300,000	225	125	6,480
3.	14 (sand)	750,000	190	130	5,300

The above results do not include shifting time of barges nor that of the dredge. At the same time, the average load and peak load for the electric alternators during one cycle are measured and the results are shown as follows:

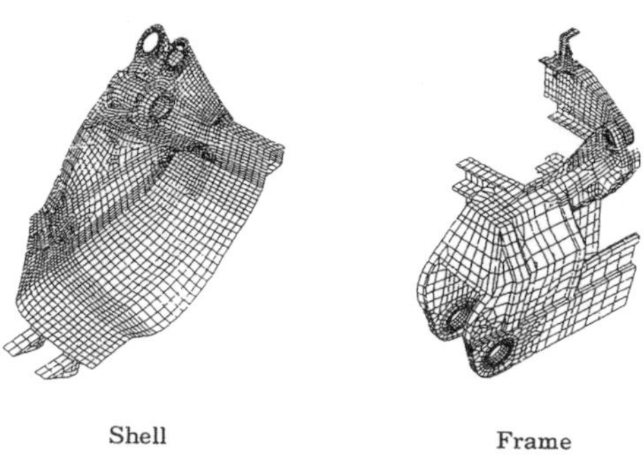

Fig. 6 Results of Analysis

Average load in a dredging cycle
1,400 kW
Peak load in a dredging cycle
3,400 kW

From the above result, total energy for dredging is about 70% of that for conventional types of dredges. When compared at dredging depth of 30 m, it is estimated that energy for dredging is 60% of conventional dredges.

Bucket Ladder Dredge (BL)

Bucket Ladder Dredge for Harbor Deepening

The problem confronting port authorities throughout the US involves deepening channels to accommodate the increased draft of modern container ships. When hard material is encountered, the challenge to deal with it using conventional dredges has produced a dilemma for the dredging industry.

At one time, the Bucket Ladder Dredge (BL) was the work horse for all types of dredging. It was advanced mechanically about the turn of the century through its application to dredge mining for gold and tin. The BL dredge is still used in other countries for navigational dredging and recently for environmental remedial applications. This paper will propose consideration be given to building BL dredges for harbor deepening in the US. A description of BL dredges that have performed in hard materials that are applicable to many harbors will be presented and illustrated with photos and comparative data.

Harbor Deepening

The hardest digging in US ports is mainly on the East Coast such as New York/New Jersey where glacial rock lines the bottom of the harbor. As deeper channels are dredged harder materials are encountered. In cases such as Port of NY/NJ, this requires drilling and blasting followed by bucket dredges; i.e., clamshell, grab, backhoe and dipper.

Bucket Ladder Dredges

BL dredges are normally classified by the size of the buckets. The bucket design can vary from cast manganese steel that is necessary for digging hard material, to fabricated plate for soft or fine material. **Table 1. BL Dredge Bucket Comparisons,** shows bucket size with corresponding production, weights and speed of the buckets.

Table 1. BL Dredge Bucket Comparisons

Bucket Size(ft^3)	Production yd^3(000)/mo.	Depth(ft) Digging	Speed bpm	Bucket wt-lbs
2-1/2	100	35	38	600
6	187	50	34	1300
9	280	70	30	2500
14	466	100	28	4000
18	560	125	25	4700
24	750	150	25	6000
30	850	164	24	8000

The buckets are linked together by steel pins moving over the upper tumbler and running on rollers. When reaching the upper tumbler the bucket is flipped over and discharges into a hopper chute. The material is then either dumped into a classifier, a conveyor, or, into barges alongside.

As can be seen in the figures cited, the buckets are driven around a vertically pivoting ladder with the digging force exerted at the lower tumbler. Approaching the lower tumbler from underneath the ladder the buckets are digging on the catenary which results in three or more of the buckets (i.e., the number depending on depth), making contact along the bottom.

It is this feature that produces a uniform surface on the bottom as the ladder reaches its planned depth and swung laterally along an arc; either constrained by a spud at the stern, or, by a span of wires fixed to anchors.

Figure 1.

BL Dredge Operation & Components

The digging function is conducted in the same manner as a cutter suction (CS) dredge. The cut begins at a depth relating to the size of the bucket and hardness of the material with the dredge swinging in a lateral arc to complete the first level.

Each subsequent cut results in the ladder being lowered then pivoting until reaching the planned depth. After raising the ladder the dredge is stepped forward to begin the next cut. Depending on the hardness of the material the final declination of the ladder may vary from 45° to 60°, but the ladder design must have been made to accommodate the greater angle.

The speed of the buckets also has a direct relationship to the wear of all moving components in the digging function; i.e., buckets, pins, lips, rollers, gears, motors and tumblers. The speed used is therefore a tradeoff between the benefits of higher produc tion with the cost and time necessary to maintain, repair and replace worn materials and components. The higher production usually wins out when management can organize personnel to perform aggressive maintenance to keep ahead of breakdowns.

The natural compensating weight of the bucket line results in a lower power requirement that has the consequence of reducing the cost of production. The dredge may be powered by shore electric or by diesel-electric on board. The largest single winch (Ladder Winch) raises and lowers the ladder carrying the combined weight of the bucket line, tumblers, rollers and its steel plate structure.

Dredge Selection

When deciding upon the configuration and design of a BL dredge to build for a range of port and harbor projects there are several governing features to be considered:

1) Maximum depth to be dredged;
2) Production rate desired;
3) Hardness and impact resistance of the material.

Figure 2

***Corozal* BL Dredge-54 ft³**

Dismantling of the dredge and transporting components for re-assembly at destination has been done frequently with BL/M dredges in past years where either long distances were involved, or, simply difficult terrain when located inland. There are some BL dredges that are self-propelled, sea-going vessels *(See **Figure 1.**)*.

Perhaps the best recommendation for the utility of BL dredges in dredging hard material in ports and waterways was in dredging of the **Panama Canal**, c.1912-1915. The largest was the BL Dredge ***Corozal*** shown in ***Figure 2***, built in Scotland and delivered in c.1912. It carried two sets of 39 buckets each: One of 34 ft^3 for digging rock, and the other of 54 ft^3 for soft soil.

The speed of the *Corozal* bucket line varied from 14-18 bpm. It produced 1,200 yd^3/hr in soft material, while digging at a depth of 50 ft. Records show that the dredge operated 414 dredge days during 1913-15, excavating a total of 1,695,107 yd^3 of which 82% was classified as rock at an average production of 4,100 yd^3 per dredge day.

There were two other BL dredges that were furnished by the French; one that was self-propelled and the other a barge-type of Belgian design. Both used 17 ft^3 buckets and achieved high production in rock and soft material.

In mining applications, a BL/M 14 ft^3 dredge has been considered as providing optimum efficiency on which to build a rugged, high production dredge. Using higher bucket speeds when dredging in sand and clay production records have been set on gold dredges in Colombia using this size dredge; i.e., as high as 550,000 yd^3 per month.

Conclusions

The BL dredge configuration has been overlooked in US civil construction dredging during much of the 20th Century. Based upon some of the facts outlined in this paper the BL dredge should receive serious consideration for port and harbor deepening projects. There is no dredge design that can effectively out-produce a BL dredge in hard material at a comparable cost. Other countries use the BL dredge for maintenance, deepening, and environmental remediation with favorable results.

Geotechnical Information & Rock Dredging; A Contractor's Perspective

Introduction

Improvement and construction dredging projects can involve both rock and general soil (or nonrock) dredging for the deepening of channels and harbors. In general, rock dredging per unit quantity is significantly more expensive than the dredging of non-rock material per unit quantity - often by an order of magnitude. Dredges capable of economically dredging rock are generally larger and more expensive than those that can dredge non-rock material. Production in rock dredging is much lower than in non-rock materials dredged by the same machine. Wear and tear on the dredge and associated pipeline or transportation equipment is considerably higher in rock dredging than in non-rock dredging. Rock often needs to be treated prior to dredging - usually fragmented by explosive blasting or sometimes by chiseling. These extra costs - more expensive plant, lower productivity, pretreatment, higher wear and tear - compound to make rock dredging significantly more costly than non-rock dredging.

Often, the owners of rock dredging projects will include geotechnical site investigation information in plans and specifications prior to taking bids for the work.

Geotechnical site investigations for dredging projects typically involve the taking of soil borings to obtain samples for soil classification and testing. Testing is done both *insitu* during the boring process and in the laboratory on samples obtained from the borings. Typical geotechnical information included in bid packages include logs from the borings indicating the soil stratigraphy and classification, the results of insitu testing of the material, and results from subsequent laboratory testing of soil samples from the borings.

Dredging contractors interpret and use this geotechnical information to select dredging equipment and methodologies, to quantify the location, types, quantities and characteristics of the soils to be excavated, to estimate production rates in the materials indicated by the borings, to assess dredging risks, and to plan the performance of the project. The geotechnical information is thus a primary driver of the cost estimate, pricing, and of ultimate project performance.

For rock dredging projects, equipment selection, methodology, productivity, and risk analyses are sensitive to variations in the location, quantity, and properties of the rock to be dredged. Hence, the risks and cost sensitivity associated with rock dredging projects are high and the quantity and quality of pre-bid geotechnical site information establishes a component of rock project risk.

Inaccurate and/or incomplete geotechnical site information for rock dredging projects can result in inappropriate application of equipment and methodologies, inaccurate production estimates, inordinate risk, and/or inaccurate cost estimates and pricing. Such situations can ultimately lead to unsuccessful projects not only for the contractor but also for the owner of rock dredging projects. Ultimately, poor geotechnical site investigation information can lead to budget overruns, late schedules, disputes and claims. Accurate, complete and comprehensive geotechnical site information can significantly lower risk levels and can result in the appropriate application of equipment and methodologies, accurate production estimates, and appropriate pricing levels providing successful projects for both contractor and owner.

This paper describes the geotechnical information required for estimating planning and performing rock dredging and the relationship of geotechnical properties to rock dredging costs and performance. General criteria, guidelines, and procedures for performing rock dredging geotechnical site investigations and providing information in plans and specifications are presented from the perspective of the contractor. General geotechnical site investigation methodologies, insitu and laboratory tests, and procedures are recommended which can help provide the information required for contractors to properly estimate, plan and perform rock dredging projects.

Geotechnical Information Required for Rock Dredging

When estimating, planning and performing rock dredging projects, the dredging contractor must have information of sufficient quality and quantity to answer the following seemingly simple questions with a reasonable level of contracting certainty:

-where is the rock to be dredged?

-how much rock is to be dredged?

-what are the properties of the rock to be dredged?

Where is it? How much of it is there? What is it? Although such questions may seem obvious and it may be reasonable that prospective bidders be provided sufficient and equal information to answer these questions prior to bidding, it can be the case that the geotechnical site information provided for rock dredging projects is inadequate or of too poor a quality to answer these questions within appropriate levels of certainty.

Rock Location & Quantity

The location of rock formation both in area and in elevation is of paramount importance in rock dredging. Rock dredging projects often involve a material profile with a soil layer overlying and sometimes underlying the rock formation. The contractor must be able to estimate the interface levels of the non-rock and rock layers and the relative quantities of rock and non-rock material.

The top of rock formation in relation to the required dredging depth, or *grade,* fixes the area of rock removal. Rock well above grade involves production rock dredging. Rock at, just above, or just below grade, even if overlaid by soft soils, must still be dredged to clear grade and can greatly influence the production of the overall material layer to be removed.

Rock dredging projects are often bid on a quantity unit rate basis without a separate unit rate or bill quantity for rock material. Such is sensible because the rock layer is often not practically re-measurable. Because the dredging of rock material can be an order of magnitude more expensive than non-rock soil dredging, the relative quantities of rock and non-rock materials are primary drivers of the overall dredging productivity and unit cost.

The contractor must be provided information concerning location of the rock, both in area and in elevation, in order to determine the rock quantity to be dredged.

The first step in rock dredging estimating and planning is to determine the area of rock dredging by establishing the top of rock elevation in relation to grade. Areas with rock above, at, or just below grade require rock dredging equipment and will yield rock dredging production rates. Areas with rock well below grade can be dredged with lighter equipment and will yield non-rock dredging production rates.

The interface of the soil and rock layers is of primary importance in dredging estimating, planning, and performance.

The thickness and area of the rock and soil layers within the dredging template is the basis for the take-off of rock and non-rock quantities. Further, the thickness of the rock layer is an important factor contributing to the dredging production. The thickness of the soil layer over the rock - the overburden - is also of importance as it may be taken at higher production or by alternate dredging equipment prior to dredging of the rock layer at reduced production. If blasting is required, blasting per unit quantity is extremely more expensive for thin rock layers than for thick rock layers.

Rock location, in terms of both area and elevation, and rock quantity in proportion to the overall dredging quantity, are primary factors driving the type and class of dredging equipment required, whether pre-treatment from blasting or chiseling will be required, and the production rates. With the direct cost of dredging equipment a function of its type and class, and the direct unit cost of dredging a function of the dredge equipment cost and productivity, rock location and quantity are direct drivers of rock project dredging costs and the eventual dredging performance. The contractor needs comprehensive and accurate information concerning rock location and quantity in order to estimate, plan, and successfully perform rock dredging projects.

Rock Quality and Strength

The geotechnical properties of the rock, specifically rock quality and rock strength, are also primary drivers of rock dredging equipment selection, methodology, productivity, project duration, and hence project cost. Rock strength and quality are the primary variables in empirical rock production estimating equations.

Rock is excavated by the dredge in layers, and it is the geotechnical properties of these rock layers that are of interest. Apart from the strength of the rock itself, the quality of the rock layer is important to its dredge-ability. Rock quality is an indication of how massive, intact, and homogenous the rock formation is, or is not. Increased stratification, fractures, fissures, weathering, and/or porosity generally lower the quality of the rock formation. Rock layers of low quality may be dredgeable with lighter equipment, with less or no pretreatment and/or at higher production rates than high quality rock formations.

Rock of lower quality is, to a degree, already broken - and requires less breakage energy from the dredge cutterhead or bucket and less blasting or chiseling to allow economic dredging. Lower quality rock has stratified soft layers and/or fracture lines available for attack by the dredge cutterhead or bucket teeth and edge. The rock pieces may be very strong, but it is the state of the rock mass that is also of importance in rock dredge-ability and rock dredging production.

Rock strength, in combination with rock quality, also governs rock dredge-ability and production rates. Rock strength is a measure of the rock's resistance to failure from compressive or tensile loading. Rocks of high strength are very difficult to break with the dredge cutterhead or bucket and may require pretreatment by blasting or chiseling or will be dredged at very low production rates. Rocks of low strength may be dredgeable without blasting or chiseling and at relatively higher production rates.

The density of the rock mass is also important to dredgeability and production. Higher density rock is often of higher strength. The hydraulic transport of the rock pieces in cutter suction dredging is dependent upon the size of the rock pieces and the rock density.

Rock strength and quality are the primary variables which govern the type and class of dredging equipment required to dredge the rock mass, whether pre-treatment from blasting or chiseling will be required, and rock dredging production. Because the cost of dredging equipment is a function of its type and class, and the direct unit cost of dredging a rock quantity a function of the dredge equipment cost and productivity, rock strength and quality are direct drivers of rock dredging costs and the eventual dredging performance. The contractor needs comprehensive and accurate information concerning rock strength and quality in order to estimate, plan, and successfully perform rock dredging projects.

Guidelines for Geotechnical Site Investigations

A number of general criteria, guidelines, and procedures for geotechnical site investigation methodologies, *insitu* and laboratory tests, and presentation of geotechnical information in plans and specifications can be applied to help provide an adequate and accurate basis for estimating, planning, and performing rock dredging projects.

General

1. Take borings in the dredge area.

Although this guideline might seem obvious, projects are often put on the street for bid with a majority of borings provided only outside the dredge area. Ship traffic can make boring in the channel difficult. In harbors, borings are often directed towards areas for civil works, such as piling, dock and foundation work, instead of in the dredge areas.

Borings may have been taken in adjacent dredge areas or on past projects, and the owner elects to provide information from these borings instead of investing in further geotechnical investigation in the dredge area.

Soils and rock layers are extremely variable in location, elevation, and in geotechnical characteristics. Representations of soils and rock conditions must be from borings taken inside the area to be dredged.

2. Perform geotechnical investigations, boring, campaigns and geotechnical testing, according to professional engineering standards.

For information from geotechnical site investigations to be of any use to owners, contractors or designers, investigations and tests must be performed to professional standards. In the United States, ASTM Standards should used. Refer to the ASTM Annual Book of Standards Section 4, Construction, Volumes 4.08 and 4.09. An excellent summary reference for dredging geotechnical investigation and classification is the ***Classification of Soils and Rocks to be Dredged*, a supplement to *Bulletin No. 47* published by the Permanent International Association of Navigation Congresses (PIANC) in 1984.**

3. Bore/core to about 5 feet below grade for non-blasting jobs and about 10 feet below grade for jobs with anticipated blasting.

For dredges to clear minimum dredging level over the dredge area, they must cut rock well below minimum dredging level. Cutter dredges must over cut due to spillage from the cutter and backhoe dredges must attack the material face from under the grade level. Blasting centers must be drilled well below grade to achieve full fragmentation below required depth. Thus, geotechnical information is needed for the material down to and below grade.

4. Take regard of area coverage and zone of influence; space borings accordingly.

Space and locate borings to provide information representative of the entire dredge area - not just concentrated areas. Do not emphasize non-representative locations. One guideline for dredging geotechnical site investigations is to space borings up to a maximum of 600 feet apart depending on the complexity and variability of the seabed and the dredging layers. ***(Verbeek, 1984)*** For rock dredging, borings spaced at about 300-400 foot centers are recommended. Use an alternating grid to provide for maximum coverage.

5. Take an adequate number of borings.

Rock location, quantity, and geotechnical properties are extremely variable even within a project area and within a given layer or formation. An adequate number of borings and subsequent *insitu* and laboratory tests are required to provide information and data that is representative of the rock to be dredged over the entire rock area. An equation, modified from that of **P.R.H. Verbeek *(Verbeek, 1984),*** providing an indication of the number of borings needed for rock dredging projects is:

$$N = 3 + (A^{0.5})/25 \quad (1)$$

where N = the number of borings to be taken and A = the dredging area in square meters. Risk due to incomplete geotechnical information in rock dredging is sometimes covered by adding cost contingencies for lower production, higher wear, longer dredging duration, more powerful plant, blasting, etc. On nearly any rock dredging project, such contingencies will be far more expensive than the cost of an comprehensive boring campaign. A complete boring campaign, with adequate coverage, will lower the risk levels in rock dredging projects. Inadequate geotechnical information that leads to unsuccessful projects is far more expensive than the costs of a comprehensive boring campaign by orders of magnitude. A few more borings will certainly cost owners less than receiving bids with excessive risk coverage or than an unsuccessful project.

Rock Location and Quantity

6. Take special regard of vertical control.

Accurate vertical control is critical to the representation and determination of the top of rock and of soil interfaces, and for determining the rock dredging area and rock quantity. Too often, tides are not correctly accounted for or inaccurate tide readings are applied to boring depths.

Take advantage of DGPS and RTK vertical positioning technologies for cost effective and practical vertical control, which together eliminate the need for tide correction. As a minimum, record tides at short intervals from a near by, calibrated, tide gauge. Take special care in measuring boring advance and corrections for offsets on the barge. Note the achieved vertical accuracy in the geotechnical summary report.

7. Supplement borings with jet probes to document top of rock elevations

Borings provide information concerning rock elevation only at the point taken, and cannot be economically or quickly taken in the numbers required to completely reflect the top of rock throughout dredge area. In many soils, jet probes are inexpensive, expedient, and can be used reliably and practically to document the top of rock formation throughout the dredge area. They should be used in abundance as a supplement to borings.

8. Correlate geophysical data with geotechnical borings.

Geophysical investigation methods such as seismic refraction can also be used to provide information concerning rock location, elevation, and interface levels. Representation of rock location and interfaces from geophysical methods require correlation with physical borings.

Rock Properties

A number of procedures and geotechnical tests both *in-situ* and laboratory, are appropriate and required to provide information on the geotechnical properties of rock formations for dredging projects.

9. Provide fully comprehensive descriptions of the rock core on the boring logs in accordance with engineering standards.

Typical descriptive information includes rock type, color, grain size and texture, strength, state of bedding, jointing, fractures and discontinuities, degree of weathering, *(PIANC, 1984)*.

10. Measure and report on the boring log the rock core recovery, rock quality designation, and rock fracture log to provide information on the rock quality.

Rock core recovery is the ratio of recovered core length to total core run and gives a subjective indication of the quality of the rock core.

A better indication of rock quality is the rock quality designation, or RQD. RQD is the sum length of pieces longer than 4 inches in the core run expressed as a percentage of the total core run length. *(Deer, et al, 1969)* RQD is 100% for strong, massive rocks and is near zero for closely jointed or for highly fragmented rock formations. ***(Franklin and Dusseault, 1989)***

Additional information can be gleaned from the rock fracture log. This log is a count of the number of natural fractures over an arbitrary length - often fractures per foot. ***(PIANC, 1984)***

11. In soft, bored rock layers, measure and report on the boring log the N value from the Standard Penetration Test to give an indication of the rock strength and quality.

Softer or weathered rock is sometimes bored as a soil layer instead of being cored. The N value from the Standard Penetration Test - the number of blows required to drive a standard split spoon 12 inches with a 140 lb. hammer - is useful in giving an indication of the *insitu* quality and strength of such rock. Use only the ASTM standard for the

Standard Penetration Test (a two inch outer diameter split spoon sampler driven with a 140 lb. hammer and a 30 inch drop). Do not use off - size hammers and drops, such as the 300 lb. hammer with an 18 inch drop and 3 inch sampler.

If such non-standard sampling is used in order to obtain a physical specimen, a standard SPT should be performed first to obtain the N value. Blow count value from non-standard tests cannot be correlated with the Standard Penetration Test N value and can provide misleading indications of material strength.

12. Test and report the unconfined compressive strength of a representative number of rock samples throughout the rock quantity to provide information concerning the rock strength.

Run tests on a sufficient number of samples to provide an adequate representation of the strength of the rock to be dredged and the variability of the rock strength throughout the rock quantity. Rock strength is highly variable even within given formations and layers. In general, test strength on several samples per boring down the rock layer. Again, the goal is to provide adequate data representative of the entire rock dredging quantity, not just isolated points.

13. Test and report the unit weight and moisture content of representative rock samples throughout the rock quantity to provide information concerning the quality, strength, and hydraulic transport characteristics of the rock to be dredged.

Refer to ASTM for standard testing procedures. Again, test a number of samples in each boring. This can often be done on the same samples that are tested for strength.

14. Store samples appropriately for testing and test only fresh samples.

Loss of *insitu* moisture can alter the geotechnical properties of rock samples. Wrap samples in plastic to avoid moisture loss.

15. Provide a color photo log of the boring run.

A picture is indeed worth a thousand words. Much qualitative information can be gleaned from color photo logs of the boring run. As a bare minimum, photos should include identification of the boring, core run designation, the running length of core, and start and stop levels of the core run.

Total core recovery and RQD measurements should ideally be labeled in the photo as well. All such data can be put on labels that are photographed along with the core. If practical, photos should be taken at the boring site so that they show the state of the core without disturbance from further handling.

16. Save boring runs for contractors' inspection.

Even better than a photo log, a physical inspection of the boring run provides valuable qualitative information for the contractor. Samples should be properly stored in sealed plastic wrap.

Conclusions

Rock location, quantity, strength and quality are the primary drivers of the type and class of dredging equipment required for rock dredging, whether pre-treatment from blasting or chiseling is required, of dredging production and costs, and of the ultimate success of rock dredging projects. Comprehensive and accurate geotechnical information is required for contractors to plan, estimate, bid and successfully perform rock dredging projects. A number of guidelines, procedures and geotechnical tests both *in-situ* and laboratory, are appropriate and are required to provide comprehensive and accurate information of the geotechnical properties of rock formations for dredging projects.

Dredge Components & Systems

Submerged Pump

Patented in 1959 in a development project under the supervision of **Prof. Jan de Koning**, The Netherlands, the Submerged Pump was first installed on the hopper dredge ***Amsterdam III*** in **1960**. This feature eliminated the suction deficiency and increased production of dredges below the 32 ft barometric pressure level. Retrofits followed and were applied to many CS and TH dredges in Europe in the 1960s increasing production levels. This modification was followed in the US by the 1970s. Until 1960 pumps were placed in the hull of suction dredges. With the advent of the submerged pump, suddenly densities of 1800 kg/m^3 were easily raised by hydraulic dredges from substantial depths (50 m).

The first submerged pump installed on a CS dredge in the US, was in 1974. Jet assisted pumps were introduced there in the 60s. However, they are less efficient since the average water flow is 40% or more (in order to keep the dimensions of the "throat" of the pump large enough). Submerged centrifugal pumps can reach a hydraulic efficiency of >80% (double curved vane design) so that the overall efficiency is >70%.

With further advances in powering submerged pumps either by hydraulic, electric motors, or, long drive shaft, suction dredging was made possible to previously unheard-of-depths. Today some Jumbo Hopper (TH) dredges, can dredge to a depth over 400 ft (122 m). This system is being used primarily for deep trenching and laying of pipelines offshore.

Water Jet Technologies

Hopper Dredge Dragheads

Water jet systems applied to dragheads include some of the following applications. *Jetflowhead* was developed to improve efficiency in pipeline landfall and offshore trenching. *Jetplough* was a development of sweep-beam technology with a large number of water jets placed along the cutting edge. This creates a "sliding surface" reducing the effort required to pull the material away. Designed as a bed-leveling device, the power and precision with which it can be deployed ensures an accurate surface level within very small tolerances.

Dracula (Dredging Assisted Cutting by Liquid Action) is a system based on high pressure jetting that allows hopper suction to work more effectively in stiff soils. Using high pressure up to 400 Bar, it has shown its effectiveness involving stiff soils such as "cemented" sand, hard or compacted in addition to stiff clays.

California Draghead. A design that was developed by the US COE it helped to pioneer the TH dredge for maintenance dredging. The draghead disintegrates the sand with a combination of the erosive system and mechanical cutting with linear blades. Because the erosive waterflow is divided in several sub-streams through the gates and only a few cm's have to be eroded per gate, Cal-Dragheads are indicated for excavation of hard packed sand. A disadvantage however is that the draghead consumes high pressure drops. Techniques are employed to minimize the deficiencies.

Dutch Draghead. Introduction of this system in the 1960s was one of the major contributions to the improvement of the TH dredge. Together with the width of the drag head, the bank cross-section per pass of the drag head varies from 0.12 m^2 to 1.6 m^2.

Jet-Ejector dredge

This system was developed in Japan in the period 1961-3 by **IHI** and built five small to large jet-ejector dredges to be able to reach depths of 100 m. These special dredges had already been in service for several years in **Kobe, Osaka** and other Japan ports. The main characteristics of this dredging system are:

1. Digs through silt and clay layers at the minimum necessary diameter to reach the sand and gravel layer underneath as quickly as possible. This minimizes the uplifting of excess silt and clay.
2. Excavation of large volumes of S&G from depths to 70 m.

A general description of a jet-ejector dredge, the ***Asia No. 1*** follows:

Maximum dredging depth:	70 m
Nominal discharge distance:	2,000 m
Dredging capacity:	250 m^3/hr
Hull size:	47.4 m X 14.5 m X 3.5 m
Ejector pump:	1,200 m^3/hr X 160 m, 1 set
Jet pump:	1,100 m^3/hr X 180 m, 1 set

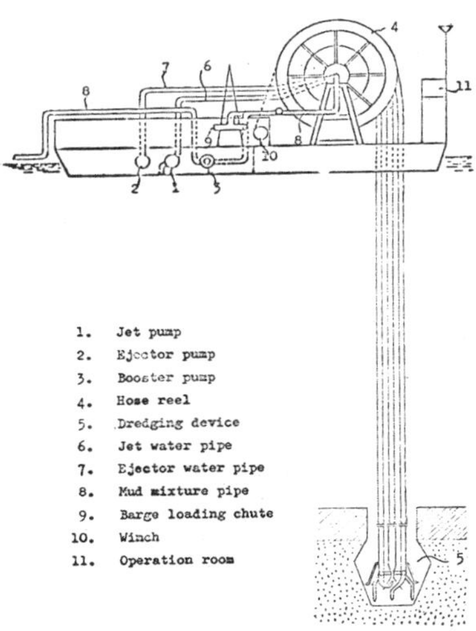

Fig-1 Dredging mechanism of "ASIA NO1"

Diesel engine for ejector & jet pump:
2,500 hp, 1 set
Booster pump: 2,200 m^3/hr X 90 m, 1 set
Diesel engine for booster pump: 2,000 hp, 1 set
Main pipe diameter:
Jet line: 300mm
Ejector water line: 300mm
Mud line: 445mm

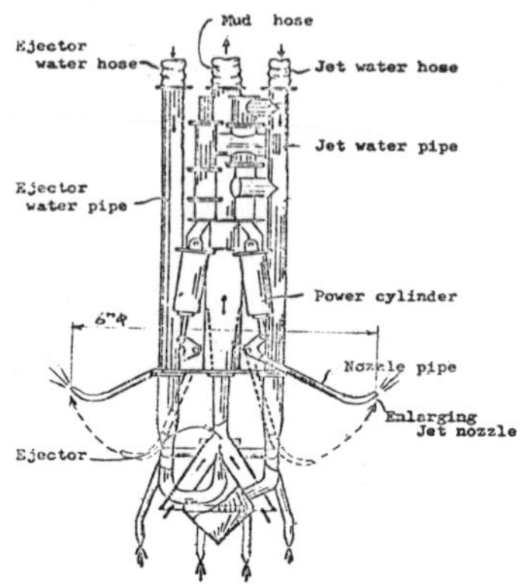

Fig-2 Dredging device of "ASIA NO1"

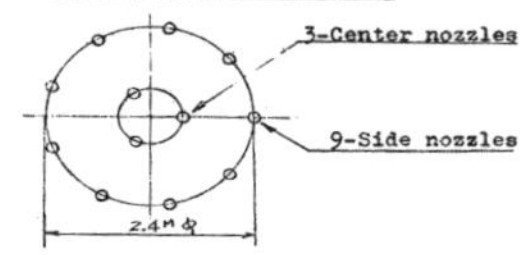

Fig-3 Arrangement of digging nozzles

These systems (photo) have a large reel, 15 m diameter installed on the fore deck on which are wound two lines of cast steel chains and three flexible rubber hoses for jet water, ejector water and mud mixture, with their tips connected to the dredging device. They are lowered and hoisted by an electric winch installed on the deck and operated in the operation room built at the bow. Pressurized water of 18 kg/cm^2 for the jet nozzles and reel shaft and rubber hoses to the dredging device by a jet pump and ejector pump respectively.

In the operation room are provided a **Co60 radioactive mud concentration indicator**, an electromagnetic flow meter and a mud-flow rate indicator which allow the operator to maintain optimum dredging efficiency.

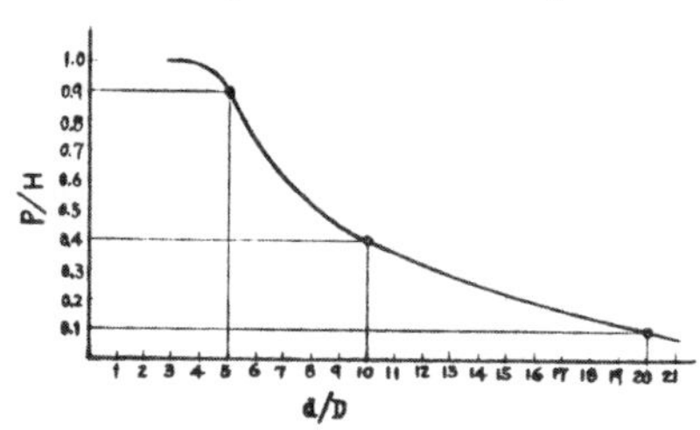

H : Jet pressure at nozzle tip
P : Jet pressure at dredging well wall
D : Inside diameter of nozzle tip
d : Distance between nozzle tip and well wall

Fig-4 Relationship between d/D and P/H

These systems were an early development to meet the increasing demand for sand and gravel for the massive reclamation project for new industrial land in Japan. After depleting the sand on the shallower harbor bottoms, the dredging companies had to resort to new methods of reaching deeper deposits.

Underwater Trenching Apparatus

Since the development of propeller driven vessels, *wheel washing* has been used as a method of removing small isolated shoals. In shallow rivers, this method is not always the most efficient, but it is often the only one available to help eliminate the trouble-some shoal areas.

In areas where the shoals are composed of silty or polluted material, the term *wheel washing*, has become controversial. In 1969 a tug boat company in Oregon, patented a *underwater trenching apparatus* which serves to deflect the wash from twin 55-inch diameter, 30-inch pitch propellers downward, to hydraulically remove bottom materials immediately astern of the vessel. This was initially installed and utilized on the tug, ***Salvage Chief*** with a hull of 192 ft. long and engine of 1800 hp with 600 rpm, to each of the two propellers.

Using a six point mooring system with the dredging door in *down* position, enables the vessel to actually pull itself astern, when under full power. This provides for quick and efficient trenching in bottom depths down to 30 ft. The vessel can operate directly in the surf and can back itself into the zero waterline while trenching. It can dig a trench about 75 ft wide and 20-30 ft deep through 1,000 ft of surf in about 24 hrs.

Adapting this concept to a smaller vessel

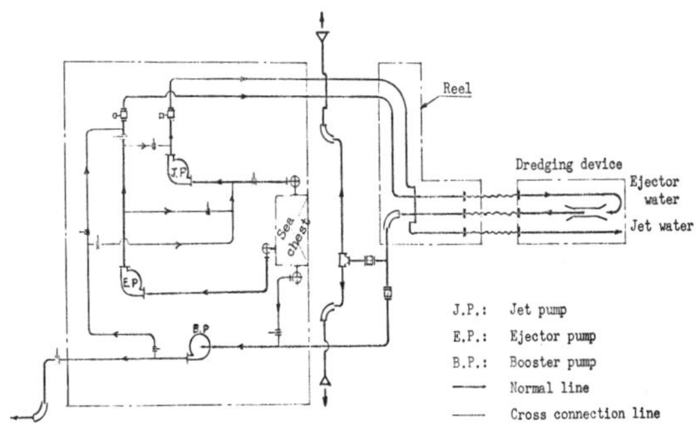

Fig-5 Piping diagram of pumps

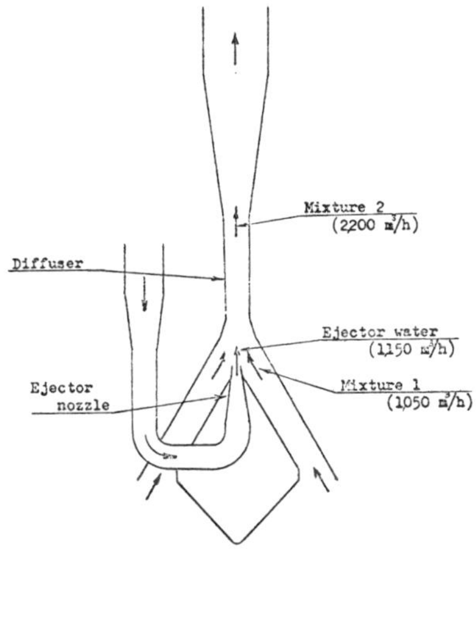

Fig-6 Relationship of ejector water, Mixture 1 and Mixture 2

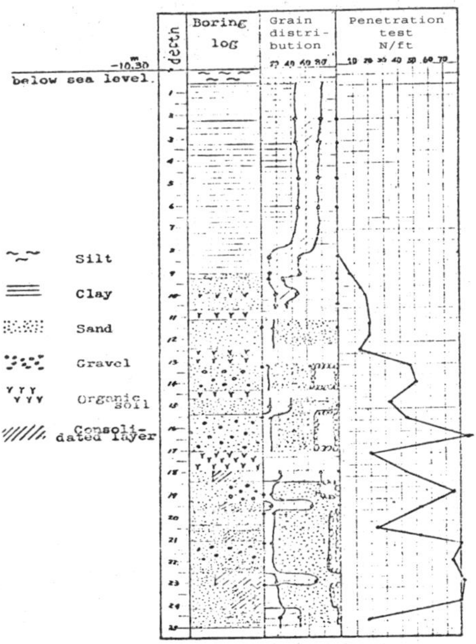

Fig-7 Nature of underground soil

for reduced cost and more flexibility in its operation, a converted LCM 8 owned by the COE in Portland was rigged with a *dredging door* (after obtaining permission from the patent holder). The door on the ***Salvage Chief*** was 22 wide x 20 ft long. The door on the LCM was 18 x 12 ft with two 34-inch diameter, 20-inch pitch propellers and 330 hp to each. The latter vessel was named ***Sandwick*** and was first utilized by the Portland District in 1972 and continued thereafter. Some production examples were: **9,000 yd³ in 24 hrs or 260 yd³** per *flushing* hour.

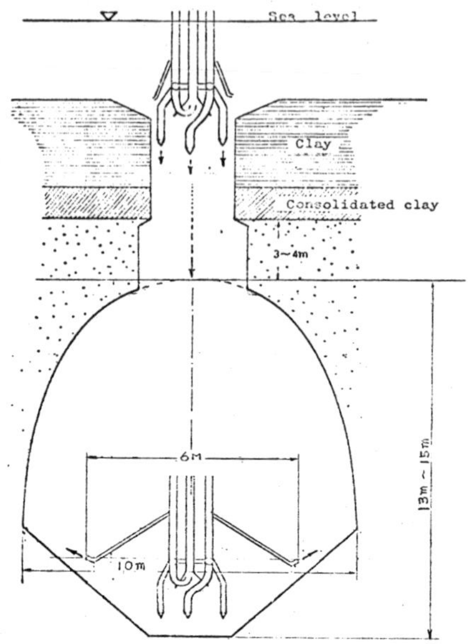

Fig-8 Standard configuration of a well formed by dredging

Hydro-jet Deep Embedment Method

Stimulated by the increasing demand for the transportation of greater quantities of material, a method was developed in the 1940s in Germany for the embedment of flex-

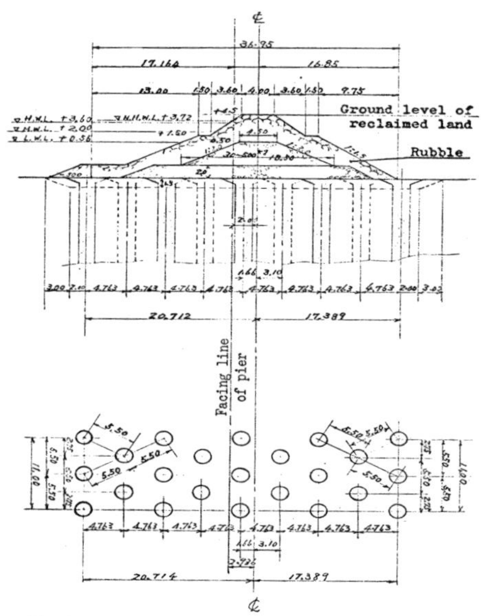

Fig-9 Foundation work employing sand columns

ible cables and pipes in the water bottom. Depending on the depth of water, three different types of machines were applied (c. 1972).

1. In deep waters, the actual hydro unit with hydro-injector is carried and guided by so-called laying vessels which also supplies the required pressure water;
2. In shallow waters, where the laying vessel cannot run, pontoons, dismountable for rail and road transport, are used as carrier and supply station of the hydro unit;
3. In extremely flat or deep waters, sledge

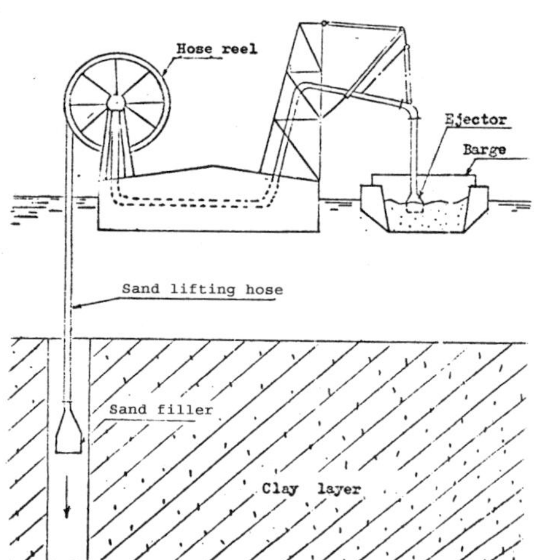

Fig-10 Sand filling operation

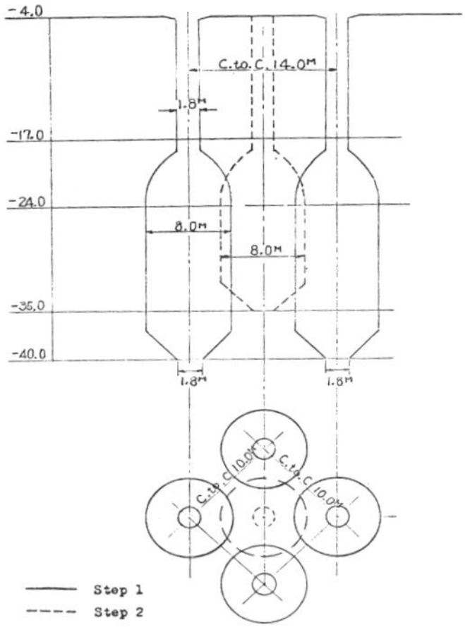

Fig-11 Wells dredged in 1st step and holes dug in 2nd step

type unit-carriers are used which may even have their own crawler drive.

Where the water jet hits the soil, the latter is loosened and thrown up. The hydro-injector can thus be pulled through the loosened soil with reduced and acceptable effort. With the same passage, the flexible cable or pipe is laid into the prepared ditch.

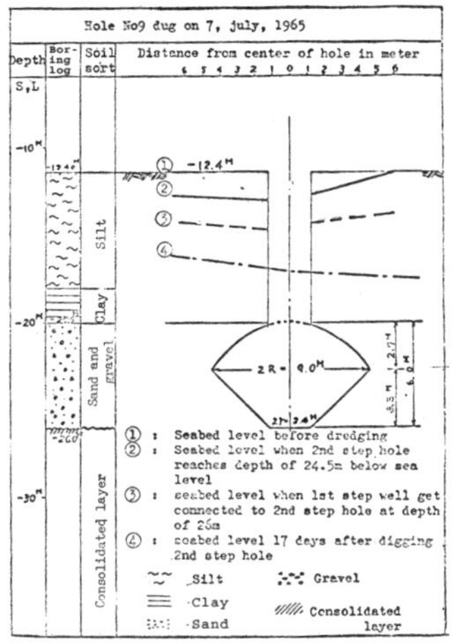

Fig-13 Condition of seabed sinking

An essential feature of the above method is the fact that no material has to be removed for the preparation of the cable or pipe ditch. The material is kept floating in the vicinity of the hydro-injector and settles again relatively soon behind the hydro-injector, into the ditch covering the cable or pipe, laid into the ditch.

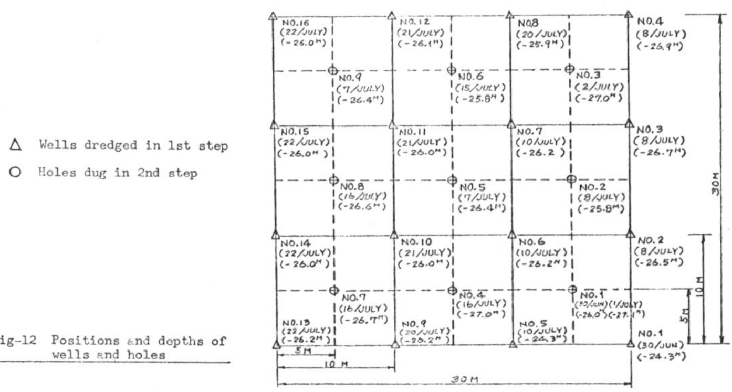

Fig-12 Positions and depths of wells and holes

Picture-1 Construction of breakwater in the port of Kobe

Picture-2 3 small jet-ejector dredgers

Picture-3 "ASIA NO. 1"

Picture-4 Small jet-ejector dredger provided with a sand filling device

Picture-5 Dredging device of ASIA NO. 1" under fabrication at IHI

Picture-6 Discharging sand and gravel into hopper barge

Picture-7 Pressurized jet water streaming out from the digging jet nozzles

Picture-8 Large gravel sucked up by ejector in the port of Kobe

Picture-9 Ejector of "ASIA NO. 1"

Dredging & Backfilling Works

New Dredging Technology for Pipeline Installation by means of Dilution

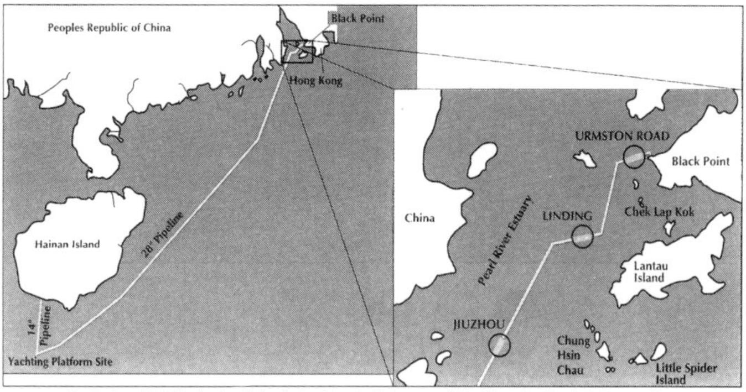

The Yacheng 13-1 Pipeline Installation Project from the Yacheng gas field offshore the island of Hainan to the Black Point Power Station at Hong Kong.

Background

Soil information for **Arco China's *Yacheng* 13-1 Gas Pipeline Installation Project**, across the **Pearl River Estuary** in the People's Republic of China, challenged the Contractors to develop new and innovative methods for the installation of the pipeline across the ***Urmston* Road Channel**. Over the last section of the **778 km** pipeline from the *Yacheng* gas field offshore the island of **Hainan** to Hong Kong, the pipeline had to be installed on a firm and leveled seabed, below a layer of soft to firm soil, up to a depth of 20 m thick in some areas.

This layer mainly consisted of previously dumped clay and silty marine deposits from many sources, including the new **Chek Lap Kok Airport**. The Hong Kong Authority requires a limited amount of soil be excavated from the trench by conventional methods and disposed of at approved dumping locations. As this quantity was far less than the amount of material required to excavate a conventional open trench, a new dredging technique was proposed by **Jan De Nul**'s Engineering Department and further developed in close cooperation with **ARCO China Inc.** and **SAIPEM-EMC J.V**.

The execution method proposed was based on *diluting* the cohesive materials in subsequent layers to such a reduced density and shear strength, that the pipeline could be installed through the diluted medium to the specified design level at the bottom of the trench. Due to dense and non-cohesive soils encountered in the deeper areas, a combination of dilution and traditional dredging methods were utilized to ultimately effect the installation of the pipeline.

The dilution works at Urmston Road, were executed during September - October 1994 by means of the trailing hopper dredges, ***J.F.J. De Nul*** (11,750 m^3) and ***Vasco da Gama*** (10,000 m^3). These vessels were respectively the largest and third largest trailing hopper dredges in the world and had been modified for the new technique. As a result of their size and capabilities, trench preparation could be executed over a minimum period of time at maximum efficiency.

After installation of the pipeline at *Urmston* Road, a rock bund was installed by means of a large, self-propelled split-barge, providing stability for the pipeline and protection against anchors. Outside Hong Kong, at the location of the *Jiuzhou* and *Linding* shipping channels in the People's Republic of China, a 3-layer engineered backfill was applied, executed by means of a side stone dumping vessel.

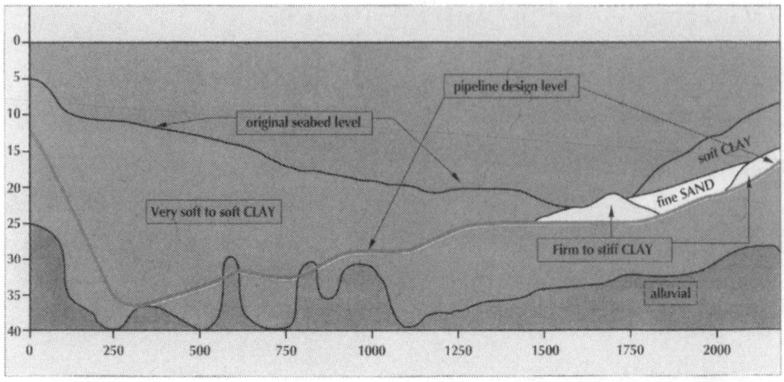

Geotechnical profile at the Urmston Road crossing.

Works executed by Jan De Nul N.V.

Gas transportation from the *Yacheng* 13-1 gas fields, offshore the island of Hainan in the People's Republic of China, to the Black Point Power Station at Hong Kong, required the installation of 778 km of 28" pipeline in the South China Sea. Over the last 71 km, the pipeline had to cross the *Urmston* Road, *Linding* and *Jiuzhou* shipping channels in the *Pearl River* Estuary.

For the crossing at the *Urmston* Road Channel, the pipeline had to be installed on a firm and leveled seabed below a layer of soft materials of up to 20 m in thickness at certain locations. This required a new and innovative method to be used, as dredging by conventional methods and disposal at approved dumping locations was limited by Authority restrictions.

After installation in this section, the pipeline had to be protected by a rock bund providing 3.5 m cover above the pipeline and with top of backfill materials always remaining –21.5 HKPD over the central section. Across the *Linding* and *Jiuzhou* channels, a trench had to be pre-dredged in order to install the pipeline below minimum future draught requirements. An engineered backfill had to be provided for sufficient stability of the pipeline and an adequate protection against shipping anchors.

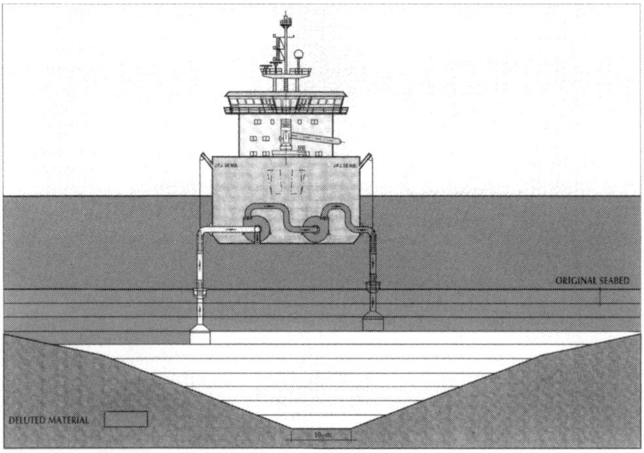

Principle of trench dilution (cross profile).

The Yacheng 13-1 Pipeline Installation Project

The 28" pipeline was required to be installed across *Urmston* Road which lies within Hong Kong's Outer Deep Bay dumping area, on the border between the Chinese and Hong Kong territorial waters. The pit at *Urmston* Road was formed by dredging methods for reclamation purposes, extraction being completed in 1991.

The dredged seabed at that time was alluvial, with the marine deposits having been removed prior to the extraction of sand. As such, the firm sea-bed in the pit was known to be changing rapidly, due to the erratic shape of the earlier dredged bottom.

Since 1991, a variety of materials had been dumped in the pit, largely comprising clayey and silty marine deposits dredged by clamshell and other methods. The sources included materials from the new Chek Lap Kok airport works and other nearby and more distant projects in Hong Kong and from the People's Republic of China.

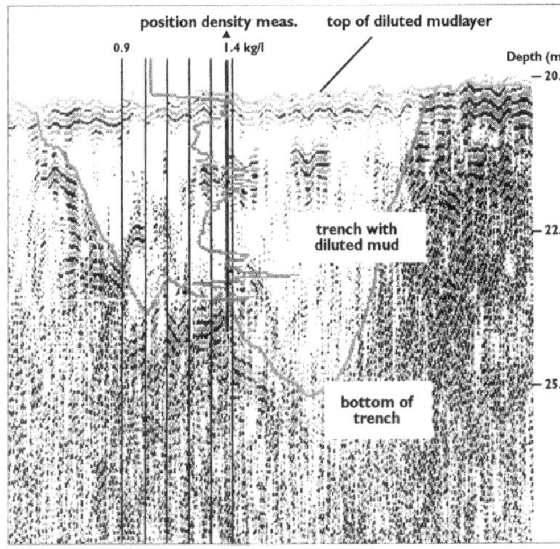

Digital record of cross section of trench during the beginning of the dilution works

In the western half of the channel the thickness of the layer of dumped mud and very soft to firm clay, varied from 10 m to over 25 m. The density of these materials was up to 1.7 t/m^3, with a shear-strength above 1 kPA. This posed two major limitations on pipeline installation. First, the highly irregular alluvium profile was unacceptable from a pipeline stress point of view, and second, the varied soil properties could not be penetrated by the pipeline during installation in order to reach its specified design level.

Due to the limited soil information available across *Urmston* Road, a detailed geotechnical and geophysical survey was undertaken to finalize the pipeline installation scope of work. From the final report, the optimum profile was analyzed, seabed preparation methodology developed and installation procedures prepared. This seabed preparation method required the utilization of two, 10,000 m^3 TH trailing suction hopper dredges. Modification of one of the 10,000 m^3 TH dredge, was made to allow high capacity water injection, via the suction pipe, into the mud medium thus reducing the *in-situ* density.

To assess the effectiveness of the proposed dilution operations, extensive trials were undertaken. These trials showed that sufficient reduction of soil shear strength to allow penetration of a test pipe could not be achieved by controlled pumping of water at varying layer depths only. As already proposed in the installation procedures, as an alternative method it appeared that optimum effectiveness could be achieved by dredging predetermined box-cut layers with one suction pipe, and simul-

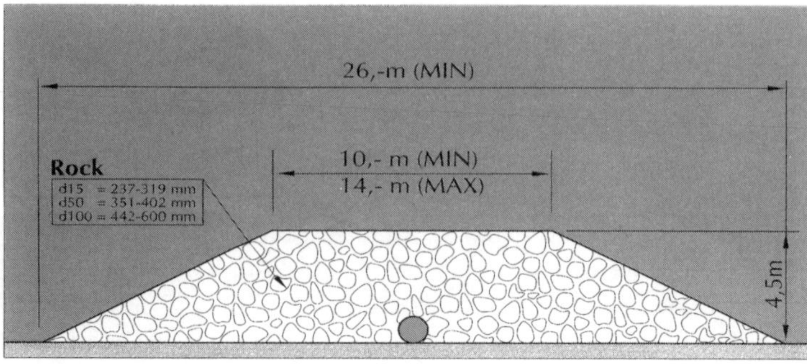

Cross section of rockbund at Urmston Road.

taneously discharging through the opposite suction pipe into the trench at the top of the nominated layer, after adding water through one of the dredge pumps. This reduced the *in-situ* shear strength and achieved a returning target density of 1.35 t/m^3.

Although the initial field trials of the dilution technique were proven to be successful, the presence of pockets of consolidated clay and granular seabed material necessitated, particularly in the deeper sections, the use of traditional dredging techniques. As part of the lighter components of the diluted soil was washed downstream by the river currents and the sand settled to the bottom, dilution was no longer effective by the time 3/4th of the required depth was achieved. Conventional dredging techniques then had to be used to reach the design depth.

At these depths, *in-situ* trench materials were disposed of at agreed dumping grounds, together with diluted soils which were loaded on top of the granular materials until overflow level was reached. As slope stability had been based on the presence of diluted material in the trench profile, dredging as a result of slope instability had to be continued at some locations in order to achieve the target design level for pipeline installation. By the combined result of these dilution and dredging operations, a mainly open trench was created into which the pipeline was ultimately installed.

The as-laid survey was performed in early October, directly upon completion of pipe-laying operations and revealed that slope failures had occurred between dredging and pipe laying in the deepest portion of the trench. However, a subsequent soil investigation revealed the soil to have higher sand content than before and now provided a suitable foundation for the pipeline.

Principles of trench preparation by means of dilution by using a large trailing hopper dredger.

Principles of the new dredging technique by means of dilution

Diluting of the trench materials is executed in accordance with detailed procedures, based on the working principles of a trailing hopper dredge (TH). The works are executed in successive layers of up to 2.0 m in thickness, where box cuts over the width of the drag head are removed. In this respect, the TH dredge sails parallel tracks at a speed of around 1 knot at average 3.5 m inter-spacing while diluting each layer. In case of limited drag head penetration as a result of more cohesive soils, the height of the box has to be reduced.

Installation of rockbund at Urmston Road by means of the 2000 m3 selfpropelled splitbarge "Verrazzano".

The *in-situ* density and shear strength of the trench materials are reduced by dredging each box measuring 2.0 m x 3.5 m, with one of the suction pipes. Before entering of the mixture in the suction pump, a quantity of water is added by controlled setting of one or more inlet valves. The density of the material is measured by the radio-active measuring devices at the outlet of the pump.

After having passed the pump, the diluted mixture is pumped down to the trench by means of the other suction pipe, which is, for this purpose, serving as a fall-pipe. The drag head of the suction pipe acts as diffuser-head and is positioned at a depth approx. 2.0 m higher than the nominated suction level.

Choice of equipment and required modifications

The dilution works at the *Urmston* Road crossing, were executed by means of Jan De Nul's trailing hopper dredges *J.F.J. De Nul* (11,750 m^3) and *Vasco Da Gama* (10,000 m^3). The following modifications had been executed prior to start of the works:

- Pump room alteration for the outlet of the starboard pump to be connected to the inlet of the port suction pipe.

- Port suction pipe adopted for reverse flow.

- Setting devices on water inlet valves for control of outgoing mixture density.

- Positioning and trucking displays and logging with hard

Seismic survey results of the as dumped profiles.

copies, to be provided if required.

- Drag head position systems installed.

Survey and monitoring

Particular survey and testing methods have been developed in order to enable monitoring of the works during the progress of dilution:

- Density measurements were carded out in the diluted layers by means of the STEMA density probe launched from the survey vessel.

- The drag head position monitoring system installed on the dredges provided on-line information of the dredge levels reached, and a longitudinal drag head survey could be added to the daily report.

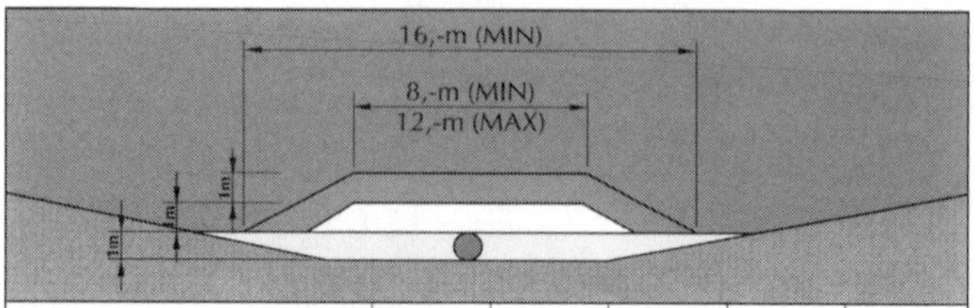

	thickness	d15	d50	d85	d100
BASE LAYER	1m	1mm	2mm	4mm	-
INTERMEDIATE LAYER	1m	10mm	20mm	35mm	-
ROCK ARMOUR LAYER	1m	102/136mm	150/172 mm	-	189/256mm

Cross section at Linding channel.

Pipe dunking tests provided information on the actual diluted levels reached. The pipe dunking was executed by lowering a sealed air-filled 12.2 m long section, of the concrete coated 28" gas pipe by means of the dredge's deck crane, while the vessel remained on position during slack tide. The reading of the underwater weight of the pipe section (both ends closed) relative to the levels reached, provided information about to what anticipated levels the pipeline would sink during installation from the lay barge. After the initial phase, a scaled 1.0 m long model of the 28" coated pipe, handled from the survey vessel was provided by the Main Contractor, in order to save time for the dredges.

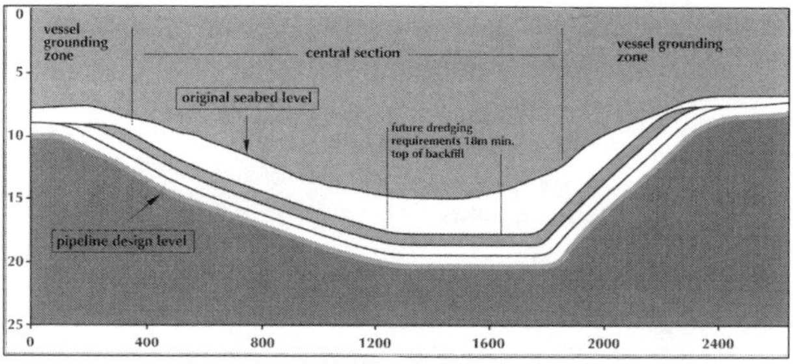

Longitudinal backfilling profile at Linding channel.

After completion of the dredging and dilution works and successful pipeline installation at *Urmston* Road, the pipeline had to be protected by a rock bund providing 3.5 m cover above the pipeline. The top of the backfill materials always remained below –21.5 HKPD over the central section. During the approx. 3 months execution period, of these back filling works, the trench profile remained remarkably stable. In view of the thickness and width of the rock bund to be installed within the central part of the *Urmston* Road section, the works were executed by means of]an De Nul's large, self-propelled seagoing split barge, ***Verrazzano***.

The self-propelled split barge was positioned longitudinally above the pipeline at slack tide by means of its twin rudder propellers and bow thruster. The load of the vessel, approx. 2400 tons of rock, was such that an average of three dumps was necessary in order to build up the theoretical profile.

In order to stabilize the pipeline as quickly as possible, the pipeline was pinned first by means of intermediate dumps of 50 m long every 100 m. Subsequent, in-between dumps, completed the first layer. Afterwards, the second and third layers were installed in accordance with the dumping sequence of the procedures. Defined *windows* were left open and continuously surveyed to monitor the position of the pipeline.

The survey data was utilized by SAIPEM-EMC JV and ARCO, to determine and optimize the dumping sequence. Intermediate and final surveys were executed after each dump by means of bathymetric and seismic surveys. Considering the method adopted, installation of rock armor materials for this kind of dumping profile, by means of a split-barge proved to give acceptable results at lower costs than in the case of using a side-dumping vessel.

In total, over 350,000 t of rock have been installed by this method, providing protection for the pipeline against 27.5 t stockless anchors over a section of 2180 m.

At the location of the Jiuzhou and Linding channels, in the territorial sea of the People's Republic of China, a conventional design of a pre-dredged open trench with engineered backfill was applied. Taking into account the fine bottom material, and in accordance with filter rules, a 3-layer backfill design has been adopted providing a cover of 2 m above the top of pipe. As such, sufficient stability for the pipeline and protection against a possible impact of 13.5 t stockless anchors, have been provided for.

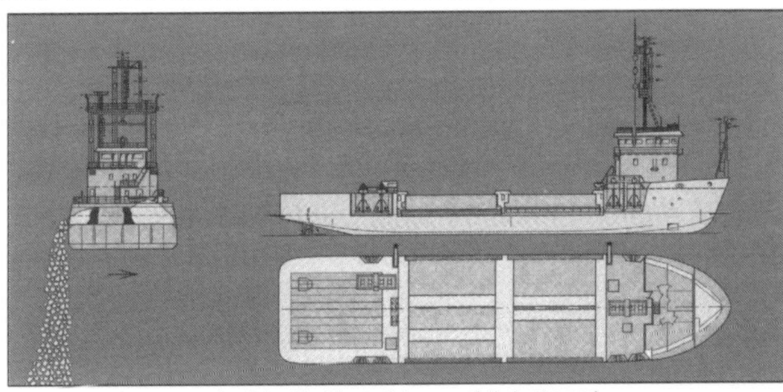

Installation of backfill materials by means of side dumping vessel.

In order to provide accurate installation of all three layers in these channels, the rock material has been placed by controlled sideward disposal by means of a side dumping vessel. The side-dumping vessel allows accurate placing of predefined quantities of backfill material at predefined locations. Up to 2000 t of backfill materials are loaded on the four cargo decks. On arrival above the dumping location, the vessel positions itself parallel above the pipeline, and installs each layer to the specified thickness and width by means of controlled side-wards dumping.

During the dumping sequence, the vessel moves side-wards in a controlled manner on its dynamic positioning system using its twin rudder propellers and bow thruster. The dumping rate can be controlled, by adjusting the speed of the hydraulic rams on the slide scrapers. By means of a combination of the dumping rate and the speed of side-wards movement, any profile of uniform or linear increasing or decreasing thickness can be achieved.

The rock was supplied by PIONEER Quarries Ltd., from local quarries located at Chung Hsin Chau and Little Spider Island in The Peoples Republic of China.

Europipe Development Project Project Overview for Europipe Landfall

The supply of additional natural gas from Norway is desirable both for energy management and for environmental considerations. For these reasons, Den norske stats oljeselskap a.s. **STATOIL** is building the *EUROPIPE* pipeline to transport natural gas from the Norwegian part of the North Sea to Germany. To satisfy the increased demand for natural gas, the Europipe pipeline - with a total length of 650 km - had to be completed by 1st October 1995.

The construction procedures are designed to minimise environmental impact while at the same time guaranteeing technical feasibility. For this purpose, various alternatives have been investigated. The EUROPIPE route applied for will enter German territory north of the East Frisian islands and will cross the Lower Saxonian Wadden Sea National Park through the Accumer Ee coastal inlet between the islands Baltrum and Langeoog. The tidal flats of the Dornumer Nacken are crossed by tunnel and the pipeline shall reach the mainland at Dornumersiel. From there the pipeline will continue to Emden.

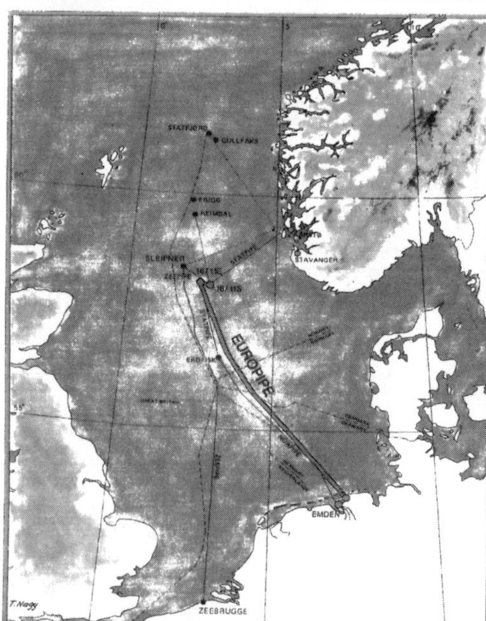

The pipe laying in the near shore area, the sand bank, the tidal inlet and in part of the Accurnersieler Balje will be done by pipe laying barges in a dredged trench. The inside tidal flats, the foreland and dike will be under-tunneled. The construction activities in the tidal flats will take approximately 15 months, and in the area of the tunnel construction site approximately 21 months.

The intervention is made in the Wadden Sea, one of the last natural landscapes of Central Europe, which has been placed under special protection because of its large area and its uniqueness. Despite the existing pressures on the environment (pollution e.g.), the still dominantly natural dyn-

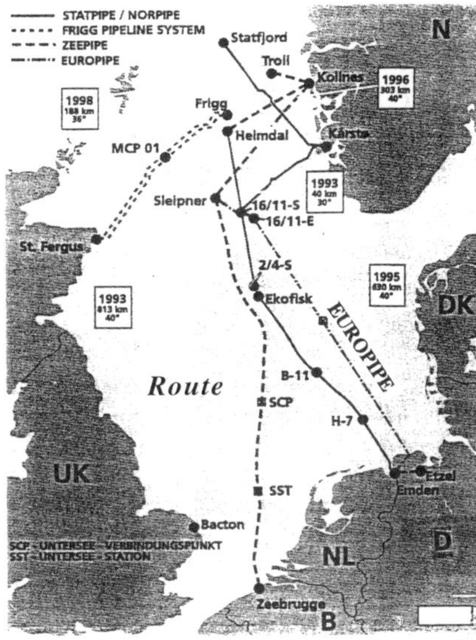

amics of the ecosystem processes are a valuable feature of the area.

The pipe laying will cause large-scale, large-range and intensive disturbances of these processes. After the pipelaying in an open trench, the original topography of the sea ground will be restored. It is probable that the characteristic processes in the area will be reestablished after a *settling down period*, since no major effects due to the existence of the pipeline and its operation are expected and since the pipe is not expected to be exposed. Thus the construction work will probably not result in any long term damage to the ecosystem processes.

Route

The pipeline landfall section with a length of ca. 12 km extends from the 3 sea mile limit north of Baltrum, between Baltrum and Langeoog, following the *Accumersieler Balje* and crossing under the *Dornumer Nacken* up to the tunnel starting shaft located inside the dyke in the *Munsterpolder.* According to its various water depths and local conditions, the route is subdivided into route sections which require different construction methods and construction equipment.

* Nearshore region

With depths between NN - 18.0 m and NN - 9.0 m, this area constitutes a natural shelf in front of the sandbank with special wave conditions.

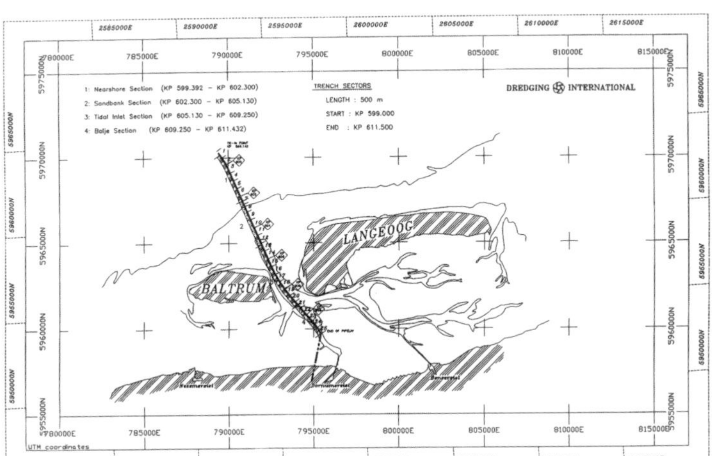

* Sandbank

With depths of less than NN-9.0m, in some parts only NN-3.0 m, this area forms a shallow water barrier between the islands. It is characterized especially by the prevailing wave climate (presence of breaking waves) and by tide-dependent restrictions on marine traffic.

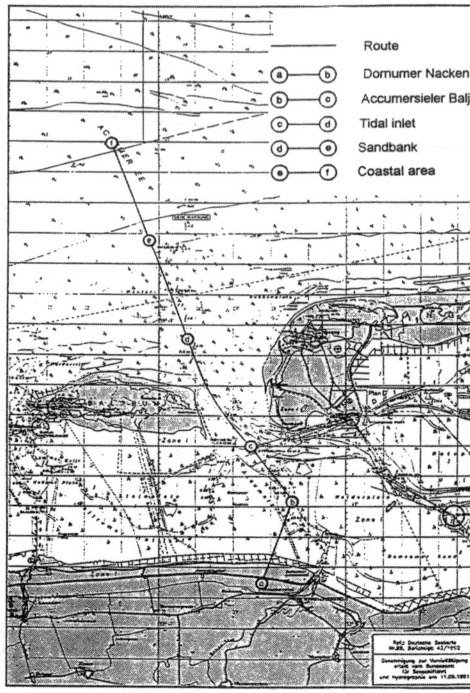

Europipe route in landfall area. Different route sections.

* Tidal channel

The tidal channel marks the most narrow passage between the islands with a water depth of up to NN - 22,0 m and high current velocities which require special construction methods and equipment.

* Accumersieler Balje

Special features of this section are its low water depth and a pronounced, directed tidal current.

* Dry-falling mudflats

From the transition point in the *Accumersieler Balje,* the dry-falling mudflats section of the *Dornumer Nacken* is under-crossed up to the *Munsterpolder.*

National Park

Lower Saxonian Wadden Sea

The *National park* covers an area from Emden to Cuxhaven of 240.000 hectares. The Wadden Sea is one of the last natural landscapes in Central Europe. It is still largely intact and represents a unique ecological area. Despite the existing pressures on the environment, the natural dynamics in the Wadden Sea represent a valuable characteristic biosphere for many flora and fauna species. (biosphere = part of the earth where life is possible).

The Wadden Sea comprises a most sensitive area with respect to flora and fauna, and breeding and resting places for birds and seals. The national park has a unique landscape; it contains a variety of islands, gullies and tidal channels, sandbanks, mudflats falling dry during ebb-tide and salt marshes. The Wadden Sea is also recognized by the European Community as an area of international importance. Special directives have been proclaimed for

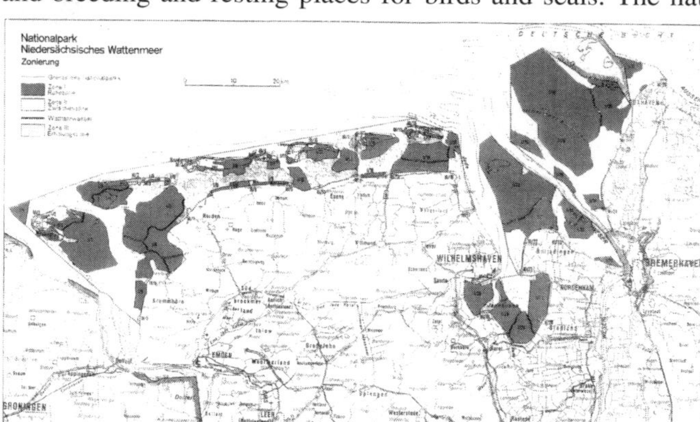

the protection of birds. In order to protect this area a *National Park* has been established. The purpose of the protection is to preserve the particular uniqueness of the nature and landscape of the tideland region and to protect it against impairment. To achieve the above objectives the National Park is divided into zones of different protection categories:

* Zone I is the Rest Zone; here all operations are forbidden which destroy, damage or alter the National Park or its individual components.
* Zone II is the Intermediate Zone; here all operations are forbidden which alter the character of the tideland area, including the islands, and in particular impair the landscape scenery or the enjoyment of its natural beauty.
* Zone III is the Recreational Zone.

A chart which shows the separate zones is attached. In light of the above, any interference to the delicate environment has to be kept to an absolute minimum.

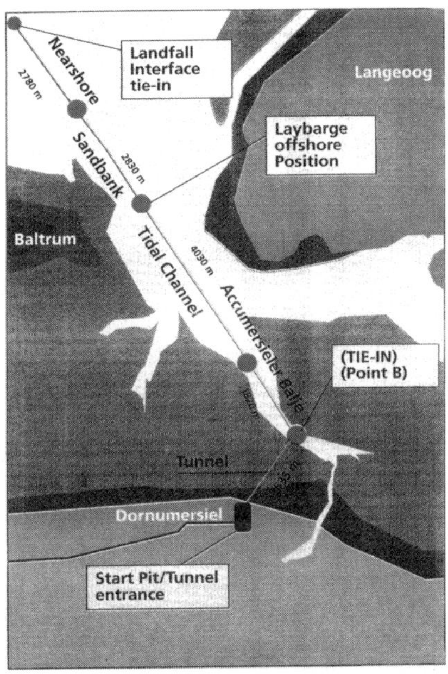

Construction Methods

In the landfall area within the territorial waters up to the Accumersieler Balje, the pipeline is to be buried permanently in the seabed. For this purpose, a trench will be dredged in the seabed into which the pipeline will be laid by a laying vessel (lay barge) moored where necessary to mooring piles. After pipe laying has been completed, the trench will be backfilled. Dry-falling mudflats section areas will be under-crossed by means of a special tunneling method. Due to the sensitive working area, all work is to be done with a view to minimizing environmental impact. The construction works have to be completed during one summer period, thus limiting the impact to one season.

The greatly differing water depths in the individual route sections require many different types of installation equipment and special construction methods. The vary-

Longitudonal Section

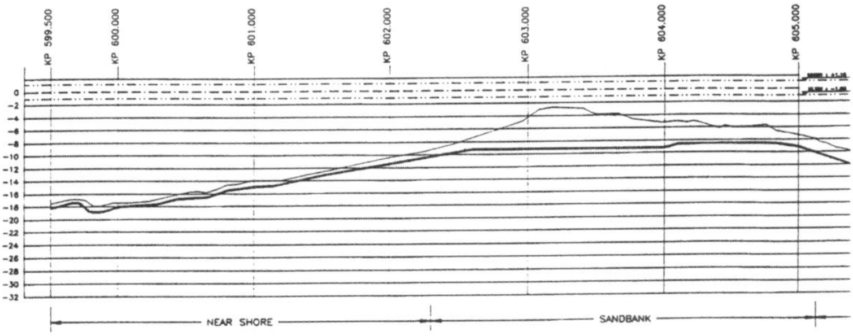

ing geological conditions along the route with different kinds of sand and clay as well as layers with peat inclusions in the soil require special handling of the various soils to be excavated for the tunnel and dredged for the pipe trench.

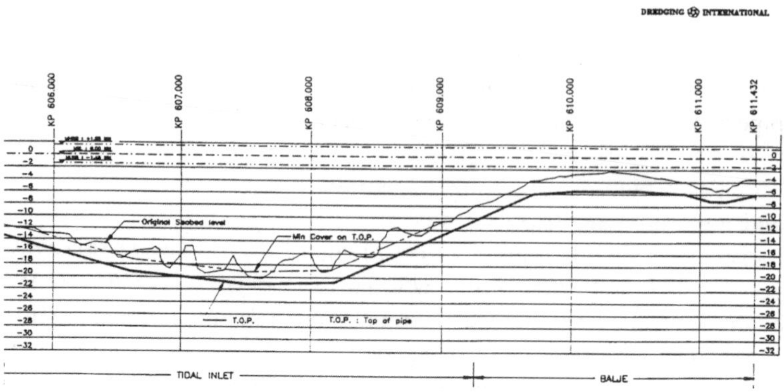

Dredging

In order to cross the offshore sand banks with the laying barge and the dredging equipment, even at low tide, and for subsequent covering of the pipeline, an access channel and a laying trench have to be dredged. The trench depth is normally determined by the required elevation of the pipeline. In the sandbank area however the trench depth is determined by the draught of the floating dredging equipment and of the lay barge respectively. This requires an access channel lying on the route of the pipeline with a bottom at NN -8.00 m. The bottom of the pipe trench within the access channel is at NN -10.25 m.

The pipe trench bottom width is determined by the required size and the operational conditions of the lay barges, as well on the basis of possible natural backfilling. It is predetermined at 15.0 m in the outer area of the coastal inlet and 10.0 m in the Balje area. For safety reasons the access channel in the sand bank will have a bottom width of 80.0 m because of the width of the equipment used and because of the flow conditions. The gradients of slope of the trench are expected to be 1:10 in the coastal section, 1:7 in the area of the sandbank and tidal channel, and 1:3 in the trench in the Balje area (which is only open for a short time). The quantities of material excavated during the dredging of the trench and channel will

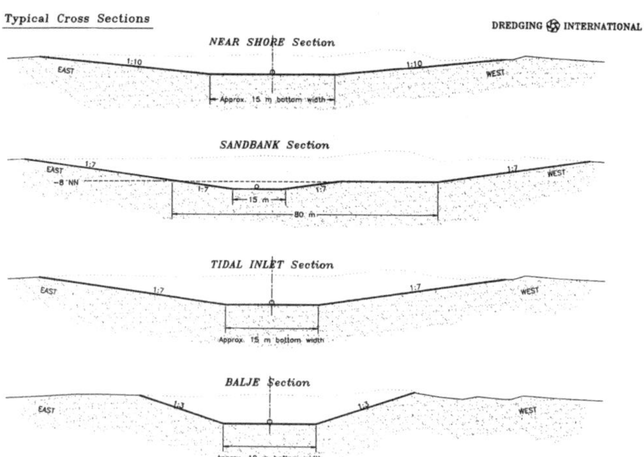

amount to a total of approx. 2.65 million m³. Taking into consideration the maintenance dredging to keep the access channel and pipe trench clear a quantity of approx. 3.17 million m³ is estimated.

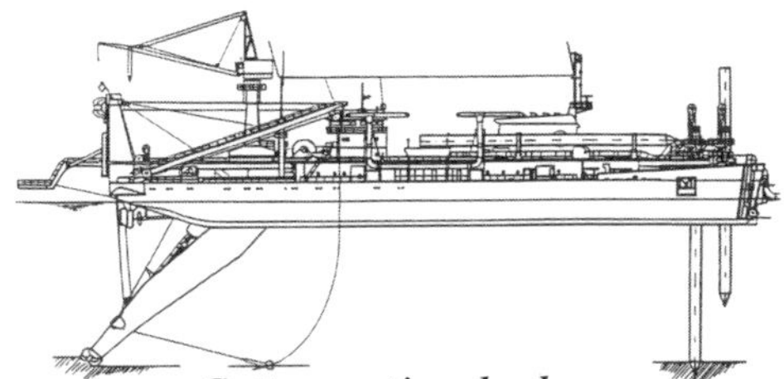

Cutter suction dredger principal sketch

The following dredging equipment and methods have been selected for the individual route sections because of the geotechnical properties of the material to be dredged, the prevailing sea conditions and the operational conditions:

Trailing suction hopper dredger principal sketch

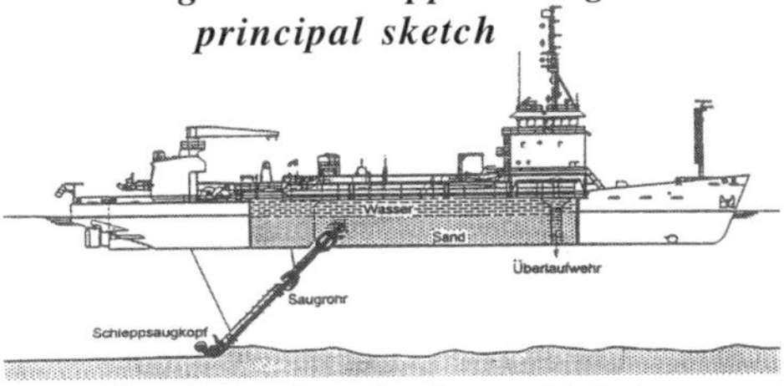

* Nearshore area

Trailer suction hopper dredger which loosens the material by waterjets or mechanically (e.g. teeth), extracts the material hydraulically and loads the material in its hopper. Any excess water is discharged via a weir.

* Sandbank

Cutter suction dredger which loosens the material by cutting, conveys it hydraulically to seagoing barges (c.q. hopper dredgers used as barges). In doing so, it provides access for the trailer suction hopper dredgers to be used afterwards.

Backhoe dredger principal sketch

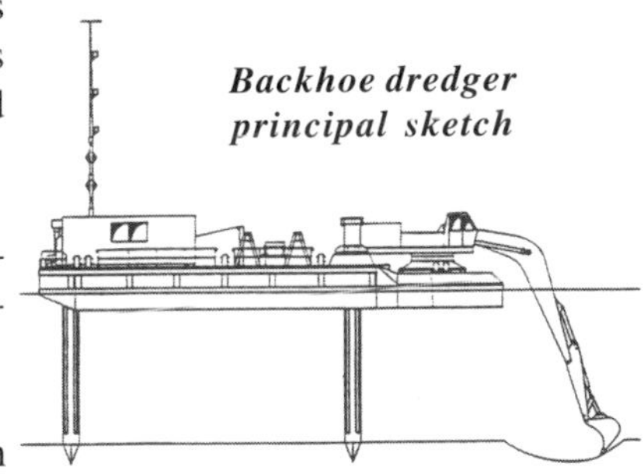

* Tidal channel

Trailer suction hopper dredger in the northern area and cutter suction dredger in the southern area, where consolidated clay is found.

* Balje

Two spud pontoon dredgers equipped with backhoes. By using this dredging technique, the dredged material remains largely compact, which means that comparatively small quantities are brought in suspension. First the top layer of sand, which is suitable for subsequent back filling, and then the clay/peat layer underneath has to be removed. The material is loaded into seagoing barges and shipped to the dumping area.

VLAANDEREN XIX supplying Hoppers

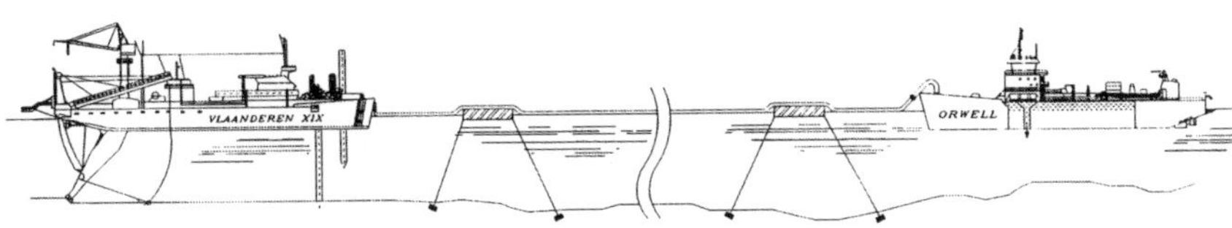

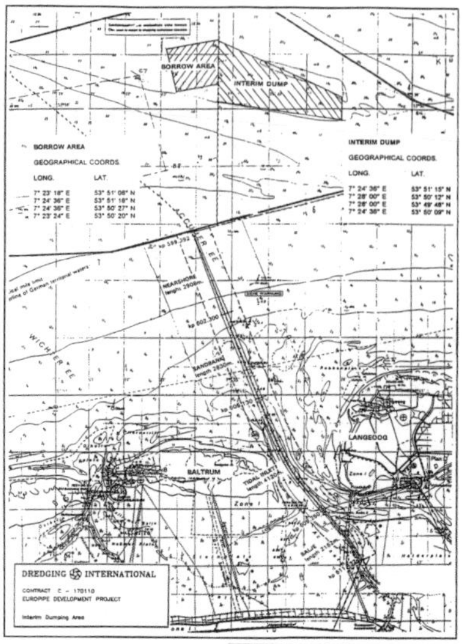

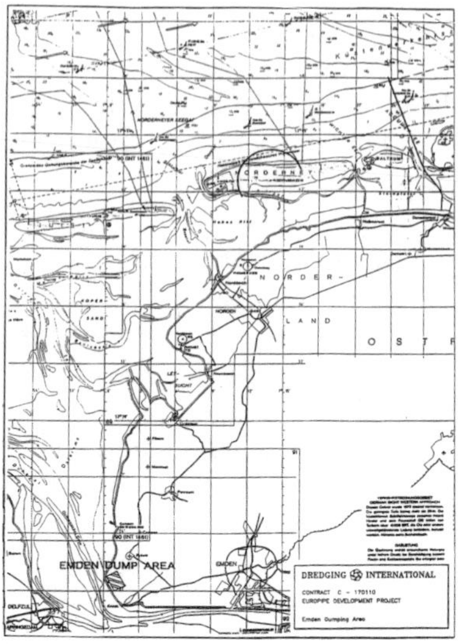

Depending on the soil characteristics of the dredged materials there are three dumping areas:

-Interim dump: Located North of the trench and to be used as interim storage area and borrow area for all sandy materials.

-Klapstelle 01: Located North of the island Wangerooge at the entrance of the Jade River. To be used as permanent dump area for all clayey materials not containing peat.

Dump Areas: - interim dump (upper L); -Emden onshore dump (upper R); - Klappstelle 01 (R)

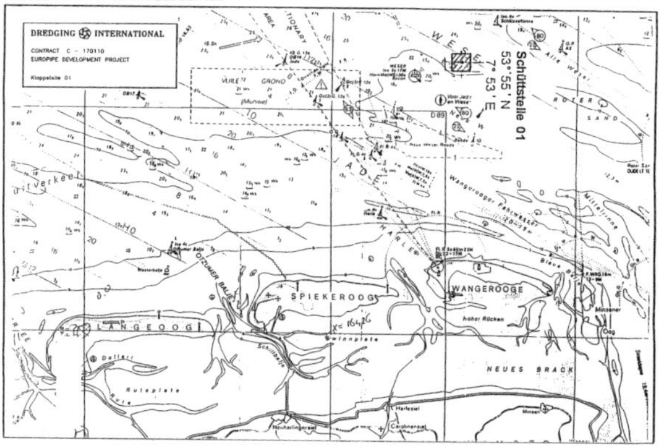

- Emden onshore dump: Located at the right bank of the river Ems close to Emden. To be used as permanent storage area for all clayey materials containing peat.

Backfilling

After pipe laying the pipe trench and the access channel will be back-filled with suitable material. In the near shore area and in the area of the sand bank the pipeline must be protected against flotation immediately after laying. The gravel material to be used for this purpose (approx. 50,000 m^3) is dumped in berms every 80.0 m by means of a sidestone dumping vessel and fall pipe vessel. The material must be brought in from outside sources because no suitable material has been found in the vicinity of the construction site. The gravel dumping vessels will be loaded in a port.

For the side stone dumping vessel, which consists of different compartments, the dumping is performed by pushing the gravel gradually over the side of the vessel into the water by the slides on the flat deck. For the fall pipe dumping vessel, the gravel is dumped right above the pipe via a fall pipe, loaded by a crane and/or conveyor belt.

Further backfilling with sand takes place by means of trailer suction hopper dredgers (TSHDs) of different sizes, depending on the available water depths. The hoppers recover interim-stored, dredged material, transport it to the site and dump it in a controlled manner. The TSHDs can discharge either directly through bottom doors or by splitting the hopper. In shallow water the material can be discharged via the bow coupling of the hopper dredger through a spreader or diffuser pontoon.

Backfilling Vessels

- Typical D. P. Rock Fall Pipe Vessel;
- Typical D.P. Side Stone Dumping Vessel;

D.P. Rock Fall Pipe Vessel

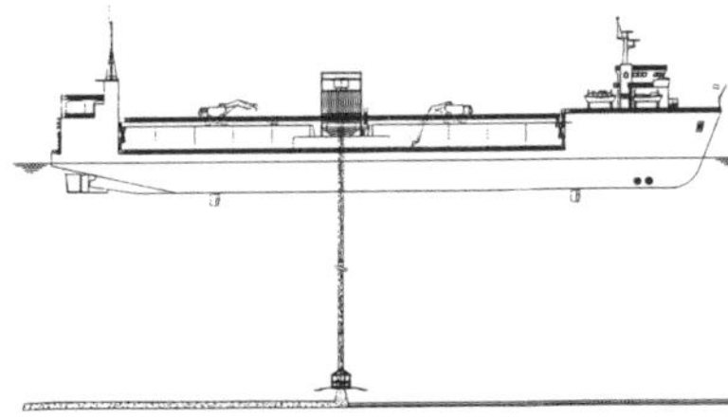

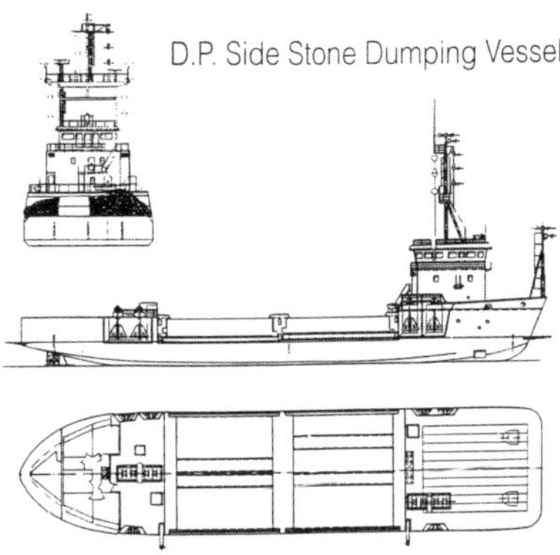

- TSHD backfilling through spreader pontoon;
- gravel uplift protection.

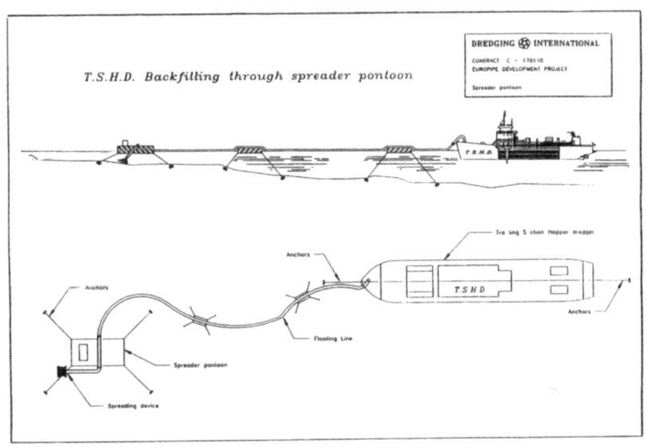

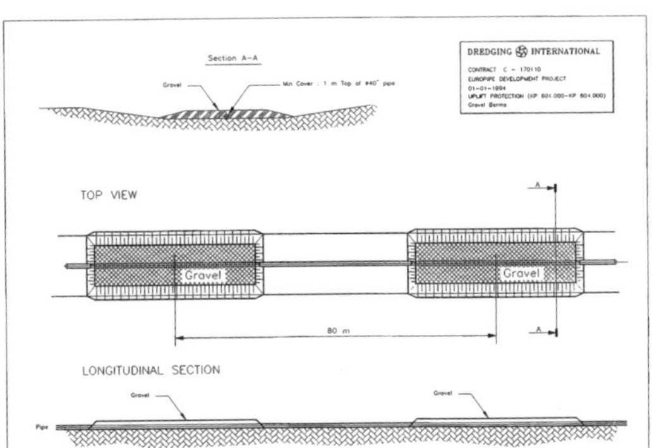

Survey Activities

The survey activities cover the dimensional control, positioning of floating equipment and the provision of project data such as charts, profiles, quantities, etc.

Horizontal position (x,y) (location)

Two positioning systems (Microfix and DGPS) will be used for the float ing equipment associated with dredging, dumping, reclamation, backfilling and survey. The use of two positioning systems is for reasons of cross checking the data and to ensure the exact position in case of failure of one of the systems. The accuracy of both systems is 1 to 3 m. The principle of the Microfix positioning system can be seen in attached figure.

Vertical position (z)

To know the water level 3 tidal stations are installed (in Balje, Tidal Inlet and Nearshore). The recorded water level is then transmitted by telemetry to the vessels. Depending on their location, the respective tidal stations are selected. The water level is recorded relative to the mean sea level (Normal Null = NN). The depth measurement will be done by echo sounder *(Atlas Deso 22).* To compensate for the survey vessel's vertical movement a heave compensator is used. The heave compensated data combined with the recorded water level gives the exact bottom level. The achieved accuracy is ± 20 cm.

Detection of obstructions

To ensure as far as possible that no obstructions or (dangerous) objects will be encountered on the sea bottom while dredging, a seabed inspection is executed prior to dredging operations. The equipment used for this purpose comprises:
- sub bottom profiler (to detect objects under the seabottom)
- side scan sonar (to detect objects on the seabottom)
- magnetometer (to detect metal objects).

Equipment

Two survey vessels provide the survey support during the dredging and backfilling operations. The vessels are selected on basis of shallow draft and reliability. Operations are expected to continue 24 hours per day, 7 days a week. Beach Nourishment survey at Langeoog is done by conventional land survey equipment.

Activities

Following survey activities are scheduled:
- Seabed inspection
Detection of (dangerous) objects on and in the sea bottom;
- pre-dredge survey
Depth of the natural seabed;
- post-dredge survey
Final level and width of the trench;
- pre-backfill

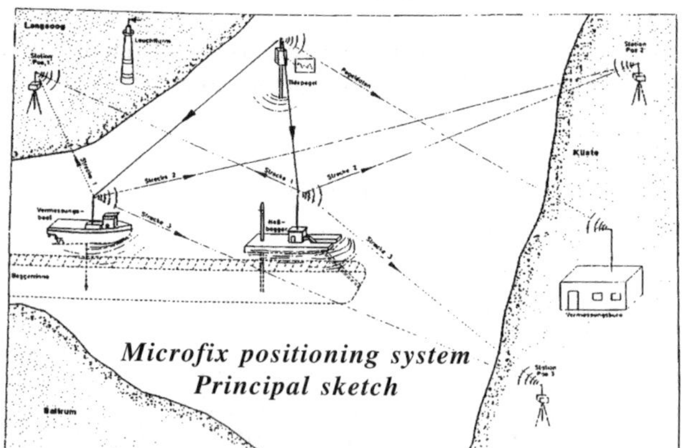

Microfix positioning system
Principal sketch

Level and width of the trench with pipeline;
- post-backfill survey Restored seabed depth;
- pre-survey

Beach Nourishment Original seabed and beach level;
- post-survey

Beach Nourishment Reclaimed seabed and beach level;
- Progress survey

The purpose of these surveys is to monitor the progress and to ensure that the required accuracies and tolerances are obtained. The data is then processed in the office to provide the necessary charts, profiles, quantities, etc.

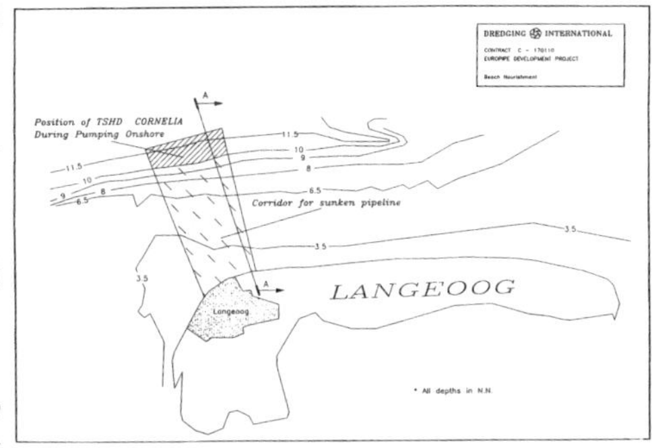

Beach Nourishment

About 50 % of the excavated material from the sand bank and material from the maintenance dredging (in total 700,000 m^3) is used for beach nourishment on the island of Langeoog.

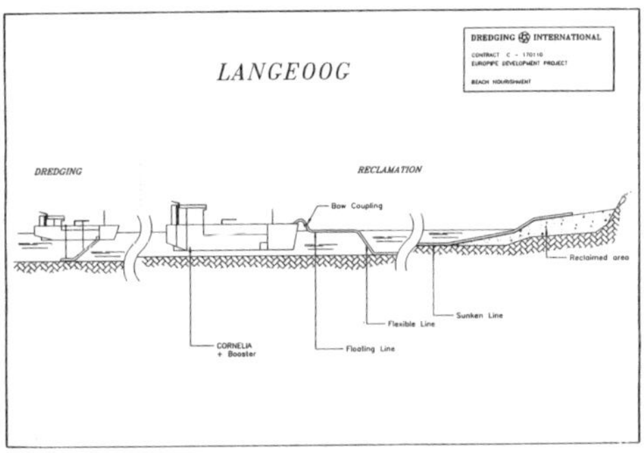

The excavated material is transported by means of trailer suction hopper dredger to a transfer point in front of the beach section to be nourished, and from there pumped onto the beach through a pipeline of approx. 3,000 m length (floating discharge line, sunken pipeline and beach line). At the transfer point off the island a heavy mooring point is placed, which consists of an 18 ton concrete block kept in position by three 10 ton anchors. For keeping the vessel in position while pumping both this heavy anchor and the vessel's own aft anchor will be used. The heavy anchor replaces the vessel's own bow anchor to obtain the required holding force. After anchoring, the coupling section of the floating pipeline is picked up by a tug, lifted aboard the trailer suction hopper dredger, and connected to the matching ball-joint. Through this pipe the dredged material is hydraulically discharged on to the beach. By means of land based earth moving equipment the sand is levelled according to the predetermined profile.

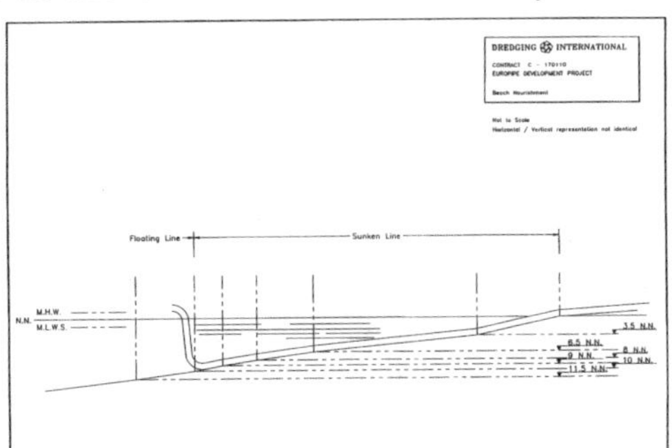

Pipe Laying

In the landfall area the pipeline is laid by floating equipment (lay barge) in the pipe trench previously dredged. Two types of laying equipment will be used in the landfall region. A small laying vessel (minibarge) will be employed in the shallow section of the Balje, while in the other sections a larger laying vessel will be used. The pipe will first be laid in the Balje section with the *minibarge*. Once the complete trench has been dredged, a larger lay barge will pick up the end of the already laid pipeline and will extend it to the offshore area.

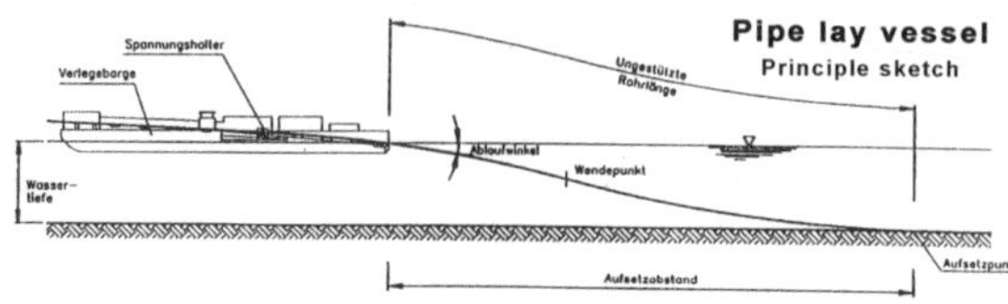

Landfall Area (Except for Balje Section)

The tidal channel section with water depths of more than 25.0 m at high water requires employment of a large laying vessel, due to the high tensile forces of approximately 800 kN on the pipeline in order to prevent buckling under its own weight. A laying vessel with a high laying rate has been chosen which will enable construction work to be completed during one summer season, thus minimising the impact to one vegetation period.

The lay barge is positioned with 10 anchors. In the near-shore area the heavy onboard anchors, moved by heavy-duty tugboats, are used for this purpose. In the sandbank and tidal channel, these tugs can only operate after extensive dredging works, as their draft prohibits sailing over the sandbank. To avoid this dredging, anchor piles arranged in a corridor of approx. 1,000 m width are used which, once set, need not be reset. The piles are placed with low noise vibrating hammers from a working barge, and are later removed.

The lay barge picks up the end of the pipeline previously laid in the Accumersieler Balje, extends it by up to 500 m per day, and let the accomplished pipeline section slide into the water under a flat angle. By successive hauling of its anchor wires it moves through the sandbank and into the near shore area. There, it lays down the pipeline at the transition zone to the offshore area. A jet prop system will be on standby during pipe laying. If the pipe level locally is too high (due to sand ingress...), it can be adjusted by means of the jet prop equipment.

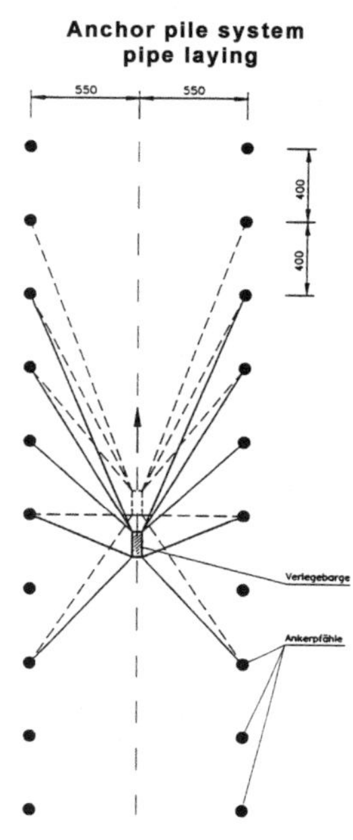

Landfall Area (Balje Section)

The lay barge cannot operate in the narrow, shallow

area of the Balje (without dredging a special access channel). On the other hand, the conditions prevailing there do not require such large equipment. Consequently, a minibarge is used, which in principle operates in the same way. It can lay about 100 m of pipeline per day. The minibarge lays the pipeline from the tidal inlet to the tie-in chamber of the tunnel section. It is positioned with 8 anchors and two on-board spud piles.

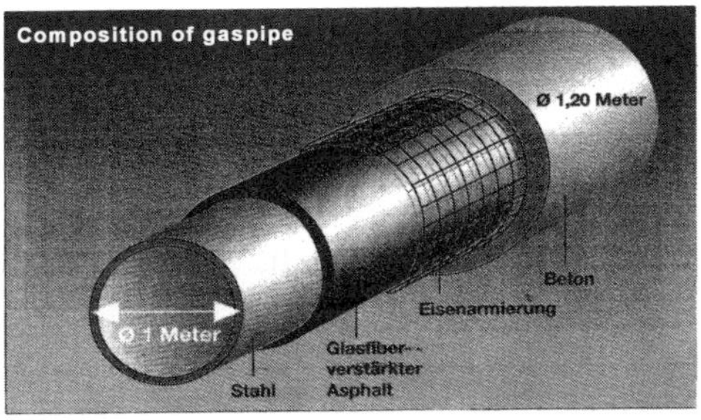

The minibarge follows a short distance behind the two spud pontoon dredgers which dredge the pipe trench. Consequently, a trench slope gradient of 1:3 will be sufficient, which will in turn minimise the quantities of dredged material. The dredged material is transported along the lay barge and, if suitable, will be dumped again immediately above the completed pipe section. For logistical reasons an interim storage is provided for a partial amount in the area of the tidal channel inside the pipeline route.

Tie-In of Sea Section and Tunnel Section (Tie-in Chamber in the Balje)

The connection between the offshore-laid pipeline and the tunnel pipeline in the Accumersieler Balje is made in a tie-in chamber situated under the sea bottom at the termination point of the tunnel. This will be constructed as a caisson structure installed by a floating unit. The pipeline laid in the Accumersieler Balje is inserted into the tie-in chamber from the laying minibarge through a lateral opening.

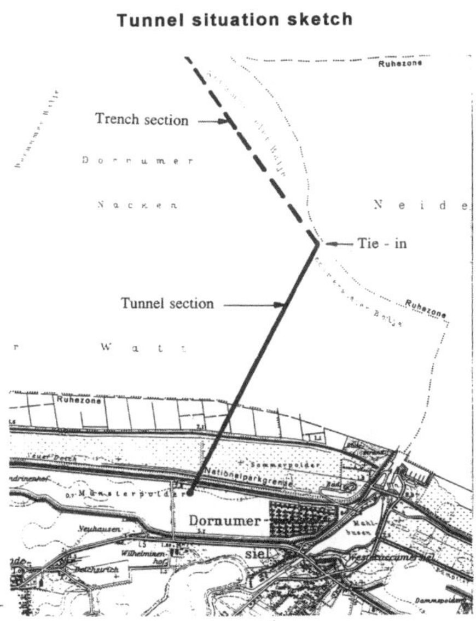

The chamber also forms the target structure for the tunnel construction. For this purpose, a special wall section is provided through which the drilling head of the tunnel drilling machine moves into the chamber. An access shaft, through which the tunnel drilling machine is recovered from the chamber, by means of a working ship, is mounted temporarily above the chamber.

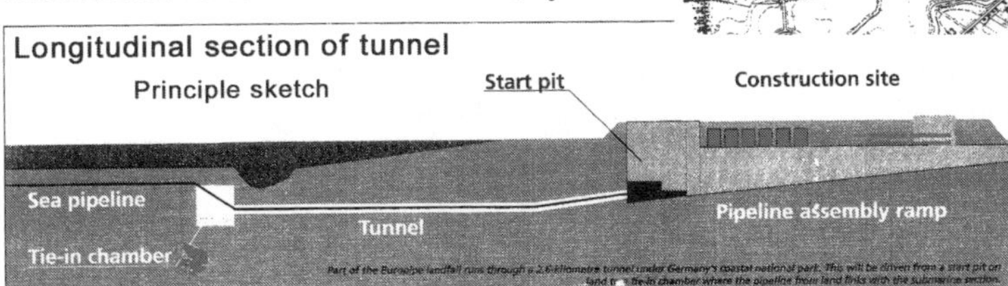

The two pipe

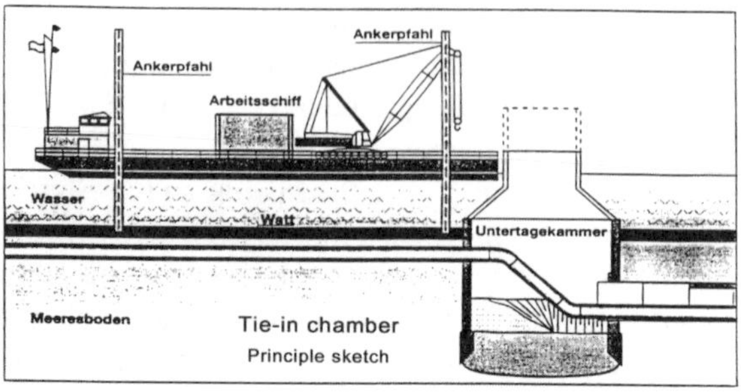

lines sections are then connected in the chamber under atmospheric conditions. After the connection is made and the pipe tested, the chamber is demolished down to the pipeline level, filled with concrete or similar material and covered with sand so that after all the work is completed the sea bottom is restored to its original status.

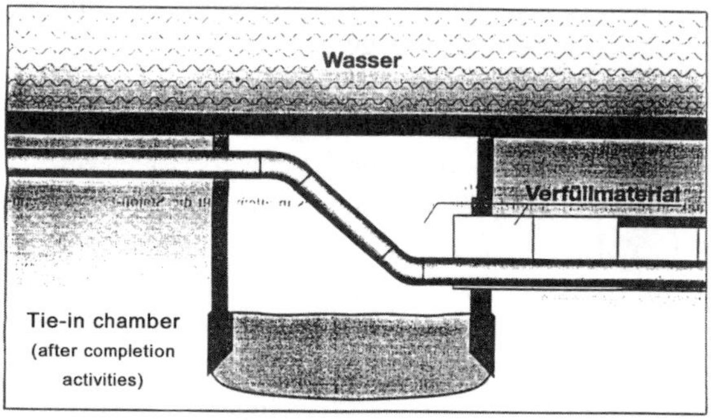

Under-crossing of the Tidal-Flats and Dike

As part of the tidal flats of the Accumersieler Balje to be crossed are designated as 'rest zone' and in order to minimise the intervention effects on the high tide protection systems, the total section of 2600 m length between the Accumersieler Balje and the protective strip of the main dike, is under-crossed at a depth of approx. NN -8 m (top of tunnel). The under-crossing is carried out from a starting shaft with reinforced concrete pipes applying the telescope jacking method. The first outer telescope pipe has an inner diameter of 3.00 m and an outer diameter of 3.80 m.

Using the telescope jacking method the first, outer, tunnel pipe section is driven on by remote control until the jacking can no longer be exerted. A second tunnel pipe section is then introduced through the existing tunnel, after removal of the drilling head of the first pipe. It is quite possible that the entire section of approx. 2,600 m can be done with the first telescopic pipe.

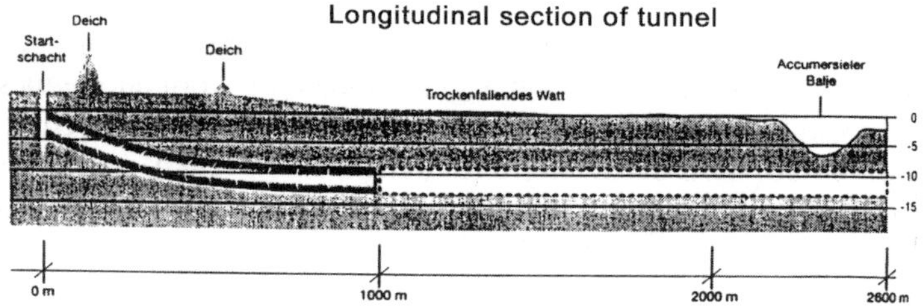

After completing the under crossing, the gas pipe (with outer diameter of 1.2 m) is pulled into the tunnel to the tie-in chamber where it later will be connected to the offshore gas pipe in the Accumersieler Balje.

Environment

As the works are executed in the National Park, any interference to the delicate environment has to be kept to an absolute minimum. Special attention has to be given to noise, exhaust gas emissions, waste handling, water quality and oil and lubricant leakages. The amount of dredged material and the duration of the construction period has to be kept to a minimum.

In order to minimise the impact on the environment, a zero dumping concept has been adopted. This means that strictly no waste (solid or liquid) can be dumped overboard. Even after treatment of the liquid waste, originating from toilets, sinks, showers, laundry and galley, the purified water cannot be discharged in the sea. Waste water will be collected in tanks, solid waste will be collected in containers and disposed of ashore. Those conditions have rather far-reaching consequences to the vessels. Keeping on board all liquid wastes (even after purification) requires large tanks.

Special measures should also be taken to reduce hydraulic oil and lubricant leakages. In order to reduce the effects on the environment it has been decided to use low sulphur, light fuels on the principal vessels. The use of lower sulphur fuels reduces the exhaust of sulphur dioxide, one of the causes of acid rain. Light fuel is used because it can be more easily removed in case of spill.

Two vessels equipped with oil spill equipment will be on site for quick response in the event of a spill. The equipment consists of oil booms for encircling the spill and an oil skimmer to remove the oil. At all times it has to be prevented that fuel, hydraulic oil and grease, spilled on deck, will be discharged in sea. Special attention is thereforerequired by all personnel involved and an environmental awareness of all personnel is essential.

Safety

It is the purpose to create by training, with people's awareness and by wearing protective equipment a healthy and safe work environment. Safety on board can be increased by paying close attention to the working environment, such as good housekeeping and wearing of proper safety clothing and by keeping all safety devices and signs in good condition.

The most common causes of injuries encountered are injuries due to boarding and unboarding operations; so a life jacket should be worn during these operations. Exercises will be held in order to train personnel in emergency procedures. An 8 hours safety training course is required for all personnel on board and visiting the vessels; this includes marine crew, superintendents, survey personnel. A person who is visiting a vessel for the first time should upon arrival be given a safety introduction. In order to achieve the highest level of safety, near misses and accidents are followed up by reporting, resulting in investigations and corrective actions to be taken. The use of alcohol and drugs is not allowed on board of the vessels according to the STATOIL safety labour regulations.

Overall Time Schedule

This Overall Operational Plan covers the period from October 1993 to July 1998 and includes both construction work and project supporting operations. The time schedule is designed for gas supplies to commence on October 1, 1995.

Milestones for Europipe construction work:

Activity	Date
Beginning of construction for tunnelling site	06.10.1993
Beginning of tunnel driving for dyke and mudflats undercrossing	16.02.1994
Beginning of dredging Laying of pipeline in the landfall area	01.03.1994
Mechanical completion	20.03.1995
Beginning of gas supplies	01.10.1995
Completion of compensation and substitution measures	01.07.1998

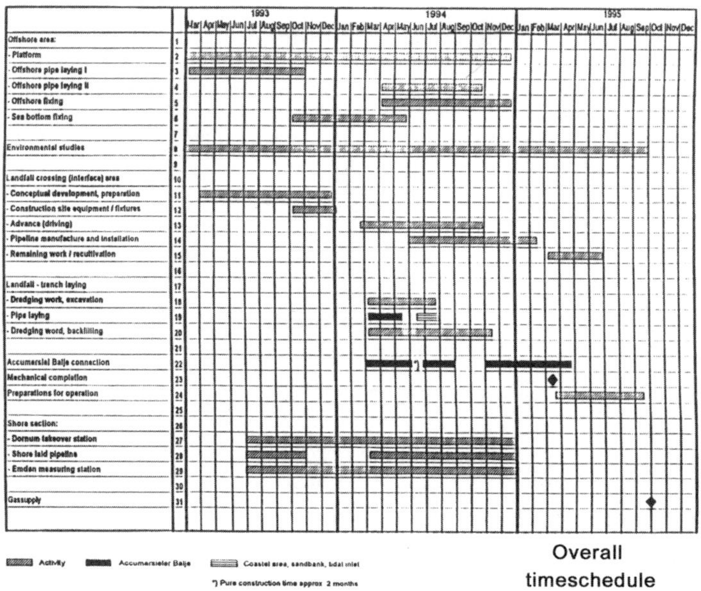

Overall timeschedule

The Europipe system is designed for an operational life of 50 years.

Planning and Progress Planning

A time schedule of all dredging and backfilling activities is found in the attached schedules. It has been established based upon expected dredging quantities, materials, working time (dependent upon weather delays...) and expected production performance (e.g. for TSHD : based upon hopper load and cycle time). The dredging activities are scheduled to start on the first of March in the Sandbank and the Balje section. Loaded hoppers (some even unloaded) cannot cross the Sandbank. So first of all an access channel has to be dredged to provide enough keel clearance. A cutter suction dredger (loading hoppers through a floating line) will make this access. It has to be remarked that hoppers cannot turn in the access channel, so they have to sail backwards once loaded by the cutter. This activity is scheduled from the first of March till the seventeenth of April.

The minibarge is scheduled to start laying pipes in the Balje section from the first of April. This implies that dredging activities in the Balje have to start on the first of March. Completion of the trench dredging is scheduled for the 15th of June. A large pipe lay barge is then picking up the pipe already laid in the Balje and completing the laying up to open sea. The backfilling activities are scheduled to start on the 15th of July and are expected to be finished between the 22nd of October and the 8th of November 1994.

Timeschedule Dredging

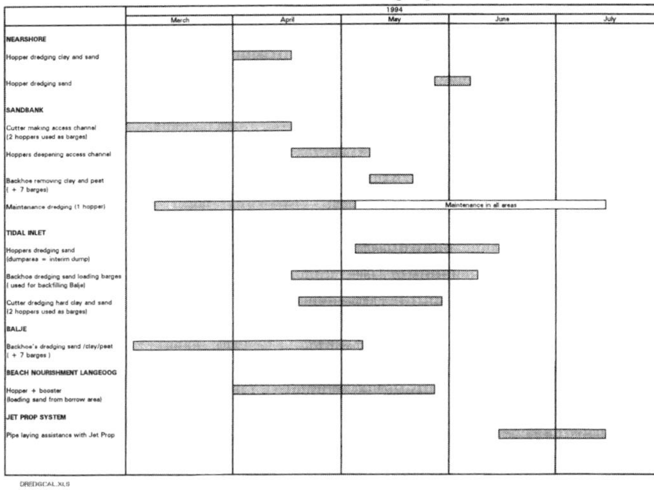

Timeschedule Backfilling

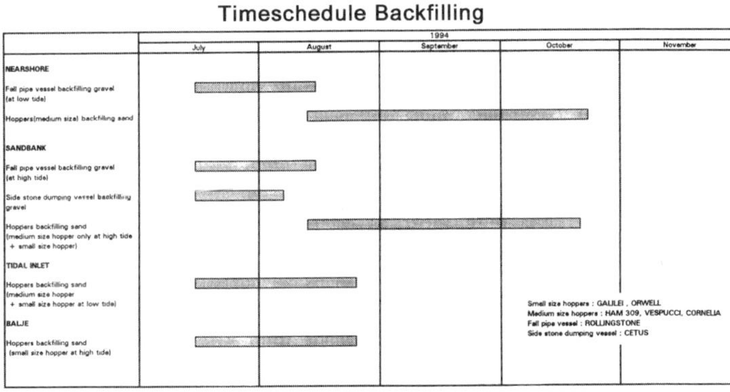

Progress Measurement

The progress of the dredging/backfilling activities has to be monitored in order to check whether the planned milestone dates will be reached. The daily reports of the vessels, which will be transmitted by telemetry to the office, are used to report the activities and production performance of the vessels. The production performance per week or per day has to be by a good average (according to a minimum and maximum performance as calculated). In case of deviation corrective actions should be taken.

The progress measurement is based on moved m³ confirmed by the survey department. If the sand ingress (e.g. due to adverse weather conditions) is greater than the dredged quantities even a negative progress is possible. So progress and production are always to be considered separately.

The UK-Continent Gas Interconnector Project

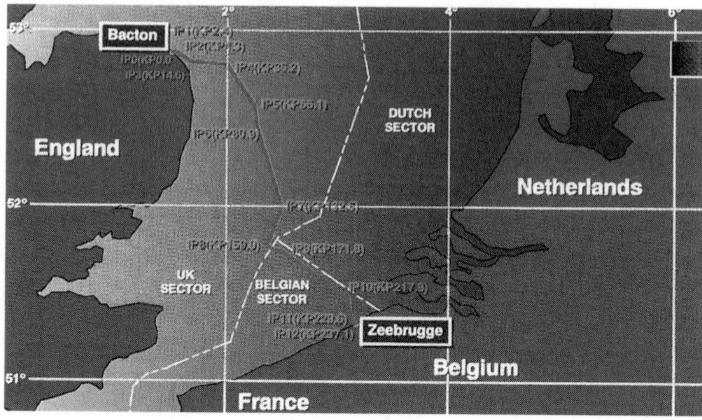

Interconnector Gas Pipeline Route

Historical Background

In early 1992, the UK Government's Department of Energy formed a Study Group comprising six companies to investigate the potential for a cross-Channel natural gas pipeline, referred to as the Interconnector Pipeline. With the UK placed as the biggest gas producer in Europe, one aim of constructing such a pipeline is for the UK to provide a secure source of gas to the rapidly growing markets of mainland Europe. As the second largest European consumer, the Interconnector pipeline could also provide the UK with an alternative source of gas supplies in the next century when indigenous gas suppliers may no longer be able to fulfill domestic demand.

The Study Group developed the basic commercial and technical concepts for such a project, and identified preferred landfalls on both sides of the Channel. In England, a site was chosen at Bacton on the east coast because of the well-developed infrastructure which was already in place there and a node connection to the *UK National Transmission System.* Similar infrastructural considerations also led to the selection of Zeebrugge as the Continental landfall point in Belgium, at which point the Belgian gas company, Distrigaz, joined the Group. Other key factors in the decision process were Zeebrugge's emergence as a gas transportation hub, together with its proximity to the French and German markets.

A project description was circulated to oil and gas companies across Europe, with a request for non-binding indications of capacity requirements. The level of response was so high that a full-time project team was established, and a commitment date of December 1994 was agreed. Nine energy companies subsequently pledged long-term capacity commitments for shipping a total of 20 billion m^3 of gas annually, and ultimately became shareholders in the newly-formed company, Interconnector (UK) Ltd. Since its inception, the Interconnector project has had the strong support of the UK and Belgian governments. This is underlined by the bilateral treaty which is being ratified in 1997.

Interconnector (UK) Ltd.

Interconnector (UK) Ltd. was formed on 16 December 1994 for the express purpose of constructing and operating the Interconnector Pipeline. The nine shareholders are:

- British Gas plc (40%)
- BP Exploration Operating Co Ltd. (10%)
- Conoco (UK) Ltd. (10%)
- Elf UK plc (10%)
- Gazprom (10%)
- Amerada Hess Ltd (5%)
- Distrigaz SA (5%)
- National Power plc (5%)
- Ruhrgas AG (5%).

The Interconnector shareholders have adopted the principles of alliance contracting, an approach which has already proved extremely successful in the UK offshore oil and gas industry. By bringing together the client and appropriate contractors and suppliers within an alliance, a project can be delivered at the lowest possible cost by sharing financial risks and rewards via an agreed formula. An integral part of such an approach is the maintenance of high safety standards and care for the environment.

In March 1996, the Interconnector Offshore Alliance was forged as a result of an agreement signed between Interconnector (UK) Ltd. and contractors for the offshore pipeline and landfall works. This Alliance comprises:

- European Marine Contractors (EMC): pipe lay and UK landfall works
- BacZee (a joint venture between Dredging International and Jan De Nul): dredging and Belgian landfall works.
- Brown and Root: engineering
- Interconnector (UK) Ltd.

For the onshore facilities, Interconnector (UK) Ltd. formed an Onshore Alliance in October 1996 with the design engineering company, Kvaerner John Brown, and the Thermodyn division of Framatome who supplied the compressor packages.

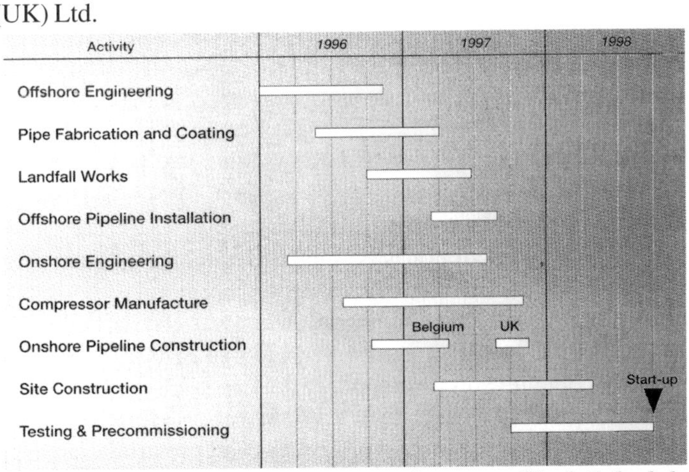

UK Continent Gas Interconnector Summary Work Schedule

Facts & Figures

UK gas will enter the Interconnector Pipeline at the Bacton compression facility, which is designed to transmit up to 20 billion m^3 of natural gas each year. Following compression to a maximum pressure of 140 bar, the gas will be transported through the

40" diameter, 235-km long pipeline to reception facilities at Zeebrugge in Belgium. Although the initial intention is to export gas from the UK, there is provision for an initial reverse flow of 8.5 billion m^3 per year to meet market demands. Orders were placed in the spring of 1996 for the purchase of line pipe, the supply of anodes, and internal and external coating of the line pipe. The contract for supply of compressor packages was awarded in August 1996.

As it leaves Bacton, the Interconnector pipeline first passes to the east of the gravel-winning area off Great Yarmouth and a restricted zone further south. It then runs southwest of the Belgian/Dutch territorial water boundary before crossing the Scheur shipping channel to the landfall west of Zeebrugge. Along this route, the pipeline has to negotiate areas of significant sandwaves and sandbanks.

Work on the Belgium onshore pipeline section began at the end of 1996, with installation activities commencing on the onshore stretch at Bacton in mid 1997. Work at the two terminals will be carried out in parallel, with construction beginning early in 1997. Offshore pipe laying activities began in March 1997. Mechanical completion of the pipeline and terminals is scheduled for the end of June 1998, allowing three months for testing and commissioning before start-up on 1 October 1998. It is estimated that investments in the scheme will total approximately £450 million. Some 130,000 tonnes of line pipe will be delivered, in 19,700 sections each measuring 12.4 m in length. Six thousand anodes will be attached to the pipe, weighing a further total of 800 tonnes.

Health, Safety and the Environment

Interconnector UK, together with its contractors and suppliers, are committed to working in accordance with the highest health and safety standards, and minimising the impact of both development activities and pipeline operation on the environment. Since as far back as 1993, discussions have been held with a wide range of authorities and third parties, including local fishermen and cable operators. The aim has been to address any concerns and problems at an early stage.

Facilities have been designed in compliance with the appropriate UK and Belgian statutory regulations, and work on both sides of the Channel will be conducted in accordance with local authority requirements. Land restoration, including the preservation of sub-surface stratification where necessary, will also be carried out in accordance with relevant planning conditions.

The Onshore Facilities

The Interconnector Onshore Alliance, formed in October 1996 to undertake the onshore works for the Interconnector Project, comprises Kvaerner John Brown (engineering), the Thermodyn Division of Framatome (compressor packages), and Interconnector (UK) Ltd. The aim of the Alliance is to design, procure, construct and commission a fit-for-purpose facility, safely and at the minimum full life cycle cost. In the endeavour to achieve an exceptional result for all parties, the Alliance has set itself targets in safety, environmental protection, noise limitation, performance and availability, in addition to sharing in capital savings. A significant proportion of the gain share will be held back until completion of one years successful operation. The value of the onshore works is estimated at some £120 million.

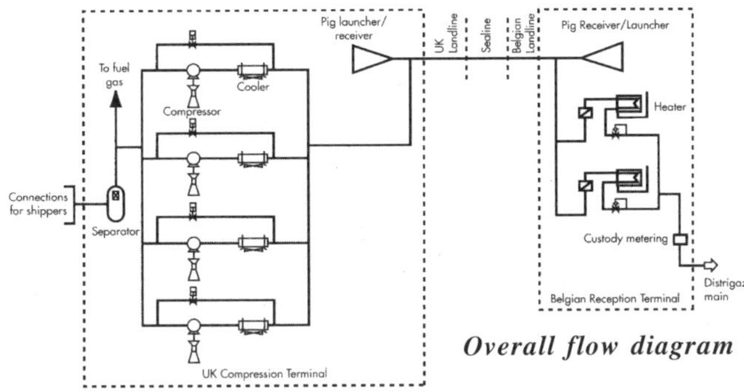

Overall flow diagram

Design and Engineering

Kvaerner John Brown's scope of work for the Interconnector project includes: engineering and project services for design and procurement; support for installation and commissioning of the compression facilities at Bacton; onshore sections of pipeline in the UK and Belgium; and the reception terminal at Zeebrugge.

The Alliance strategy enabled the Kvaerner John Brown project team to propose an initiative which suspended engineering activities whilst opportunities to optimise both the design and cost of the onshore works were being investigated. This initiative resulted in reorganisation of the conceptual design layout at both locations and developed a design which not only reduced material and equipment, but also shortened the construction timeframes. This alone resulted in an approximate 10% cost reduction for materials, with further benefits being identified during construction.

A further bonus of the Alliance strategy was the development of construction sub-alliances for the major on site construction activities. For the Zeebrugge facility, a sub-alliance was set up with Kvaerner John Brown's sister company, Kvaerner John Brown NV of Antwerp, to utilise its extensive Belgian construction expertise. For the much larger Bacton construction works, a construction sub-alliance is to be awarded following selection of the contractor tendering the most attractive construction alliance cost estimate, coupled with a suitable alliance strategy and a commitment to safety and care for the environment.

Thermodyn skid mounted barrel compressors at the Le Creusot factory

The Bacton Compression Facility

The five hectare site for the Interconnector compression facility will be located within the existing Transco gas terminal at Bacton. Gas received from the UK National

Cofferdam construction at Bacton

Transmission System or direct from independent suppliers will be compressed to a maximum pressure of 140 bar prior to shipment to the beach, and from there subsea to Zeebrugge. Two transmission lines, previously used to transport gas from Bacton to the Midlands, will be used to bring gas to Bacton from elsewhere in Britain. This will join North Sea gas landing at Bacton, both from the National Transmission System and from any Direct Access connections, upstream of the new compression plant. During normal operations, the Interconnector facilities will be controlled from the Bacton Control Room.

Kvaerner John Brown was responsible for the detail design of the terminal, and for purchase of all material and equipment following a series of competitive tender enquiry issues. Work at the Bacton facility then began early in 1997 with site preparation and civil works, while the main mechanical, electrical and instrumentation installation work will start in the summer. Mechanical completion is scheduled for June 1998, ready for operation at the start of October. The onshore pipeline will be installed in time to tie-in to the offshore section, after which the whole system will be dried and prepared for accepting gas.

The Bacton terminal includes the following major plant:

• compressor house containing the four 28 MW gas turbines;

• four banks of after-coolers incorporating electrically-driven fans and heat exchanger pipes to cool the gas following compression;

• suction scrubbers to prevent solids or liquids entering the compressors;

• connection points from the existing Transco National Transmission System feeder and for Direct Access supplies;

• 40 m vent stack to depressurise the terminal pipework and equipment for maintenance or in emergency;

• utility requirements include 11 KV electrical supplies with standby generator and battery backup, communication including a Pipeline Integrity Monitoring System, fuel gas system, fire water and sewerage;

• pig receiver/launcher for cleaning and on-line inspection of the pipeline during operation;

• control facility to monitor safe and economic operation of the Interconnector system;

• metering

The Compression System

The main equipment for the Bacton terminal comprises four turbo compressor sets supplied by Thermodyn. Each train consists of an RD (6-5)B centrifugal compressor driven directly (without an intermediate gearbox) by a General Electric LM 2500+ Gas Generator and Nuovo Pignone Power Turbine.

The centrifugal compressor is being manufactured at Thermodyn's works located in Le Creusot, France. Thermodyn selected a 5-wheel RD barrel compressor rated at a flow of 5 billion m^3 of gas per year, with a pressure lift from 43 bar to 143 bar. The machine is fitted with Pacific tandem dry gas seals and is protected by a Thermodyn patented anti-surge algorithm. The compressor is designed to accommodate six impellers for possible future re-wheeling.

The gas turbine is packaged by Nuovo Pignone in Florence, Italy. The gas generator is the new model of the established LM2500, and will be the first unit sold for a mechanical drive application. Power is extracted from the gas generator by the new power turbine running at 6,100 RPM. This machine develops some 25% more power than the LM2500 design on which it is based. The gas turbine is equipped with a dry low emission system designed to meet current environmental legislation. The first train will be assembled in Le Creusot in order to undertake a full load string test and verify the integrity of the compression train, the control software and the power delivered.

The compression trains will be delivered to Bacton between September 1997 and January 1998. Installation and commissioning will follow ready for full operations commencing in October 1998.

The Zeebrugge Reception Facility

As in the UK, Kvaerner John Brown's scope of work in Zeebrugge includes engineering and procurement for both the onshore section of pipeline from the nature reserve as well as the terminal reception facilities. After the Interconnector pipeline comes ashore at the Belgian landfall, an underground directionally-drilled pipeline passes under the dunes and the Fonteintjes Nature Reserve. The pipeline then passes through the polder before connecting to a pig receiver/launcher within the reception terminal.

Thrust borr equipment installed in Bacton landfall shaft

Work at the Zeebrugge site is being carried out in parallel with the Bacton compression facility, with construction beginning in the Spring of 1997 and mechanical equipment scheduled for delivery in the late summer. The onshore pipeline will be laid concurrently with the landfall of the offshore pipeline. A hydro test of the whole pipeline will then be conducted. The Zeebrugge terminal site covers 4 hectares and has enough space to enable installation of compression facilities should the UK become a gas importer. If and when this occurs depends on future UK gas field developments.

A number of heating trains will be installed in parallel to maintain the gas temperature to Belgian grid specifications. This is necessary as a result of cooling caused by a pressure drop in the pipeline (83-105 bar depending on flow rates) and during flow regulation. Four 4.0 MW water-bath heaters will keep the gas temperature above 2°C. Valves will be installed to protect the Distrigaz grid from depressurising if the Interconnector pressure should fall for any reason, as well as to isolate the terminal from the Distrigaz high pressure grid. A control room will be included. Metering will be carried out at the adjacent Distrigaz facility.

Interconnector Landfall Zeebrugge

Design and Engineering of the Offshore Pipeline

The initial technical feasibility study was performed by British Gas during the second half of 1993. The feasibility report was presented to the study partners later that year when approval was given for a Technical Study Group that included representation from the partners to undertake a conceptual design.

In early 1994, the Study Group appointed EMC (European Marine Contractors Ltd.) to head the contractor team to support the conceptual design. Other members of the contractor team included Brown & Root and Dredging International. Prominent amongst the various consulting groups providing technical support were the Belgian Engineering Consultancy, HAECON N.V. They played a vital role in providing coordination of Belgian Authority and Environmental Engineering issues in support of the many consent and permit applications submitted by Interconnector (UK) Ltd.

A comprehensive route survey campaign was executed by Geoteam in 1994. The survey included retrieval of data giving the depth of water along the route, seabed profile and texture. In addition, in situ solid testing was conducted using a cone penetrometer device and further samples were taken onshore for laboratory analysis. Analysis of the results helped to lend definition to the preliminary design, and therefore confidence in the cost estimate projections to be approved by the partners. In addition, HAECON provided expertise which they had acquired through their previous support to Statoil on the Zeepipe project and various harbour and marine engineering works.

Several advanced techniques were used in support of the Interconnector studies:

- Differential bathymetry maps of the Belgian coastline in the vicinity of the pipeline corridor.
- Mathematical modelling to demonstrate evolution of the coastline.
- A combination of geoelectric and geotechnical surveys to help delineate the freshwater and saltwater layers in the Fonteintjes Nature Reserve in the Zeebrugge coastline dune belt. The information obtained contributed to an environmental assessment undertaken by AIB (Belgian Certifying and Engineering Authority), and later in the selection of the directional drilling technique used for crossing the reserve.

ERT, connected to Edinburgh Herriot Watt University, completed the environmental evaluation for the offshore pipeline. The University of East Anglia completed the environmental assessment for the Bacton landfall. Whilst this work was underway, the offshore route was also being shaped through discussion with authorities and third parties. The route had to skirt sand and gravel winning areas in the UK and Belgium, and a restricted military exercise zone. Information provided by fishing interests also proved influential.

Detail Design

As engineering contractor for the Offshore Alliance, Brown & Root (B&R) was responsible for design of the offshore pipeline and the onshore sections at the Bacton and Zeebrugge landfalls. As a pioneer of *Alliancing* from industry acknowledged projects such as ANDREW & ETAP, Brown & Root was able to identify key success factors and thus rapidly facilitate the early formation of the offshore alliance. One such factor was the encouragement of initiatives that *add value* to the project, through improved safety, environmental impact or cost effectiveness. In excess of 70 opportunities were identified for Alliance management consideration, and have directly contributed to the overall project success.

Automatic UST of weld

Plate in the 'U' shape entering the 50.000 tonne 'O' press at British Steel's 42"

B&R first prepared a preliminary design for the pipeline during the conceptual phase, which established the route between Bacton and Zeebrugge. Once the Alliance had been established, B&R was responsible for final detailed design. This involved careful assessment of pipeline material requirements, together with preparation of specifications for these materials, anodes, coatings, pipeline welding and field joints. The final wall thickness for the offshore section was based on a calculation using the internal diameter which achieved a 2% saving of materials. B&R was also involved in enquiry bid packages for materials, and technical evaluation of the bids. The Interconnector route had largely been determined by information obtained from the survey campaign undertaken in 1994. During the design phase, advice was sought from experts in the fields of geotechnics and sand wave behaviour and it emerged that the pipeline route would need to be adjusted in the sand wave areas to further limit the potential for scour. A survey carried out by Seateam in 1996 helped to provide the information for the optimised route.

The Interconnector pipeline is trenched nominally flush with the seabed along its length. The division of post-lay trenching work between use of ploughing and jet sled was optimised as much as possible, in conjunction with determining the pre-lay dredge and weight coating requirements. During design of the pipeline, much effort was invested in selecting an optimal concrete weight coating. A concrete density of 3400 kg/m^3 first used in UK was adopted to provide stability and resistance to up-heaval buckling.

The internal surface of the pipeline is coated with a layer of epoxy to improve the flow characteristics and maximise throughput. Much effort has been expended in developing an environmentally friendly approach to pigging and testing the pipeline. The volume of the pipeline, approx. 175,000 m^3 means that the normal dosage rates for biocide and corrosion inhibitor would create a disposal problem in coastal regions. In order to avoid this situation a discharge model was set up to assess the extent of the affected area and a test programme was initiated to investigate the feasibility of avoiding the use of such chemicals without corrosion risk to the pipeline.

The test programme successfully achieved this objective and is considered to be a major break through for the industry. B&R was also responsible for design of the crossings where the pipeline route encounters live fibre-optic telecommunications cables, as well as the future NorFra pipeline.

Pipe Supply

The line pipe specification developed by the Alliance was particularly tight on production tolerances, resulting in reduced fitting up times on the lay barges. The 40" pipe for the Interconnector gas line was supplied by two manufacturers: Nippon Steel Corporation in Japan, and British Steel in the UK.

Nippon Steel had been awarded the contract to manufacture and supply 60% of the 40" gas pipeline in March 1996, as part of an alliance with the leading multinational company, Marubeni Corporation. Marubeni, with an annual turnover of more than US$146 billion (1996 figures), had pre-qualified for the Interconnector project on behalf of Nippon Steel in August the previous year. Such an alliancing arrangement is standard practise within the Japanese steel industry, where a client's contact with the steel mill takes place through an intermediate trading company such as Marubeni.

As contract principal, Marubeni was charged with overseeing all negotiations between the Interconnector Project Team and Nippon Steel, guaranteeing that stringent specifications and documentation requirements were met. Marubeni also played a key role in ensuring that the pipeline was shipped on schedule from Nippon Steel's Plant in Japan for delivery to the coating plant at Immingham in the UK, and dealt with all financial aspects of the contract.

Pipe end X-ray radiographic examination system

Nippon Steel manufactured some 77,000 tonnes of pipe at its fully automated Kimitsu Plant, responsible for initial steel-making through to the plate rolling processes. Equipped with the most up-to-date facilities and the very latest in pipe-making technology, Nippon Steel's Kimitsu Plant was ideally suited to supply the high quality pipe demanded for the Interconnector Project.

British Steel supplied 94 km of line pipe for the Interconnector pipeline, and was required to meet particularly tight production tolerances. The supply of bends also formed part of the contract, and this was sub-contracted to Fabricom of Belgium using mother pipe supplied from the UK. Of the total order, 88 km of pipeline was supplied with an outside diameter (OD) of 1016 mm (40"), and an OD of 1027 mm (40.45") for 6 km of landfall pipe. Three thicknesses were manufactured to achieve a constant internal diameter of 972.48 mm, ranging from 21.76 to 29.86 mm.

Manufactured at British Steel Tubes & Pipes' 42" U.O.E. pipe mill at Hartlepool in the UK, the Interconnector piping significantly exceeded the target mark for both ovality and straightness. Of the 1016 mm x 21.76 mm pipe supplied, for instance, all of the 2,000 pipes measured met the specified ovality limit of 5 mm and 99% were better or equal to 3 mm. The ability to manufacture line pipe which meets the most consistent tolerance of any existing U.O.E. mill is achieved by the exceptional strength and power of Hartlepool's forming press, together with good control and monitoring of mechanical expansion.

Technical specifications for the project were representative of a demanding submarine line pipe, with tight restrictions on the yield/tensile ratio. These were coupled with a stringent toughness requirement involving impact testing at-30°C and a Battelle drop weight test at -10°C. A material grade of EP450 was specified in accordance with EEMUA Standard 166 (equivalent to API 5LGrade X65). To meet requirements, British Steel supplied a low carbon micro alloyed steel with close control on chemistry consistency to ensure that subsequent pipe properties met the specification.

Coating

The 235-km long, 40" diameter Interconnector line is constructed of carbon steel, internally coated to enhance flow. The pipeline is also externally protected with an anti-corrosion coating of asphalt enamel, together with a concrete weight coat for stability on the seabed. Aluminium sacrificial anodes are fitted to provide cathodic protection. Universal Pipe Coaters Ltd was awarded the contract for internal, anti-corrosion and concrete weight coating of the Interconnector piping. Some 130,000 tonnes of bare pipe were delivered to the company's Immingham facility in the North of England. Of this approximately 80,000 tonnes was received by sea from Nippon Steel via the Port of Immingham, while the remainder was delivered by rail from British Steel. The final load out weight of coated pipe totalled 365,000 tonnes.

Contract specifications called for the application of an internal flow coat using Copon EP2306 HF supplied by E. Wood of the UK, and based on a minimum thickness of 65 microns. An external anti-corrosion coating of asphalt enamel was applied to the subsea section, with a thickness ranging between 5 and 7 mm. The enamel was manufactured at site by Phoenix Contractors of Denmark, who also supplied the primer and outer wrap. The inner wrap was supplied by Isover of Holland.

An anti-corrosion layer of 3-layer polypropylene coating was applied to the Bacton and Zeebrugge landfall sections by Isotub Coatings, under subcontract to Universal Pipe Coaters. Due to the negative buoyancy requirements, the specification called for three different concrete thicknesses of 65, 85, and 105 mm, with a concrete density of 3400kg/m^3 and individual pipe weights of up to 23 tonnes. Non-galvanised reinforcement mesh within the concrete coating was custom-made by Bekaert N.V. of Belgium, and was applied in two layers for each coating thickness. Cement was supplied by Titan of Greece, and the bulk aggregates were supplied by Tarmac of the UK and Universal

Aggregates. Sacrificial anodes were attached to over 6,000 pipes with the concrete coating chamfered down to a standard anode thickness of 50 mm.

Offshore Pipeline Construction

Dredging Activities

The BacZee joint venture, comprising the Belgian companies Dredging International as sponsor and Jan De Nul as co-sponsor was established to undertake all dredging and backfilling works on behalf of the Offshore Alliance, as well as construction of the Zeebrugge landfall.

The seagoing cutter dredger 'Marco Polo' (21,913 hp) loading the 3,500 m³ split barge 'Nina'

Seabed preparation works started at the end of 1996 in the Belgian Scheur channel, followed by the Zeebrugge approach in January 1997, the Bacton landfall in February and pre-sweep works (shaving the peaks off the sand waves) along a 120-km-long section of the route in March. The Zeebrugge works consisted of bulk dredging to give the pipeline adequate protection whilst crossing the busy Scheur shipping lane, the main access channel to the ports of Antwerp, Flushing and Zeebrugge. The top layers of sand and soft materials were removed by Dredging International's 5,000 m^3 *Vlaanderen XX* and 8,500 m^3 *Antigoon* and by Jan De Nul's 7,000 m^3 *Christoforo Colombo* trailing hopper, whilst the bottom hard layers of very stiff clay were excavated by Jan De Nul's cutter suction dredger *Marco Polo*. Interference with navigation in the busy shipping channel was kept to a minimum through continuous dialogue with the port authorities and the use of this barge loading cutter dredger without the need of floating discharge lines.

An access channel had to be created in the foreshore to bring the lay barge close to the shore (1,200 m), and so allow a safe pipe pull. Dredging International's *Vlaanderen XX*, 2,500 m^3 *Orwell*, 1,800 m^3 *Vlaanderen XXI* and Jan De Nul's 2,000 m^3 *Galilei* were used as trailers, while Dredging International's *Vlaanderen XIX* was deployed as cutter suction dredger.

'De Zeebouwer' executes offshore installation of the cofferdam

The *Gerardus Mercator* and *Pearl River* assisted in stabilising the pipeline in the Scheur trench, after which the *Antigoon* and *Orwell* completed the works over a 10-week period.

The pre-sweep area between the UK and Belgian landfalls contains a succession of sand waves and sand banks which lie on the pipeline route. These sand ridges have permanent cores with upper layers which are subject to constant movement. Some 600 of these sand waves had to be dredged to reduce pipeline stress during pipe lay. As dredging depths were up to 65 m, Jan De Nul's newly-built 18,000 m^3 *Gerardus Mercator* (equipped with an underwater suction pump for increased production at greater depths) was the key vessel for this part of the works. Also Dredging International's 17000 m^3 *Pearl River* was recalled from the Far East to perform part of the Presweeping together with the *Antigoon*. Equipped with sophisticated dynamic vessel- and draghead positioning systems, these vessels were able to dredge the cut lines accurately. Multibeam survey equipment further enhanced their efficiency.

The Bacton Landfall

The Bacton landfall works were carried out by Smit Land & Marine, under subcontract to EMC. Excavation of the 3.2 m deep trench required at the Bacton shore approach was performed by Dredging International's backhoe dredger *Big Boss* in February 1997. The removal of sand above the chalk layers was carried out by Jan De Nul's *Galilei.*

Construction of the shaft, tunnel and cofferdam, commenced in December 1996. A 15-m shaft was sunk from the cliff top down to the foot of the cliff, followed by construction of a tunnel from the shaft-base towards the sea. In March 1997, EMC's semisubmersible laybarge *SEMAC 1* commenced pipe lay at Bacton using two beach stationed linear winches to pull pipe ashore. The 24" hydro test intake and discharge line was installed by pulling on to the beach. The 40" pipeline was then pulled into the cofferdam and the *SEMAC 1* began laying towards Belgium. After EMC had completed the shore pulls, the onshore landfall section of the pipeline was installed in the tunnel, with pre-fabricated risers installed in the shaft and tied-in to the pipe string.

Shore approach dredging for the installation of the Interconnector Gas Pipeline of Bacton and Zeebrugge

The Zeebrugge Landfall

The offshore works at the Zeebrugge landfall were undertaken by Jan De Nul's civil department, with pipe laying work beginning in April 1997. A 445 m long cofferdam was

The trailing hopper dredger 'Gerardus Mercator' (18,000 m³) was at time of execution of the project the largest dredger of the world and has been used as the key vessel for the offshore pre-sweeping

constructed close to the beach, and a tie-in cofferdam was constructed to tie the onshore and offshore pipe together under dry conditions. Construction, excavation, backfill and removal of the seaward end of the cofferdam were performed by the jack-up platforms *Zeebouwer* and *Halewyn*, owned by Dredging International's sister company, Hydro Soil Services (HSS).

The onshore element of the Zeebrugge landfall work was also performed by HSS, comprising the most environmentally sensitive part of the project. The pipeline route at this point crosses the Fonteintjes Nature Reserve, an area containing a freshwater lake with rare orchids. A significant body of work was involved in evaluating the ground conditions (see previous section) and after an intensive selection process, direction drilling was chosen as the most appropriate technique.

After welding of the 500 m pipe string had started in the polder area, a protected sand platform was created on the beach from where a hole was drilled some 17 m under the Nature Reserve. After reaming the hole with the drilling rig in four passes up to 56", the pre-constructed pipe string was pulled-in using a catenary created by six telescopic cranes. Between the exit point of the directional drilling in the polder and the new terminal, approximately 1500 m of land line has been installed by Jan De Nul's civil department.

Pipelay

In addition to shore approach works at Bacton, EMC was responsible for installing the Interconnector offshore pipeline between Bacton and Zeebrugge, including management of permanent material procurement, ploughing and trenching, rock dumping, diving construction and pre-commissioning. Two lay barges were required to lay the offshore pipeline: a large semi-submersible vessel for the bulk of the pipe laying, and a flat-bottomed barge for the shore approach and first 16 km out from Zeebrugge where the water is shallow.

Since the Interconnector gas line crosses several fibre-optic cables at various locations along the route, EMC has installed mattresses prior to pipe lay to provide separation and protection at these crossing locations. After pipe lay is complete, rock will be dumped over these crossings to provide complete protection. The Interconnector pipeline also crosses the route of Statoil's NorFra pipeline. At the crossing point, the pipeline will be trenched to 0.6 m below the level of the seabed and covered by mat

Advanced pipeline plough - PL2

tresses ready for the NorFra line to be laid.

SEMAC 1 laid approximately 215 km of the 40" pipe to the tie-in point some 16 km from the Zeebrugge land fall. In April 1997, EMC employed the flat-bottom lay barge *CASTORO 2*, belonging to its parent company Saipem, to install the pipe in the shallow waters from the Zeebrugge cofferdam to the offshore tie-in point. The pipeline was subsequently tied-in by means of the above water technique using EMC's jet trench/utility barge *CASTORO 10.*

In view of the long design life and the harsh subsea environment, the decision was taken to trench the pipeline to a level just below the seabed. This will be achieved by a combination of dredging, ploughing and jetting. EMC will carry out the post-lay work using the jet-trench barge *CASTORO 10* and the diving support vessel *Bar Protector,* with the advanced diver-less pipeline plough *PL2.* Completion of all offshore works is scheduled for August 1997, and EMC will then pre-commission the pipeline ready for planned first gas in October 1998.

The Semac 1

Sedgwick Energy & Marine Limited, a specialist arm of Sedgwick Group plc, was appointed as insurance broker to Interconnector (UK) Ltd. in December 1995. The appointment was for placement of construction risks and third part liability insurances, and insurance consultancy services during the project period.

Sedgwick Energy & Marine is

part of Sedgwick Group plc, Europe's largest risk services consultant and insurance broker. Sedgwick Energy is the world's leading oil, gas and petrochemical insurance broker and risk adviser. Operating from offices in London and Norwich in the UK, with linked resources in Australia, Azerbaijan, Canada, Kazakhstan, Malaysia, Norway, Russia and the US, Sedgwick Energy has its own engineering services team - qualified petrochemical engineers, who combine practical industry experience with insurance and risk management expertise.

Excavation of landfall cofferdam

The programme was placed on time and on budget, despite the high project values to be covered. With its wide experience of construction insurance, Sedgwick was able to predict the market response to the project accurately and place the necessary coverage with recognised construction insurance markets.

The bulk of the Interconnector project values consisted of pipelines, regarded as many insurers as the least attractive of offshore construction risk exposures. Positive presentation of the risk and of its high industry profile was essential to ensure that underwriters did not see the project as yet another pipeline risk.

Sedgwick Energy & Marine's technical approach enabled Interconnector to receive broad and cost-effective insurance protection. The insurance programme provides significant flexibility for the project, particularly in the area of overall project values. An example of this is the provision of automatic coverage for increases in values of up to 25% of overall project values. Significant limits were also required for third party liabilities, because of the project's shareholder arrangement. Cover was once again placed on time and on budget.

Dredging International

Dredging International is a partner in the BacZee joint venture, which is the Dredging Contractor for the Interconnector Project and a member of the Offshore Alliance. BacZee's scope of work included all dredging and back filling activities, together with construction of the Zeebrugge landfall.

At the Bacton land fall, excavation of the required 3.2 m deep trench was performed by Dredging International's backhoe dredger *Big Boss* in February 1997. The company's trailing suction hopper dredgers *PearlRiver* and *Antigoon* were selected for dredging activities in the pre-sweep area between the UK and Belgian landfalls where hundreds of sand dunes lie perpendicular to the pipeline route. Equipped with sophisticated dynamic vessel positioning and draghead positioning systems, these two vessels were able to accurately and efficiently dredge the cut lines at a dredged depth of up to 55 m, allowing the lay barge to continue its programme on schedule.

The *Antigoon* also performed large volume dredging in the busy Scheur shipping lane near the Zeebrugge landfall, together with another trailing suction hopper dredger, the *Vlaanderen XX*. Initial gravel backfill to stabilise the pipeline will be carried out by *Pearl River*, which has a carrying capacity of 25,000 tonnes, and will be completed by *Antigoon*.

To ensure the stability of the Interconnector pipeline in the Belgian foreshore, it will be necessary to perform backfilling over its entire length using various types of

gravel and sand. These will all be supplied from the gravel market, or dredged in defined borrow areas by the *Antigoon* and the *Orwell.*

The shallowness of the foreshore at the Zeebrugge landfall means that an access channel had to be created so that the lay barge could approach within 1,200 m of the shore to allow a safe pipe pull. Dredging International used a range of shallow-draught vessels to dredge this section, including the trailer *Orwell*. The seagoing cutter suction dredger, *Vlaanderen XIX* was mobilised for the area close to the cofferdam.

To bring the pipe onshore at Zeebrugge and connect it to the land section on the beach, a 445 m-long cofferdam was installed, excavated, backfilled and partly removed by the jack-up platforms *Zeebouwer* and *Halewyn*, owned by Dredging International's sister company, Hydro Soil Services (HSS). Winches to produce a 1,000 tonnes pulling force were anchored on the beach to pull the pipe from the lay barge to the shore, through the pre-installed cofferdam. The cofferdam was then backfilled and removed. The beach tie-in to the land section of the pipe was performed at the same time, ready for tourist use of the beach in mid-June 1997.

The final part of the Zeebrugge landfall was the most environmentally sensitive part of the project. The pipeline crosses a dune area which is a recognized nature reserve containing a fresh water lake with rare orchids. As the crossing has to be executed in subsoil, horizontal directional drilling was selected as the most appropriate technique.

HSS commenced preparation of the 40" diameter pipeline string in the polder area. An offshore platform was then created on the beach for installation of the drilling equipment. After the first 10 cm subsoil hole had been completed to an accuracy of +/-20 cm, the hole was reamed in four passes to 56". The 500-m pipe string was then pulled from the polder to the beach in just 12 hours.

GPS-Satellite Guidance & Reference System

The GPS system was developed by the U.S. Army and Navy in the late 1960s to early '70s. The system was initiated to develop programs for timing and space-based navigation. GPS is the first and only global, 3-dimensional radio navigation system providing continuous operational service today. The US Air Force currently finances and operates the basic system of over 24 satellites and associated ground monitoring stations.

GPS service is provided for the benefit of users in countries around the world as well as for the US government; essentially without charge. GPS is generally considered an invention rivaling the compass for its broad advantages to both land and water bound industries. The dredging industry and its customers have gained greatly by improving accuracy in all aspects relating to excavation and reclamation. [13]

There are a variety of electronic components that are interfaced with the GPS, providing accurate means of controlling the dredging function by automation. Such systems that are produced by IHC include what they term, *Accurate Dredging*. This includes the following subsystems.

Dredged Profile Calculator

Knowing exactly the position of the cutter head that determines dredging depth and width, is imperative for efficient and accurate dredging with stationary equipment. The Dredged Profile Calculator calculates and displays the position of the cutter or wheel, related to the set profile and the waterline by means of LCD-display. Automatic adapting possibilities to the tide and the craft's draught are available.

Dredged Profile Monitor-DPM

The DPM enables the dredging master to verify and control the position of cutter or wheel, related to the channel's centerline, the waterline and the working depth and profile set-points. The DPM is an independent system but can also be incorporated in larger systems especially as a signal source for the automation of the dredging process, such as the Automatic Cutter (or Wheel) Controller. The system can be equipped with automatic profile/depth dredging, communication with survey equipment and geographical positioning.

Ladder Winch Controller

The main efficiency factor for the production of a sand-winning dredge of the plain suction, cutter or wheel type is a constant suction process. The Ladder Winch Controller is designed as a stand-alone system for controlling the ladder winch in order to keep the vacuum and density at the desired values.

Hydrographic Cutter Dredge System

The system is based on an unique new developed hydrographic dredge system, which enables the cutter dredge operator to see on-line, the actual updates from the cutter dredge on the bathymetric electronic chart in absolute positions. The system is able to log all relevant data and to present on distance on-line (remotely) the same hydrographic dredge data and views.

The **Hydrographic Cutter Dredge System** communicates with:

- **Positioning systems (D) GPS (LR/SR/RTK)**
- **Gyro compasses**
- **Echo-sounders**
- **Tidal receivers**
- **Dredge depth indicators (DPM/DPC)**
- **Automatic cutter controllers (ACC)**
- **Plotter/printers**

Artificial Intelligence (AI) for Dredging Control

The control of dredging operations such as on TH dredges, using AI or *fuzzy logic* computer programs has been applied to the drag-arm and drag-head operation. This has been a decided advancement for the goal of dredge automation .

A *fuzzy* **dredging control system** was developed in the 1990s in Japan. Engineers from the Ministry of Transport, Mitsubishi Heavy Industry Ltd, and MEC Engineering Service Co. Ltd, cooperated in the development. An intelligent dredging system was developed in which a statistical PLC at a sampling interval of every two

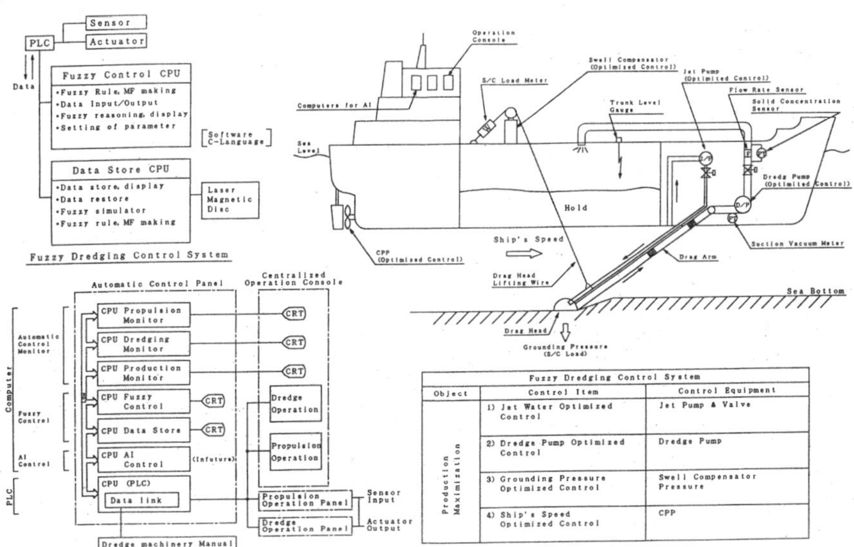

Fig.1 Configlation of Fuzzy Dredging Control

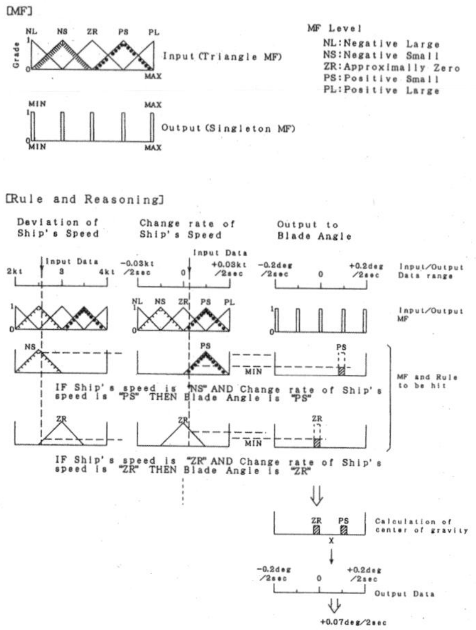

Fig.3 Procedure of Fuzzy Reasoning

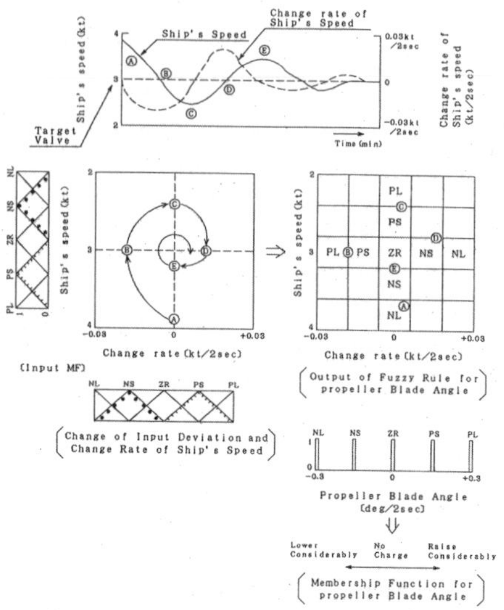

Fig.4 Example of Rule for Fuzzy Control

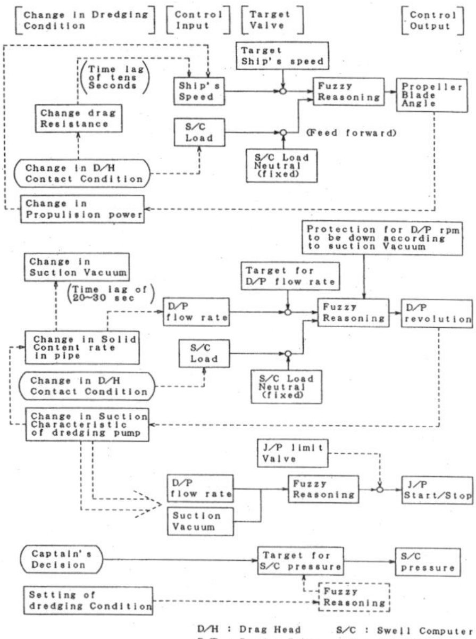

Fig.2 Characteristic of Fuzzy Control

seconds and carries out reasoning according to pre-tuned fuzzy rules and membership functions.

A conceptual diagram of the fuzzy dredging control system, is constructed with eight industrial computers, to make data communication by a **LAN** with a sequencer (PLC). The program is developed in the **language C** and loaded into two CPUs. The CPU handles the fuzzy rules, setting of the membership function (MF), setting of the control parameters, online fuzzy reasoning, data input and output with the sequencer, and screen display. The data storage CPU handles data storage during dredging, offline data generation, and fuzzy control simulation.

A control algorithm of the fuzzy dredging control system is shown in

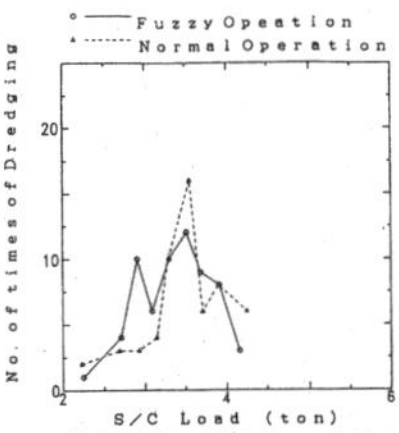

Fig5 S/C Load VS No. of times of Dredging (1992)

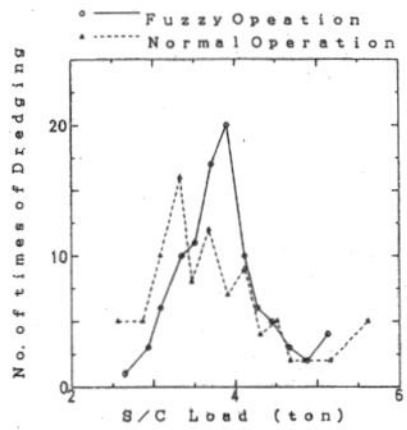

Fig6 S/C Load VS No. of times of Dredging (1993)

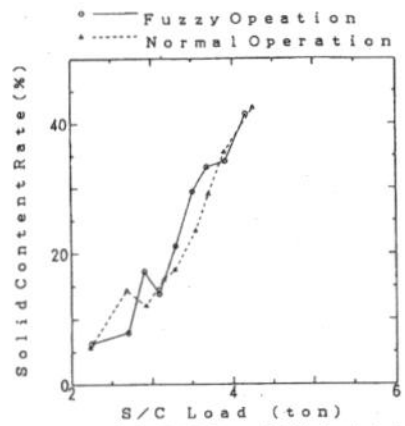

Fig11 S/C Load VS Solid Content Rate (1992)

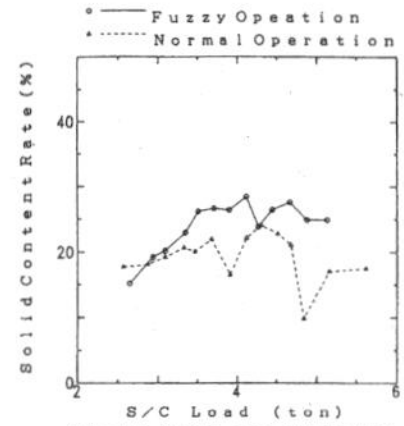

Fig12 S/C Load VS Solid Content Rate (1993)

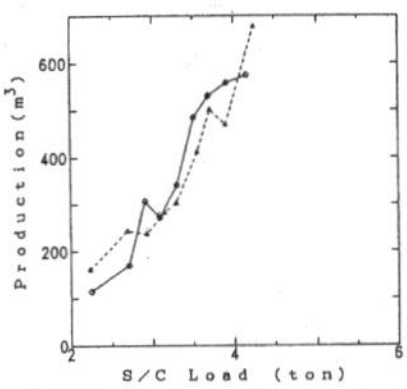

Fig 7 S/C Load VS Production (1992)

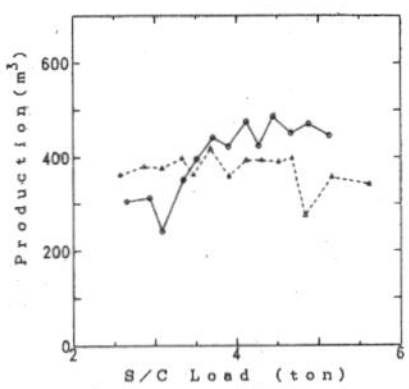

Fig 8 S/C Load VS Production (1993)

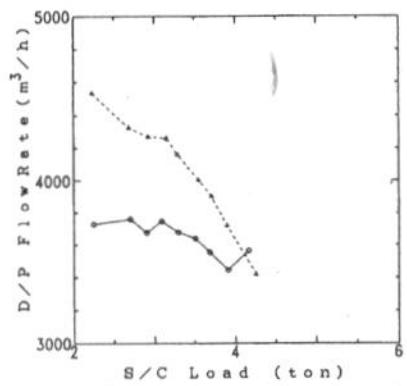

Fig13 S/C Load VS Dredge Pump Flow Rate (1992)

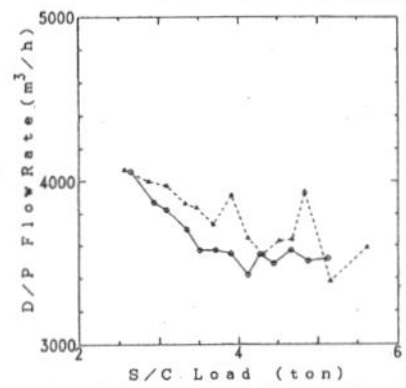

Fig14 S/C Load VS Dredge Pump Flow Rate (1993)

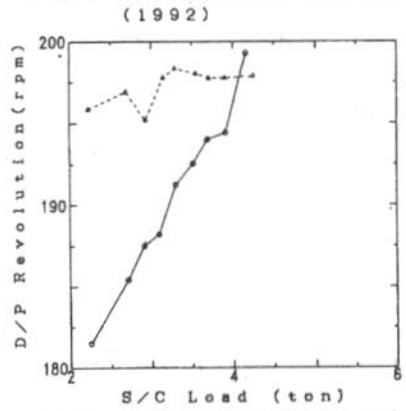

Fig 9 S/C Load VS Dredge Pump Revolution (1992)

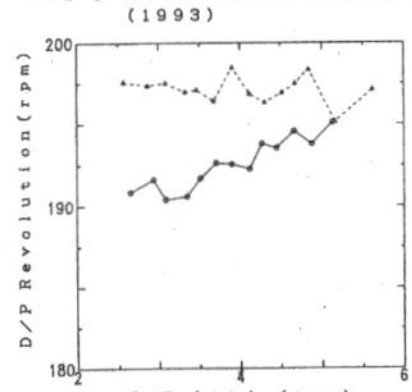

Fig 10 S/C Load VS Dredge Pum Revolution (1993)

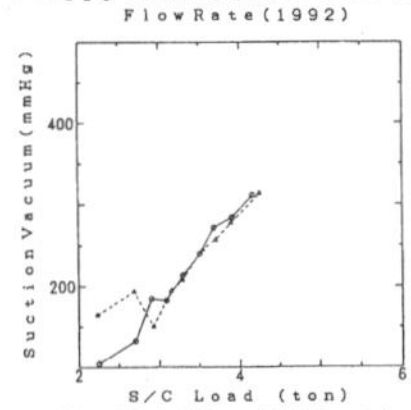

Fig15 S/C Load VS Suction Vacuum (1992)

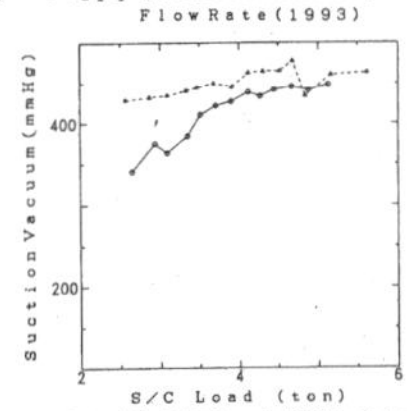

Fig16 S/C Load VS Suction Vacuum (1993)

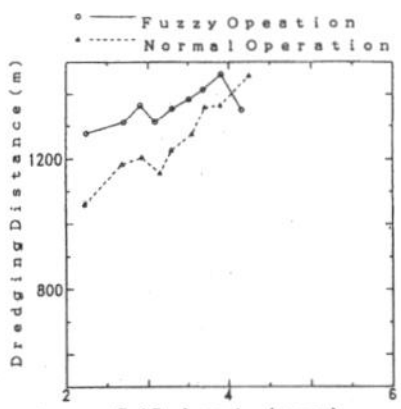

Fig17 S/C Load VS Dredging Distance (1992)

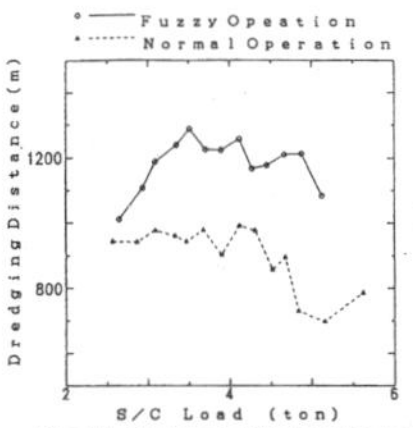

Fig18 S/C Load VS Dredging Distance (1993)

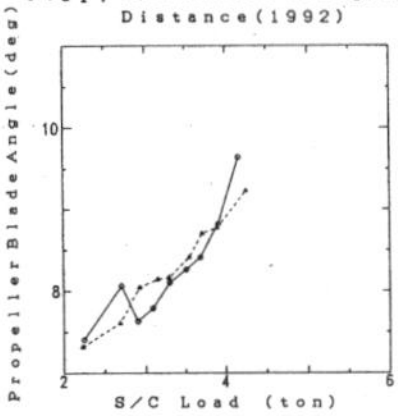

Fig19 S/C Load VS Propeller Blade Angle (1992)

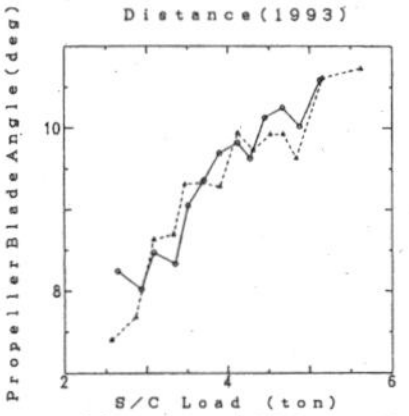

Fig20 S/C Load VS Propeller Blade Angle (1993)

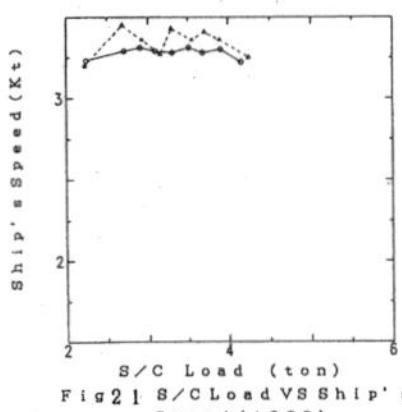

Fig21 S/C Load VS Ship's Speed (1992)

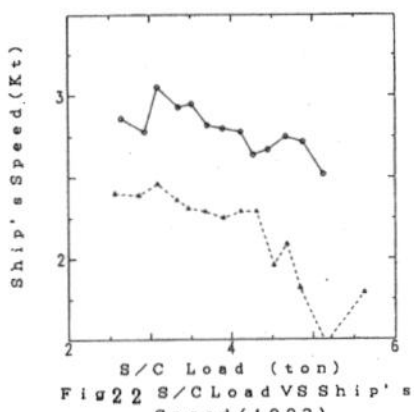

Fig22 S/C Load VS Ship's Speed (1993)

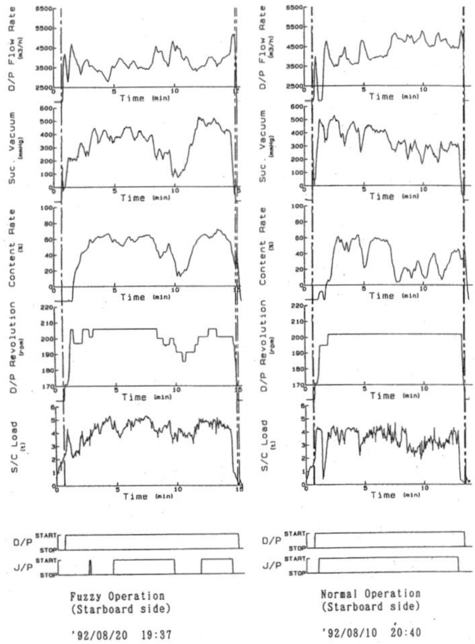

Fig23 Example of Dredge Pump RPM Control

Fig. 2. This model is prepared based on data recorded and analyzed for normal dredging done previously and on knowledge gained from a questionnaire filled out by the captain. The controls are applied to the following functions:

Hull part		Machinery	
Length a.o.	94.90 m	Main generator	2,000 kw x 2
Length p.p.	88.00 m	Main gene. engine	3,000 ps x 2
Breadth (mld)	16.00 m	Propulsion motor	1,300 kw x 2
Depth (mld)	7.20 m		
Draft (mld)	5.60 m		
Gross tonnage	3,526 t		
Trial speed	13.29 kt		
Service speed	11.50 kt		
Dredging part		**Oil recovery part**	
Type	Side drag	Oil recovery scoop	Cyclone type & Inclined type
Dredging depth	22.00 m		
Dredging pump	4,100 m^3/h x 2		
Jet pump	500 m^3/h x 2	Oily water pump	500 m^3/h x 2
Mud hold dumping	Conical valve	Discharge pump	1,000 m^3/h x 2
Hold capacity	1,700 m^3	Recovery oil tank	1,450 m^3

Table 1 Principal particulars of the dredger

1- **Propeller pitch**
2- **Dredge pump speed**
3- **Start and stop of jet pump**
4- **Swell compensator pressure**

Development of the intelligent dredging system started in 1994. The fuzzy control algorithms based on the actual-ship experiments will form the framework of future intelligent control. The target is to analyze the dredging data statistically, add a learning function, and have a system that accurately judges the nature of the soil on the seabed, to optimize the fuzzy dredging control parameters in accordance with the state of the seabed.

Content	Number of points
(1) Number of input data points	18 Data
(2) Number of output data points	8 Data
(3) Number of fuzzy rules	
Relating to propeller pitch	68 Rule
Relating to dredge pump	64 Rule
Relating to jet pump	30 Rule
	Total 162 Rule
(4) MF setting	5 functions per each input/output

Table 2 Fuzzy rules, etc. actually used

When the intelligent dredging system is completed, the dredging operation will be optimally controlled based on working conditions and will produce more efficient dredges.

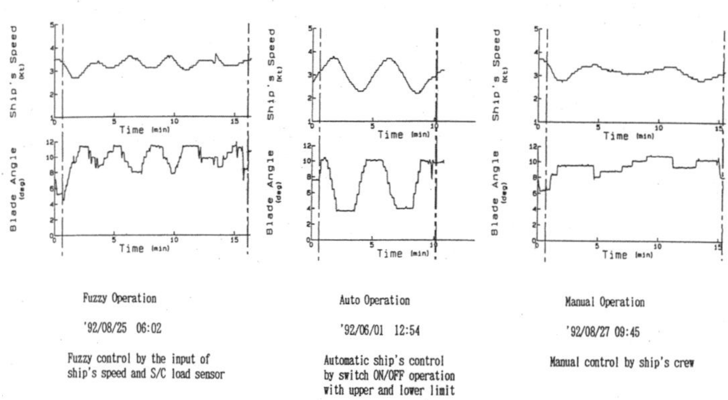

Fig2 4 Example of Constant Ship's speed control during Dredging

No	Item recorder	No	Item recorder
1	Ship's speed	25	Drag arm position
2	Ship's position	26	Drag pump start / stop
3	Propeller pitch	27	Jet pump start / stop
4	Propeller power	28	Dredge - discharge valve open / close
5	Ship's draft	29	Dredge - load-in valve open / close
6	Amount of water discharged	30	Dredge - overboard valve open / close
7	Loaded amount	31	Dredge - suction valve open / close
8	Mud hold level	32	Jet - mud hold valve open / close
9	Specific gravity of seawater (set value)	33	Jet - arm valve open / close
10	Specific gravity of mud (set value)	34	Conical valve open / close, half open
11	Tide level	35	Dredging start / stop
12	Trim	36	Dump . Dewater start / stop
13	Heel	37	Swell compensator load
14	Drag head depth	38	Fuzzy operation propeller pitch output
15	Swell compensator air pressure		
16	Swell compensator stroke	39	Fuzzy operation D/P rpm speed output
17	Dredge pump flow		
18	Dredge pum suction vaccum pressure	40	Fuzzy operation S/C pressure output
19	Dredge pump rotation speed	41	Fuzzy operation J/P degree of necessity output
20	Solid content rate		
21	Drag head deflection angle	42	Fuzzy operation state of computer
22	Drag head winch start / stop	43	Fuzzy operation J/P start / stop output
23	Intermediate winch start / stop	44	State of operation manual, automatic, artificial intelligence (AI)
24	Trunnion winch start / stop		

Table 3 Items of data collected

Item	Fuzzy operation	Normal operation
(1) Ship's speed target	3.3 kt (1992)	3.0 kt (1993)
(2) Swell compensator pressure	constant at about 12 kg/cm^2	
(3) Dredging depth	around - 14 m	
(4) Mud hold overflow	none	
(5) Propeller pitch control	Fuzzy control	Manual operation by crew
(6) D/P speed control	Fuzzy control	Automatic operation (fixed at about 200 rpm during dredging)
(7) J/P start / stop control	Fuzzy control	Automatic operation During dredging on (1992) off (1993)
(8) Drag head window	Full open	Full open (1992) Partly (1993)

Table 4 Operating condition of actual - ship experiments

Item of data collect	Unit	1992		1993	
		Fuzzy operation	Normal operation	Fuzzy operation	Normal operation
Ship's speed	Kt	3.29	3.36	2.82	2.25
Propeller blade angle	deg	8.22	8.41	9.37	9.17
Dredge pump speed	rpm	191.27	197.68	192.40	197.10
Loaded amount of soil	m^3	407.61	401.78	411.85	381.67
Dredge pump flow	m^3 /h	3880.49	4237.71	3584.43	3800.38
Solid content rate	%	28.42	28.19	25.38	19.94
Suction vacuum pressure	mmHg	226.05	237.49	417.59	447.41
Swell compensator load	ton	3.38	3.49	3.84	3.75
Swell compensator operating time	min	1.76	1.85	4.41	4.48
Swell compensator stroke	m	0.08	0.07	0.18	0.18
Dredging time	min	13.55	12.38	14.00	13.47
Dredging distance	m	1375.55	1281.51	1218.62	935.49
Number of dredging cycles	case	63	58	98	92
Drag head depth	m	14.18	14.18	13.83	13.86

Table 5 Comparison of experimental data, average values in fuzzy and normal operation

Computer Applications

On Kruse Control

Kruse Controls, Inc. was contracted to automate many of the features found on a cutter/suction (CS) dredge. The customer's goals for a new 100 ft sand and gravel dredge were to improve production efficiency, reduce downtime and automatically create daily reports detailing production, depths and locations dredged.

In order to meet the customer's goals, the shore powered electric dredge was equipped with a ***PLC (Programmable Logic Controller)***, several ***VFD's (Variable Frequency Drives)***, one computer for use as an operator interface and one computer for dredge position/depth control and logging.

The dredge has a *750KVA* on-board transformer, which develops the 480v AC power necessary for the dredge's forward and aft winches and ladder motors. Each motor is equipped with a *VFD* to allow speed and torque control for swinging the dredge or raising and lowering the ladder.

The *PLC* is the **on-board industrial computer** which controls the entire dredge. Through it's input and output modules, it is connected to all of the field devices and *VFD*s which allows the *PLC* to monitor and control the dredge operation based on the operator entered parameters. Field devices typically found on manually controlled dredges, such as pressure transmitters, nuclear density meter, flow meter, inclinometer and linear displacement sensors are monitored by the *PLC* program to allow quick and automatic response to adverse conditions without operator intervention.

This dredge is equipped with a graphical touch screen, rather than levers, pushbuttons, switches and foot pedals, which is attached to the operator interface computer. This computer communicates with the *PLC*.

The operator interface displays all of the dredge information including operational status, speeds, GPM, TPH, water pressure, oil pressure, ladder depth, discharge pressure, suction relief valve position, motor torques and suction vacuum. Additionally, all fault conditions and alarms are immediately displayed and annunciated on the operator interface screen, and then archived on the computer with a time and date stamp.

The dredge operator can navigate the dredge and operate its components manually via the touch screen, or he can enter automatic settings for production rate, suction vacuum and ladder depth, which allow the *PLC* program to control and optimize the process.

When in the fully automatic mode, the dredge swing speed, cutter speed and suction pump speed are controlled by the *PLC* program to maintain the operator specified TPH (Tons Per Hour) set-point. The *PLC* program through automatic suction relief valve positioning maintains the suction vacuum.

Semi-automatic mode allows operator control of the swing speed. While swing speed is controlled in the operator-selected direction, motor torque is controlled for the opposite swing winch to maintain a taught cable and eliminate dredge walking from the cutter.

Regardless of the operating mode, conditions are continuously monitored by the *PLC* to automatically switch the controls back to manual mode under certain conditions. In critical situations, the swing winches are stopped.

A separate screen on the operator interface computer displays historical trends of Tons per Hour, discharge pressure, oil pressure and other critical parameters.

A second on-board computer allows the dredge operator to visually monitor the dredge location and ladder depth in reference to the customer's surveyed boundary lines. This computer runs an application of ***HyPack Max / DredgePack***$^{(TM)}$ software for dredge depth and location logging and reporting. A GPS system interfaces with the *DredgePack* to allow coordinates, speed, location and heading to be displayed and logged on the computer.

A scaled symbol of the dredge is super-imposed in real-time on the screen. The color graphical depiction of dredging locations and depths are printed daily for the customer's records.

If necessary, with the press of a computer key, the dredge operator can drop a virtual buoy on the screen to indicate trouble spots in the pond or to relocate to the same spot in the future. A customized software communication driver was developed to allow the *DredgePack* software and the *PLC* to share dredge ladder depth, draft and tide corrections, and other parameters. This allows for automatic control of the ladder depth and dredge location as required to satisfy permit constraints or topographical model tracking. The next phase will be to add an on-shore booster pump station to the *PLC* control system, using radio-telemetry to bring the data back to the main *PLC*.

CHAPTER 12

DREDGING RESEARCH

In recent years, R&D on dredge design and operations has increased and been conducted in various locations. This has included USA, The Netherlands, Belgium, Germany, UK and Japan. Some of the research projects are summarized here.

The Netherlands

The beginning of the **Dutch Research Association (CSB)** was in 1963, when the **Rijkswaterstaat** (Government) and the **Dutch dredging industry,** formed a partnership aimed at *pre-competitive restricted research*. Research was concentrated in **Delft University of Technology (Geotechnics and Hydraulics departments),** where certain aspects had been studied beginning in 1924.

In the years of 1963-69, efforts were concentrated on hydraulics; slurry transport and the suction process. From this activity, a special group was formed called, **Cutting and Dredging Technology Group,** to carry out research into the physical processes. A separate R&D effort was carried out by the commercial organization of **IHC HOLLAND**, through its **MTI Research Institute** but kept separate from the CSB group.

The research into the physical processes was directed towards the mechanical loosening (cutting) of soils. The pore-water phenomena was seen to be predominant; for the first time cavitation was even observed in the pore-water in a cutting experiment.

The **Dredging Flume,** also called the cutter and trailing suction flume, was completed in 1974. This was a further development starting from the smaller pilot flumes built by **Professor Bos** in 1969 and '72 at **Delft University of Technology**. The Flume is 50 m long, 9 m wide and 2.5 m deep. It has a trailing carriage (with cutter drive, dredge pump and associated equipment), mounted over it. In 1987 a large movable glass wall was installed on the sides for studies of hydraulic fill and slope erosion (breaching) processes using the pump system.

Prof. Bos: ***You have to start from physical insight and then design or select the proper tool (dredging machinery).***

Conclusions

"Searching for dredging physics means to a large extent walking in the *no-mans-land* between hydraulics and soil mechanics. In the fields of both soil mechanics and hydro-morphology new areas of knowledge have been established.

It is interesting to note that hydraulic engineers were the first to grasp the role of pore-water in the soil-dynamics of dredging. They were familiar with water and with dynamics. At that time soil mechanics was still focused on quasi-static problems. Eventually, a new branch of soil mechanics emerged."

The CSB program has been funded by the participants themselves for many years. The partners as of 1989 were: Amsterdam Ballast Dredging; Royal Boskalis Westminster BV; HAM; Van Oord Utrecht BV; Volker Stevin Dredging; Zanen Verstoep NV; and the Dutch government (Rijkswaterstaat).

United States of America

Dredging Research in Transition

Passage of the ***National Environmental Policy Act (NEPA)*** in the 1960s, followed by the ***Marine Protection, Research and Sanctuaries Act (MPRSA)*** and the ***Clean Water Act (CWA)*** in the early 1970s, established environmental rules dealing with navigation dredging and disposal of dredged material.

These landmark legislations set the environmental direction for **the U.S. Army Corps of Engineers (USACE)** navigation dredging and dredged material disposal into the 21st century. These environmental laws required radical changes in the skills and knowledge, which the Corps needs in carrying out its navigation mission. This mission is inexorably linked to our economy, defense needs, and leadership in world trade.

The Corps' environmental mission (pursuant to the MPRSA and the CWA) mandated a leadership role in not only achieving appropriate environmental standards but in the innovation and technology necessary to be a world leader in dredging and dredged material management.

The fact that dredging often occurs in highly urbanized and industrialized areas with their associated contaminant load, only made the challenge more difficult. Linked with the knowledge that huge quantities of dredged material, can have numerous beneficial uses, interfaced with the environment stewardship requirements and mandated innovation and high tech development to meet their mandate.

Dredged Material Research Program (DMRP)

The Corps' research and development expertise centered at the **U.S. Army Engineers Waterways Experiment Station**, Vicksburg, MS in partnership with field offices, was called to develop the technology to achieve the least costly environmentally - acceptable dredging and disposal alternatives to maintain a world class navigation infrastructure.

In 1972, the **Rivers and Harbors Act** authorized the Dredged Material Research Program (DMRP) to ensure a technology base to meet the mandates of the **NEPA**, the MPRSA, and the CWA. The DMRP was a five-year program initiated in 1973 to develop the Corps' technology base for dredged material management in the areas of aquatic disposal, upland placement, wetlands management, and beneficial uses.

The key conclusion of the DMRP was that there were no disposal alternatives that were appropriate in all cases, and that all possible disposal management alternatives should remain in consideration until the economically and environmentally suitable alternatives are selected.

The DMRP formed the technology base to serve as the foundation for dredged material testing and assessment required by domestic legislation and international treaty (the **London Convention**) regulating ocean disposal. The DMRP also established the technology necessary to develop techniques for identifying, assessing, and managing contaminated dredged material, and as the national lead in wetlands management and restoration.

Unfortunately, the engineering aspects of dredging and placement lagged behind the environmental emphasis. Correspondingly, the Dredging Research Program (DRP) was initiated in the late eighties to establish the systematic efficiency necessary for cost effective dredging transport and placement.

Dredging Research Program (DRP)

The seven-year dredging Research Program (DRP), initiated in 1987, had as major objectives the development of equipment, instrumentation, software, and operational monitoring and management procedures to reduce the cost of dredging to a minimum consistent with Corps mission requirements and environmental responsibility.

To better understand the dynamics of dredged material disposed in open water, objectives included:

(a) Acquisition of field data sets for developing simulation methods and computational techniques to evaluate boundary layer properties;

(b) Forcing function for formulating short- and long-term dredged material fate models, and

(c) Development of improved site monitoring techniques.

Investigations of material properties related to navigation and dredging included:

(a) Development of instrumentation and operating procedures for analyzing properties of consolidated sediments;

(b) Definition of navigable depth in fine-grained sediments and rapid surveys of fluid mud, and (c) establishment of dredging -related soil and rock descriptors.

Dredge plant equipment and systems processes research:

(a) Improved draghead designs and eductors for sand bypassing;

(b) Designed a portable single-point mooring buoy for hopper dredge direct pumpout,

(c) Developed methods to increase dredge payload for fine grain sediments, and

(d) Prepared information for dredging manuals incorporating state-of-the art technology.

Vessel positioning, survey controls, and dredge monitoring studies developed:

(a) A 3-D, 10-cm accuracy, GPS positioning system for surveying and dredging;

(b) A real-time system for determining offshore tide and wave conditions;

(c) Better production meter knowledge, and

(d) An unmanned inspection monitoring and standardized reporting methodology for hopper dredging activities.

To enhance management of dredging projects, a comprehensive model of dredging project activities to evaluate effects of decisions, project changes and provide guidance for optimizing use of open water disposal sites, was developed.

The DRP performed the first detailed study in the U.S., to accurately quantify and document the economic benefits of a federal R&D program. Based on certainty simulation analyses, there exists a 90% certainty that direct and indirect annual and one-time benefits from the use of DRP products by the Corps' national dredging program, will exceed **US$109M** through 1998 in **FY94** dollars.

Dredging Operations and Environmental Research (DOER) Program

Accomplishing the navigation dredging mission, while balancing environmental protection is the major challenge of the 1990s. Integration of operational and environmental aspects of dredging and disposal must be accomplished within a climate of increased dredging workload need and environmental constraints, with decreasing fiscal and manpower resources.

The Environmental Research (DOER) Program, initiated in 1996, was formulated toward addressing documented deficiencies of the primary Corps users (**field operating division and district offices). Objectives of the DOER program** are to develop technologies, methodologies, and techniques to assure the operational and environmental issues of the USACE dredging program are adequately and efficiently met.

The DOER program investigates dredging issues in six focus areas:

(1) **Nearshore and Offshore Placement of Dredged Material,** will predict the time-dependent movement of non-contaminated sand and sand/silt mixture of dredged materials placed in the near-shore zone, and all materials placed in the offshore region;

(2) **Environmental Windows for Dredging Operations,** will work to reduce cost of dredging operations, attributable to compliance with environmental windows that are determined to be over-restrictive, inconsistent, or technically unjustified;

(3) **Contaminated Sediment Characterization and Management,** will reduce costs and enhance the reliability and acceptability of methods associated with the assessment, dredging, placement, management, and control of contaminated sediments from Corps navigation projects;

(4) **Instrumentation for Dredge and Site Monitoring,** will implement automated dredge inspection and innovative dredge contract payment methods, characterize and delin-

eate possible contaminated areas of Corps dredging projects, and provide a cost-effective method to accurately monitor cap status with improved timeliness, spatial extent, and reduced effort;
(5) **Demonstration of Innovative Equipment and Process Technologies,** will conduct demonstrations of emerging dredging technologies and operations in cooperation with the USACE field operating offices and other resource agencies;
(6) **Ecological Risk Management for Dredging and Disposal Projects,** will develop a technically sound approach, for characterizing and managing risk. This makes use of existing guidance (the dioxin management strategy and the technical framework) and proven tools for conducting risk-based evaluations of dredged material, that are consistent with the **U.S. National Academy of Science/US Environmental Protection Agency (EPA)** paradigms for risk assessment.

Benefits of the DOER program include:

Cost-effective practices for dredging and dredge material disposal:
Environmental protection enhancement through application of effective environmental windows;
Compliance with environmental statutes, for identifying and managing contaminated sediments;
Reduction of costs of disposal of dredged material by beneficial placement in the nearshore zone;
Greater flexibility in dredging in sensitive ecological areas; and
Expanded options for beneficial uses of contaminated and non-contaminated dredged materials.

This research is being conducted with full coordination and cooperation of other appropriate agencies including the EPA, the U.S. National Marine Fisheries Service and the U.S. Fish and Wildlife Service. The DOER program includes an aggressive technology transfer mechanism to ensure rapid implementation of research products into Corps projects.

Education & Research

There are some leading schools who prepare students for a profession in the Dredging Industry. Efforts are made through the dredging societies of WODA (World Organization of Dredging Associations), to encourage students to learn about the industry and to locate jobs; from a trade level to graduate university programs.

The following institutions have contributed through their graduates to the advancement of Dredging Technology.

Universities

United States of America

Dept. of Marine Science, Louisiana State University, LA
Civil Engineering Dept., MIT, MA
Civil Engineering Dept., Northwestern University, Evanston. IL
Texas A&M University, Center for Dredging Studies, College Station, TX
Institute of Marine Science, University of Texas, Austin, TX

Europe

Delft University of Technology, Delft, **The Netherlands**
University of Strathclyde, Glasgow, **Scotland**
Institute fur Schiffshydrodynamik, Technische Universitat, Berlin, **Germany**
Civil Engineering Dept., University of Adelaide, **Australia**
Kyushu University, **Japan**

Laboratories

WES-USACOE, Waterways Experiment Station, Vicksburg, TN, USA
USACOE, Coastal Engineering Research Center
HR Wallingford, UK
Port & Harbour Research Institute, Ministry of Transport, Japan
MTI Research Institute, IHC HOLLAND

APPENDIX

CONTENTS

APPENDIX A

Beneficial Effects of Dredging Turbidity, Supposed Toxicities of Marine Sediments

Dr. Joel F. Gustafson

In contrast to the usual procedure in presenting a topic, I shall place the major thoughts and conclusions first because I want you to have these in your minds as I develop the following topics. It will take some time to reprogram our thinking about turbidity.

For decades, we have regarded turbidity as being dangerous (biologically), and undesirable (esthetically), and now controllable (politically). Recently, regulatory provisions of state and national origin have been created or are being proposed to reduce man-caused turbidity to a minimum level, and to prohibit it wherever possible.

These actions have been taken in almost complete ignorance of the degree of damage, if any, due to turbidity, and in spite of the fact that wind and tide-created turbidity dwarf those of man's activities (Gustafson 1972).

Our state and federal agencies concerned with environmental issues are products of our own culture. The technical staffs of these agencies operate on the bases of information, assumption, and political expediency.

This statement is not unduly harsh. If the actions were not taken under the protective cover of environmental concern, then it would be necessary to develop facts prior to the initiation of regulations. Despite the hundreds of millions of dollars now being spent by many industries to clean up their air and water discharges, too many still resist such programs and the resentment of the public is high. It has been the intent of regulatory agencies to force industry to develop pollution control devices, practices and procedures. If the regulations are excessive, industry must prove its case for lesser degrees of control.

Do not misunderstand me. As an ecologist I believe that the need to protect and enhance the quality of our environment is paramount. In many areas of air and water pollution it may take a decade or more to significantly reduce the pollutants in the environment even after the discharges cease ***(Meadows et al, 1972).***

The costs of cleaning up our environment will be many billions of dollars. It behooves us, then, to spend wisely because we will not be able to do it all at once, and some programs may be delayed for many years.

The present trend of regulations to control turbidity in our estuaries and harbors is based upon a ***fear syndrome***. The teachings of the evils of turbidity, has created a *knowledge* that all turbidity in all places is bad. Acting under this fear, and in a determination to protect the environment, agencies have placed the dredging industry in a hazardous position.

Let me first point out the natural role of fine sediments in fresh water and marine environments, and then let me offer some indications that the toxicity part of the *fear syndrome* is without basis in fact.

Beneficial Roles of Clays

Clay sediments consist of particles of mineral matter which are near colloidal sizes. Each clay particle attaches to itself a shell or a halo of water molecules. Due to the orientation of the water molecules, the clay + water shell assumes a negative charge. Each of these units (clay + water shell) travels as a unit and acts as a unit. The many chemicals attached to the clay mineral particles dissociate from the clay surface and extend out into the water halo. These chemicals, although they are physically bound to the clay particles, are also available for exchange with other chemicals in the water beyond the halo. The chemicals that are exchanged, are predominantly positively charged radicals (cations). Among the cations which are exchanged from the surrounding water into the water halo are the plus two charged metals such as **arsenic, cadmium, chromium, copper, lead, mercury and zinc.** These, and other metals, and other chemical substances, are drawn out of the surrounding water and are quite firmly attached to the clay particle.

In fresh water situations, the clay particles remain suspended until all water movement ceases and then they slowly fall out of the water column and accumulate as sediments on the bottom. As fresh waters bring clays into marine areas, the clay units aggregate into small clumps as the salinity approaches 14%. This is due to the fact that the abundant numbers of positive ions in salt water change the surface charge of some of the clay particles to a positive charge. As more and more clay masses assume positive charges, they join with negatively charged clay masses. Aggregations of such combinations, join together to form flocculants that fall out of the marine waters during periods of slack tide. Those areas within an estuary or a harbor, in which this phenomenon of aggregation and flocculation is most evident, are nature's construction terraces commonly known as mud flats.

The ability of clays to remove materials out of the water column is astounding! The greatest removal of material is accomplished within the first 20 minutes of suspension of the clay within the water. ***(Marshall 1962).*** The longer the period of suspension, the greater will be the success of removal. Oils, pesticides, sewage pollutants (with the exception of nitrates), and metals are all stripped or scrubbed from the water by the clay sediments. In the case of the large organic molecules associated with sewage, the molecules from layers on the outside of the water shell of the clay particle. We might compare this grossly with an onion in which the leaves surround the plant embryo. Within a period of a few minutes, each particle may attach two and a half times its own bulk of organic sewage molecules to its water halo.

We know that bacteria attack these attached sewage substances much more readily when they are attached to the clay than when they are dispersed within the water. In this sense, then, the adsorption of sewage to clay makes the subsequent release by bacteria of the nutrients from that sewage more rapid. However, this is true only so long as the clays remain suspended. As the clays in the marine environment fall to the bottom to form mud sediments, the lack of, oxygen greatly reduces the liberation of organic nutrients from the sewage molecules attached to the clay halo.

In organically rich sediments, oxygen becomes deficient at the surface, becomes a negligible 1/8th of an inch within the sediment, and typically, is absent completely at depths greater than ½-inch. As the clays fall and become bottom sediments, the materials which have been stripped out of the water column are almost completely isolated. The great majority of the organic sewage molecules and of toxic substance such as oil, metals and pesticides, have been removed from the water column and placed in semi-permanent storage. The activity of the burrowing bottom organisms, such as worms, clams, and crustaceans does very little to increase the turnover of substances stored on the clay particles. As clay sediments form, bacteria and diatoms produce copious amounts of mucous, which assist in the binding together of the clay particles. Consequently it requires considerable energy on the part of winds or tides to lift and suspend muddy sediments. Such re-suspension is limited to the uppermost portion of the bottom sediments. Therefore the toxic substances and the sewage particles which are adsorbed to the clay particles are literally placed in deep storage.

Concentrations

Table I portrays concentrations of toxic substances in bay sediments and the bay water. It also presents some data on the toxicity (poisonous) levels of some metals. At this time I have no further data from the Environmental Protection Agency (except that listed on the acceptable maximum limits of certain metals (5th column).

Reference No. 1 illustrates the very much larger concentrations of the metals at tached to the sediments versus those concentrations suspended in the water. The difference in concentrations between sediment and water speaks to the adsorptive capacity of the sediments and proves their important role in the removal of these, undesirable chemicals from the water.

The reader may note that all of the references cited in this paper are recent. This illustrates the fact that we have not been doing the kind of work that we should have been doing until recently. Unfortunately the eye of most investigators seems drawn to the closeness of the concentrations between the toxic levels and the normal levels of bay water. It is my opinion that most toxicity levels have been determined in aquaria using filtered sea water.

Table I
CONCENTRATIONS OF TOXIC METALS IN SEDIMENTS AND WATER

Metal	Dredge Spoils	Sea Water		S.F. BayWater		Toxicity		F.W.Q.A. Criteria
Cadmium	130	0.08	(1)	0.06	(3)	0.01 - 10	(1)	(4)
Chromium	150	0.00005		0.02		1.0		
Lead	310	0.00003		0.02		0.1		0.005 = 50 ppm
Nickel	610	0.0054		0.06		0.1		
Mercury		0.00003	(2)	0.0002				0.0001 = 1 ppm
Zinc		0.01		0.02				0.005 = 50 ppm
Copper		0.003		0.02				

Concentrations in Mg/1 = ppm

Concentration in % on a dry wt. basis

(1) = Ocean Dumping 1970
(2) = Water Quality 1968
(3) = Gustafson, February 1972
(4) = F.W.Q.A. Criteria

Concentrations of Toxic Metals in Sediments and Water

Table II presents the results of a pilot study designed to test the effectiveness of adsorption and the permanence of that adsorption when subjected to vigorous agitation. (One hundred grams of sediment in 1 liter of bay water was agitated with a propeller for 20 minutes. Following the drop of the sediments after an additional 30 minutes, sample water was removed, carefully decanting so as not to re-suspend the sediments.)

It should be noted that in four of the five instances the re-suspension of bay sediments caused an increased removal of the metals from the water. These rates of increased removal ranged from 69.8% to 97.4%. In the fifth case there was a 45% increase in the concentration of cadmium following the re-suspension of the muds. This single increase might have been an error in analysis. I must emphasize that this simple experiment was nothing more than a pilot study. The evidence indicates that sediments upon re-suspension typically do not liberate metals but rather further remove metals from the water in which they are suspended. Pesticides and PCBs followed the action of the metals.

Table II

Adsorption of Toxic Metals from San Francisco Bay Water Following re-suspension of San Francisco Bay Muds

In ppm	Bay Water	Water +Sediments	Removal-%
Mercury	0.0014	0.0000361	97.4
Lead	<0.02	<0.0024	88.0
Zinc	<0.02	0.0024	88.0
Copper	<0.04	0.0121	69.8
Cadmium	<0.02	0.0290	45.0 (gain)

The experiment just cited contains no replicates. It therefore lacks both substance and reliability. It is presented at this time as an indication of the kind of experiment that should be done many times in many locations.

The dredging industry must take a role in developing those kinds of information which are sadly lacking at the present time. This experiment consists of the determination of concentrations of the water before and after bay sediments are re-suspended. Although not necessary, it would be informative to also analyze the concentrations of these metals in the sediments. Sediment analysis requires additional laboratory work, and therefore expense. If expense is a criterion in some locations, the sediment analysis can be omitted, but no area of the United States, and presumably elsewhere, can afford to omit the expense of the simple experiment indicated above.

If the dredging industry were to develop an overwhelming mass of information on this point, it would produce an almost immediate improvement in their present low status.

Nature Dwarfs Man

What about the sewage organics that are adsorbed to the clays? The short-time re-suspension of sediments such as might occur during dredging or even during disposal in confirmed areas, will not produce an excessive growth of algae. The *fear syndrome* assumes that a sudden liberation of sewage products is accomplished by the re-suspension of sediments. Previously *(Gustafson 1972)* we detailed evidence showing that in waters subject to wind action and to tidal scouring that the re-suspension by nature dwarfs that of man's activities. We do not see algal blooms or extensive growths of plankton following strong or persistent winds, nor do we in fact see algal blooms following the very great introduction of new sediments and agricultural wastes that are brought into a bay by flood. At this time we do not know enough to explain why algal blooms occur or do not occur. But, for our immediate purposes, we know that they are not obviously associated with the re-suspension of bay sediments.

There are some additional beneficial roles that I shall mention in conclusion, even though we have previously demonstrated the very important beneficial roles of turbidity particles. Turbid waters offer shelter and protection to the multitude of larval and immature marine life that utilize bay waters as nursery grounds. Turbidity prevents the penetration of light and perhaps this is one of the several factors involved in the failure of rich bay waters to produce excessive algal growths. The increased turbidity levels would reduce the depth of the water in which algae could engage in photosynthesis. The food substances attached to clays are digested by many of the marine organisms that live within the bottom sediments and by many organisms that are filter and particle feeders.

Clam Experiments

In another pilot study *(Gustafson 1972b)* we explored the critical question of whether marine organisms could or would digest the metals and pesticides off of clay particles. We selected a common bay clam, *Mya arenaria*, a filter feeder, as our test organism. Three 25 gallon aquaria were used in our experiments. Two of these aquaria were test aquaria into which bay sediments were added and kept suspended for twenty minutes. The third aquarium was a control for the other two. The clams were carefully washed, measured and paired so as to get clams of equivalent sizes in each of the aquaria. About three-four inches of sand was placed in the bottom of each aquarium and 10 clams were carefully placed in an upright position in the sand.

Four times each day, 15 gallons of San Leandro Bay salt water were added or removed so as to duplicate the tidal regime of the bay. The *tides* were added at 0800 (8 am), removed at 1300 (1 pm), added at 1800 (6 pm) and removed at 2200 (10 pm), thus providing nine hours per day of feeding time. Each aquarium was aerated vigorously for the first hour after the *tide* water was added. To test the effects of a high turbidity level, 100 cc of bay channel sediment was added to 100 cc of bay water and stirred rapidly for several minutes.

Then the thick slurry was slowly added to each of the test aquaria. This addition of sediment created a turbidity level of approximately 1500-2000 *Jackson Turbidity Units*. A small electric motor and propeller kept the sediment in suspension for 20 minutes. The actual levels of turbidity were measured with a Jackson Turbidity Meter, and much more accurately by a Beckman Spectronic 20 Meter, and much more inaccurately with a Secchi Disk. The vigorous addition of oxygen for the first hour of each new *tide* (addition of fresh bay water and the addition of the sediment) was made in order to simplify the number of stress situations presented to the clams. When large amounts of sediments are suddenly resuspended, there is a marked reduction in the oxygen level in the water. We were concerned in this experiment with the ability of the clams to remove metals and pesticides from the sediments. Aeration (oxygen) was added to the water to remove the additional factor of a depressed oxygen level.

Turbidity in Dumping

We emphasize that in dredging operations, turbidity levels of *350-400 Jackson Turbidity Units (JTU's)* would be reached in the disposal (dumping) of bucket dredge operations and that this condition would clear in a matter of minutes. There would be almost no increase in the turbidity units by the removal of sediments by a bucket or clamshell operation. Hydraulic dredging operations produce a low level of turbidity on removal and about 2000 JTU's on disposal, and this level of turbidity would persist for several minutes and then rapidly clear to the normal level of turbidity *(Anon 1970).*

The normal background level of turbidity in San Francisco Bay is about 50 JTU's. This small amount of turbidity is almost below the ability to measure with a *Jackson Turbidity Meter.* The probable turbidity caused by a bucket dredge removal of bay sediments would not raise the turbidity level above that of the normal background level. In our test and control aquaria, the, roiling action of winds did not occur and consequently our experimental levels of turbidity eventually became lower than that would be normal levels in San Francisco Bay.

Our experiments subjected the clams to an average turbidity of 1,000 for 240 minutes for 10 days! This experiment then increased the possible turbidity level by 20 times and the period of exposure by 12.0 times. Surely if the clams were to remove metals or pesticides from the sediments, the experiment gave them ample opportunity to accumulate the materials. The high turbidity was not conducted for longer periods of time for the reason that clams will cease feeding for a time if the water quality is objectionable. This indeed did happen. *(See the data under five days in the following table).* Our timing of high turbidity exposure periods was designed to provide the opportunity of feeding during high turbidity periods but most of the time the clams were at or below normal background levels of turbidity.

All chemical determinations were conducted by Pacific Environmental Laboratory at San Francisco, California. The clam turbidity experiments commenced on June 28, 1972 at 0800 and terminated on July 8, 1972 at 0800 hours.

Table III
Concentration of Toxic Metals and Pesticides in Clams

	721922	721923	721924	721925
in ppm	**Control**	**5 days**	**10 days**	**10 days**
Mercury	**0.06**	**0.11'**	**0.09**	**0.12**
Lead	**1.7**	**1.3**	**1.8**	**1.8**
Zinc	**15**	**6.7**	**11**	**7.1**
Copper	**19**	**13**	**17**	**16**
Cadmium	**10**	**8.0**	**14**	**12**
Polychlorinated Biphenyl	**0.11**	**0.07**	**0.17**	**0.16**
Chlorinated Pesticides	**<0.01**	**<0.01**	**<0.01**	**<0.01**

Mercury analyses were determined by flameless atomic absorption. Other metal analyses were determined by AAS spectrophotometry. Analyses for pesticides were by *GLC, Tracor Electron Capture* and *Coulson Electrolytic Conductivity.*

Prior to making any analyses of the data we remind the reader, that the levels of turbidity were five times that which could be produced by bucket dredge removal. That is 20 times that of disposal and for a period of time, 120 times that which could be seen on both removal or disposal. It is not possible scientifically to divide the data cited by 600 or 2400 to equate the degree of increase or loss of each chemical that would be developed during a bucket dredge operation, since we have no knowledge as to whether these actions are linear or not.

Forced Feeding

An examination and interpretation of the data of *Table III* should be made with great caution. The level of turbidity selected in this pilot study was much too high. The clams were force-fed excessive amounts of clay particles. The drop, after five days of exposure in concentration of five of the six chemicals, demonstrated that the clams ceased feeding, and in a forced diet digested some of their own tissues and liberated metals. Mercury increases make even this conclusion suspect. The irregularity of increases and decreases suggests that the experiment should be redesigned at more appropriate levels of turbidity. I suggest that experiments be conducted at turbidity levels of 100, 250 and 500 JTUs, and that replicate analyses be made. At this moment, I suspect that clams and presumably other surface feeding organisms would cease feeding when turbidity levels are excessive. This would diminish the possibility of any significant increases of the toxic substances.

Studies Urged

This paper has presented evidence detailing the many beneficial roles of clay particles and the sediments that they form. Incomplete understanding by many of the nature of adsorption and the relative permanence of this attachment has contributed to the very serious position in which the dredging industry finds itself.

Even though one normally does not publish data from pilot studies in which determinations are made on single events, we have done so because we think the indications are important and wish that many others will get involved in these studies. Our preliminary investigations have demonstrated the probability that the toxic metals adsorbed to clays are not released when the clays are resuspended.

Our discussion has indicated that organic molecules are not liberated when resuspended in amounts sufficient to cause concern on the part of marine ecologists. The clam experiment did not prove whether or not metals are digested off from clays. I suspect that they are not. Unfortunately the lack of sufficient understanding on the part of many has nurtured a level of apprehension which I have titled the *fear syndrome.* In consideration of this dilemma, the dredging industry is urged to take part in studies to demonstrate those beneficial roles of clay particles which we have described.

Cord grass in the Mill Valley salt marsh protects shorelines from wave action and increases sediment fall-out. Note the black sediment in the lower foreground.

As tidal waters enter the edge of the salt marsh, sediments fall out quickly. The water clears and the mudflat at the face of the marsh increases in elevation.

Pipe and hydraulic discharges into bay waters produce a turbidity of about 2,000 JTU's. These very high levels persist a few minutes after discharge reaches the bottom.

Looking toward the San Mateo Bridge, winds of 16 to 20 miles per hour suspend sediments from the mudflats but not from the channels.

Two test and one control aquaria were established with a three-inch layer of coarse sand. Ten Mya arenaria clams were placed in each aquaria along with 15 gallons of bay water and sufficient bay mud which was suspended to bring the sediment level up to 1300-1700 JTU's.

Note: Mercury analyses by Flameless Atomic, Absorption spectrophotometry. Lead, zinc, copper, cadmium analysis by Atomic Absorption Spectrophotometry. GLC instruments were Tracos electron Capture and Coulson Electrolytic Conductivity. Chlorinated Pesticides-Lindane, Heptachlorepoxide, DDE, DDT, DDD, Dieldrin, Aldrin, BHC, Chlordane and Enfrin-were all less than 0.02 PPB (GLC determination) before and after suspension of sediments.

After six hours, water was removed to simulate a tidal cycle. For 10 days clams were subjected to alternating six-hour periods of no water (low tide) and six hours of high turbidity (high tide). At this rate of suspension the siphon tubes of the clams could not be seen in the aquaria one hour later.

Joel F. Gustafson, Ph.D., president of Resources and Ecology Projects, Inc., a consulting firm specializing in marine pollution and environmental impact studies. He held appointments in the Departments of Marine Biology and Ecology & Systematics at the California State University of San Francisco.

This paper was presented at the Pacific Chapter meeting of WODA, October, 1972. Published in World Dredging & Marine Construction, Dec. 1972.

APPENDIX B

Cleansing of Sludge & Dredged Material

ir. H.De Vlieger, ing. B. Nevejans

Maintenance dredging of rivers and canals is an economical and environmental necessity. Where the dredged material formerly was disposed on land or under water or where it was used as soil for example in farming, nowadays most of the sediment is contaminated and needs a proper treatment preceding any further use. In addition to this, as the number of water treatment facilities continues to grow, the volume of contaminated sludge originating from those water purification plants grows equally and needs to be cleansed.

Pure disposal of the contaminated sludge and dredged material is no longer acceptable. Furthermore, the total number of organised disposal sites and particularly those accepting sewage sludge and contaminated dredged material is decreasing.

Therefore **DEC NV** has invested the last few years in developing new techniques for the stabilisation, sanitation and enlargement of existing disposal sites for fine-grained contaminated sediments. Besides this a great effort was put in new techniques to minimise the volumes to be disposed off and to maximise the beneficial usage of sludge and contaminated dredged material.

The Research Project *Zandwinningsput*

From October 1st, 1997 to September 30th, 1999, a large-scale research project supported by IWT (= Institute for the Promotion of Innovation by Science and Technology in Flanders) was conducted with the aim to develop, optimize and evaluate three techniques for the stabilization and sanitation of existing disposal sites for fine-grained (contaminated) sediments. The research project, including a large-scale application test on an existing sludge disposal site in the Antwerp harbour area, was co-ordinated by DEC NV. The University of Gent (the Laboratory of Soil Mechanics and the Laboratory of Microbial Ecology) was responsible for the scientific research and evaluation.

The three innovating techniques, evaluated in this research project, are *Soft Soil Improvement (SSI)*, an in situ sanitation and stabilization technique, Funsol, a vacuum consolidation technique and the Microbial Liner, an alternative for conventional synthetic liners as sealing system for sludge disposal sites.

Soft Soil Improvement

The properties of soft soil, such as large deformation, low strength, high water content and recently, contamination by human activity, are often responsible for the problems that occur in modern building projects. Soft Soil Improvement is an in situ solidification and stabilization technique that can help to overcome these problems.

The basic principle of SSI is a combination of the better known hydraulic jet grouting and the mechanical deep soil mixing methods.

Jet grouting involves injection of the soil with grout at very high pressure (VHP grouting) from a rotating lance (a double wall drilling rod). The technique works effectively in sand, which can be liquefied, but is insufficient in clay or sludge, where the soft soil is not liquefied by the jets.

Soil mixing *(Deep Chemical Mixing or DCM)* is a technique developed by the Japanese contractors. They mix in situ soft soil with a cement grout using a large-scale multi-shafted agitator. The technique is effective in soft clay, but the required plant is large and expensive because the power needed to mix the clay without pre-cutting is extremely high.

The SSI technique and plant are based on a jet-grouting rig with a special mixing blade on the bottom end of the drilling rod. High-pressure injection nozzles are positioned along the blade. These nozzles help the rotation and penetration of the mixing blade into the soft soil. Low-pressure grout is applied from small diameter openings to fill the void behind the blade as it rotates. Thus SSI combines the advantages of both methods. The jetting process allows the use of a much lighter rig than used for DCM, whilst the rotating blade gives much better mixing of clay or sludge than ordinary jet grouting techniques.

SSI is developed to strengthen and stiffen soft soil, but by adding specific binding agents to the injected cement grout the same technique is suitable to immobilise any possible contaminants present in the treated soil.

The Mixing Blade

The mixing blade can be seen as a large mechanical mixer, equipped with high-pressure injection nozzles. These nozzles are positioned along the blade and oriented, under a certain angle, towards the direction of rotation. The high-pressure grout injection cuts a path for the following blade and makes that a rather low torque is experienced while rotating the blade. The high-pressure injection is also responsible for an intense hydraulic mixing of the injected grout and the soft soil.

The entire needed volume of grout can be injected through the high-pressure injection nozzles. However such a high pressure (300 to 400 bar) demands much energy and therefore the flow-rate through the high-pressure injection nozzles is kept to a minimum. To inject additional amounts of grout, a parallel circuit is installed in which grout is pumped under *normal* pressure. Through small diameter openings, positioned along the blade and oriented opposite towards the direction of rotation, the additional amount of grout is injected into the soft soil. The rotation of the blade itself produces, next to the intense hydraulic mixing of the high-pressure injection, an additional mechanical mixing of sludge and cement.

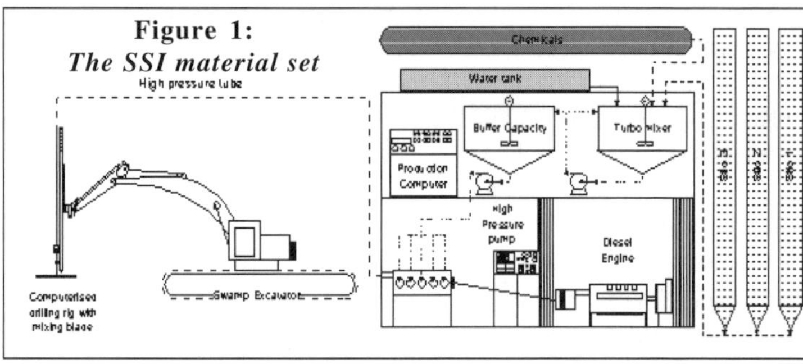

Figure 1:
The SSI material set

The Material Set

The SSI set incorporates three main units: a grout mixing unit, with storage facilities for water, cement, binding agents, etc., a pump unit and a computerised injection unit, mounted on a swamp crane (the *Swamp Excavator*). For the entire SSI material set conventional equipment was used, only the mixing blade is purpose-build for the technique.

The three units of the SSI set and the storage facilities are fully transportable by 20 ft and 40 ft containers. In three days time, with a minimum of auxiliary equipment, the mobile plant is installed. The SSI set can be erected and operated by a maximum of three workers.

The injection unit is mounted on a so-called *Swamp Excavator*. This is a hydraulic excavator mounted on a swamp undercarriage, which allows the excavator to ride on top of the untreated soft soil. The *Swamp Excavator* is equipped with a K.A.R.T.-D.G.P.S.

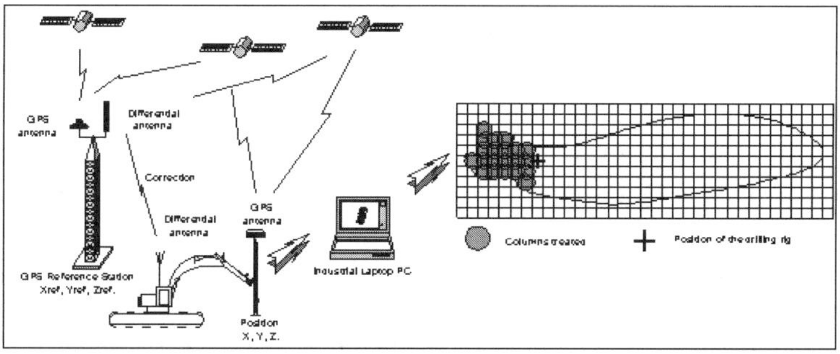

Figure 2: *The K.A.R.T.-D.G.P.S.*

(Kinematic Accuracy Real Time Differential Global Positioning System). This satellite positioning system allows the operator to position the drilling rig with a very high accuracy. A computer registers the precise position of the column by using the data from the K.A.R.T.-D.G.P.S. Each column realised is marked on a laptop screen to distinguish the performed work.

The Process

The mixing unit prepares the grout mixture to be injected. Up to four different products can be mixed in predefined quantities through a controlled measuring system. According to the required object, different types of agents can be used, such as cement, bentonite, fly ashes, lime polymers, etc. The mixing process is executed according to a preconceived recipe, after which the grout mixture is delivered ready for use to the production line. The control of the mixing unit is fully automated and computerized but it can also be used manually.

After that a high-pressure pump sends the prepared mixture with a pressure of up to 400 bar to the injection unit. Besides the high-pressure pump, a low-pressure pump can send an additional amount of grout to the low-pressure section of the injection unit. The computerised injection unit controls and registers all the injection parameters: injection time, operational depth of the mixing blade, torque for the rotation of the mixing blade, down/up speed, rotation speed and the flow rate of the injected mixture. All those data are collected on a memory card (PCMCIA standard) and at the end of the week they are transmitted to a standard PC with memory card reader to write the production report.

Data for every Single Executed Column is Recorded Separately.

A homogeneous mixture of grout, binding agents and the soft soil is created in a mechanical way by the mixing blades and in a hydraulic way by the turbulence of the high-pressure injection. The mixing blade itself is prepared in relation to the type of soil to be treated. The blades can have a different length according to the consistency of the soft soil. They are equipped with injection nozzles of different diameters that can be oriented in different directions. The injectors fulfil two specific goals. When put in a vertical position, they will literally cut the soil, and thus facilitate the penetration of the mixing blades. Secondly in a more horizontal position, they will create a deeper penetration of the injected grout mixture into the soft soil and thus create a more homogeneous mixture. By orienting the injection nozzles more horizontal, a wider spread of the grout mixture is achieved and thus a column with a bigger diameter (up to 60 %, according to the consistency of the soil) than the length of the mixing blade is realized. The mixing blade also can be divided in high- and low-pressure sections. Where high-pressure sections fulfil above mentioned specific goals and low-pressure sections allow us to vary the volume of the injected grout in a wide range according to the requirements of the contract.

Once the soft soil is injected with the grout mixture a certain period (several weeks) for hardening of the cement must be taken into account before further application of the site.

Large-scale Application Test

The process as discussed above was tested on an existing sludge disposal site in the Antwerp harbor area. The main object was to accumulate practical know-how and to evaluate SSI as a possible stabilization technique; in order to create a suitable foundation for raised surrounding dikes.

On a limited area (16 m x 100 m) along the existing surrounding dike SSI was executed. With a mixing blade of 1.5 m, about 100 columns with a diameter of 2.2 m and a length of up to 14 m, were created.

During the installation of the columns several injection parameters were varied and evaluated:

- the water/cement factor of the injected grout
- the number, diameter and orientation of the high-pressure injection nozzles
- the injection pressure (high and/or low)
- the down/up speed
- the injected cement concentration

After a period of 28 days, samples of the treated SSI-columns were taken and the compressive strength was determined.

Results

The main object of the large-scale application test laid in the accumulation of know-how. Most of the injection parameters were tested in order to define the limits of the material set (in relation to the site characteristics).

Besides this testing of the injection parameters as such, the influence on the production was evaluated. Under normal conditions, according to the down/up speed, a production of 10 to 30 meter column per hour is reasonable. For columns with a diameter of 2.2 m this equals a production of 40 to 120 m^3/h. Notice that this production involves only the installation of the columns. The positioning of the *Swamp Excavator* and the drilling rig, the start up and shut down (cleaning) of the material, etc. are additional factors, depending on the project, that need to be taken into account when calculating the overall production.

As a possible stabilisation technique, the injected cement concentrations lie between 250 and 350 kg dry cement per cubic meter in situ sludge. In this concentration, sludge with practically no core resistance before treatment, shows a compressive strength of over 2 MPa after treatment.

Fields of Application

By varying the binding agents, their concentrations and the column configuration, different objects can be achieved.

To stabilize soft soil, a mixture, consisting mainly in cement, is injected. The columns can be placed following a predefined configuration according to the expected load on the soft soil.

For the immobilization of heavy metals or other contaminants in the soft soil, specific binding agents can be added to the grout mixture. The diameter of each column is calculated so that each of them interacts with its neighbors. By this way, the entire volume of soft soil is treated and thus the method prevents any lixiviation of contaminants.

The method described before can also be used to create a vertical shield to isolate contaminated sediments from the environment. The columns are placed in one line and each of them intersects with its neighbors.

For yet an other field of application, the unit can easily be used to insert a consortium of bacteria into contaminated soft soil. The technique stays the same, but instead of using a mixture of cement and binding agents, bacteria are implemented to enhance bio-activities for degradation or immobilization of contaminants *(see also Microbial Liner).*

Features and Advantages

Soft Soil Improvement is an in situ technique, so contaminated material mustn't be replaced to other locations, in contrast to classic sand replacement methods. This results in a reduction of time and money! Furthermore, the environment is preserved because no disposal site is required for excavated silt.

Due to the easy transportation and installation of the equipment and the mobility of the swamp crane on the very soft soil, no site is inaccessible for the technique. Due to the automated control systems, the execution of the technique is not complicated. Furthermore, the fact that the whole installation consists of conventional, easy to install equipment, makes the technique, even in very small projects, profitable.

By using primarily the cutting force of the fluid injected at a high pressure the wearing of cutting tools and total installed power is reduced.

In varying the agents and the column configuration every object can be achieved. This makes the technique far more flexible than the classic methods for stabilizing soft soil. The insitu sanitation of contaminated soft soil, using the described technique, offers a durable and stable solution.

Funsol

The disposal of fine-grained (contaminated) sediments, inherent to maintenance dredging activities, is an increasing problem. New disposal sites are, for different reasons, no longer the obvious solution. So, the yet available storage capacity must be used in the most efficient way.

One of the most important problems in the storage of fine-grained material and sediments in disposal sites is the very long natural dewatering and consolidation period. *Funsol* is a vacuum consolidation technique, which facilitates and accelerates this natural dewatering and consolidation process.

Funsol is developed to extract water and/or gas from fine-grained sediments, stored under water. By extracting water (and gas), the volume of the disposed sludge decreases, which results in a gain in storage capacity.

Principle Vertical PVC-drains are inserted into the disposed sludge layer. A surrounding sand layer around the PVC-drain can be installed. Inside each drain a small collector tube is installed and connected to a network of larger collector tubes. At a certain distance below the sludge surface, a mechanical seal is placed around this small collector tube inside the drain. On top of this seal a cement-bentonite mixture makes the sealing airtight. The network of collector tubes is connected to a vacuum pump and thus an under-pressure is created inside each drain.

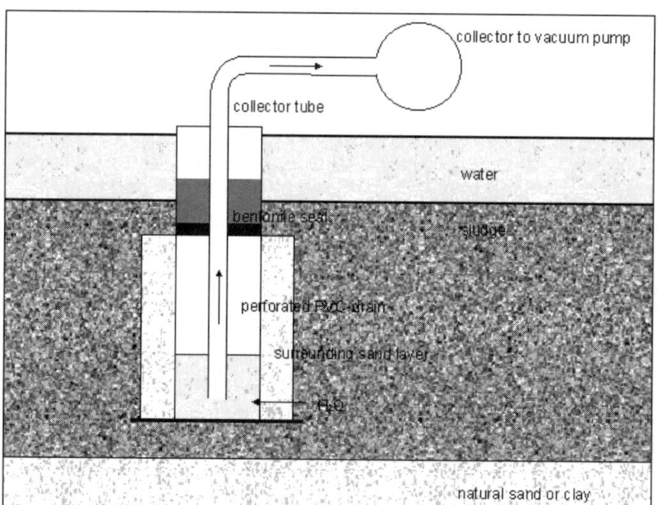

Figure 3: ***The Funsol drain***

Special attention is paid to the position of the seals inside each drain. These seals need to be installed at least 1 m underneath the top of the sludge layer to assure a minimum of leaks (= water from above the sludge, drawn into the drain tube).

The difference in pressure between a drain and the sludge makes the water flow towards the drain. This way a large amount of water is gradually drawn out of the silt, in an area around the drain tube.

The material set

A specially designed machine (the *Funsol Drain Installer*) inserts the drains in the sludge layer. The FDI-machine was built in such a way it can easily be installed on conventional coupling pontoons. A system of coupling pontoons was selected for its transportability over road and because of the possibility to alter the layout of the pontoon depending on the needs.

On the water the pontoon is moved by means of a fixed boat. A water depth of 1 m is sufficient to keep the whole installation floating. A combination of two, fixed spud piles and two horizontal movable spud piles make the pontoon manoeuverable with great preciseness.

The FDI-machine itself is basically a drill mast that can push down and pull up a fixed hollow pipe. It is conceived in a way that a hydraulic aggregate controls all the possible movements of the machine.

The FDI-machine is fixed near the edge of the pontoon on a rotating connection. This way, by rotating the machine around its connection point, several drains can be installed from one position. This means an important gain in production time.

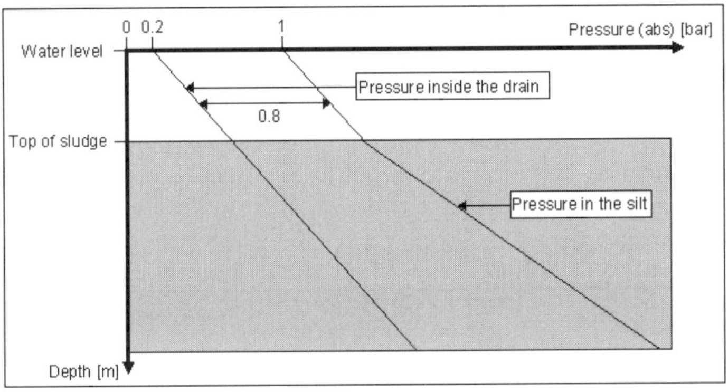

Figure 4:
Principle of vacuum consolidation

Installing the drains

First the pontoon is positioned on the starting line. From one position several drains are installed. The installation of a drain is realised by pushing down a hollow pipe, that guides the drain, into the sludge. Depending on the needs, the hollow pipe can be filled with sand around the PVC-drain. After that, the pipe is pulled back and a bottom plate, fixed at the end of the drain, makes that the drain, with or without a surrounding sand layer, stays in the sludge.

Once the drains are inserted into the sludge, the bentonite sealing is installed and the small collector tubes are connected. The pontoon is moved to the following position by means of the spud piles and the next series of drains is installed. Once a line is completed, the whole installation sails to the next line. At the end the different collector tubes are all connected to a vacuum pump and the consolidation starts.

Large Scale Application Test

The first eight months of the research project were preserved for the engineering and construction of the *Funsol Drain Installer.* After that a large-scale application test was conducted on the ***Zandwinningsput.***

Five different test fields were installed, varying the distance between the drains and the presence of a surrounding sand layer. In all the different fields an under-pressure of 0.9 bar was installed for over several months.

Field A distance between the drains: 2.5 m surrounding sand layer installed.
Field B distance between the drains: 1.5 m no surrounding sand layer.
Field C distance between the drains: 2.5 m no surrounding sand layer.
Field D distance between the drains 3.5 m no surrounding sand layer.
Field E distance between the drains 3.5 m surrounding sand layer installed.

During the consolidation period several process parameters were monitored, such as the installed vacuum, the extracted volume of water, the pore pressure in the sludge and the consolidation of the treated sludge layer.

Results

The differences in consolidation in the several test fields illustrated the importance of a surrounding sand layer. The best results were gained in the two fields where the vertical PVC-drains were installed with a surrounding sand layer. The reason for those better results lies in the fact that due to the surrounding sand layer the draining surface is drastically enlarged.

Starting from the results of the different test fields, completed with several laboratory tests, the Laboratory of Soil Mechanics from the University of Gent could develop a theoretical model describing the consolidation of sludge.

Know-how

Depending on the location and needs, different layouts are possible. The distance between the installed drains determines the speed and effectiveness of the consolidation. A surrounding sand layer around the PVC-drains enlarges the draining surface more than six times, prevents the drains from clogging and thus can improve the consolidation process. A cost-benefit analysis must determine the appropriate layout for each project.

Features and Advantages

The use of the vacuum consolidation technique to realise a fast dewatering of sludge has several advantages. The dewatering results in a reduction of the volume and an increase of the density and stability of the soft soil. No surcharge is needed, as the water is extracted from the sludge, creating a vacuum in the drains. Neither a sealing foil is needed, as the seal is created by the sludge itself.

The technique can be used in almost every location. Funsol has the possibility to insert a drain-network in a sludge disposal site after the storage has been filled. This means that every existing storage facility for fine-grained sediments can maximize its capacity, using this vacuum consolidation technique.

Microbial Liner

Probably the most innovating of the three techniques evaluated in the research project *Zandwinningsput* is the development of a Microbial Liner as an alternative for conventional synthetic liners as sealing system for sludge disposal sites.

Conventional synthetic liners are relatively expensive and can be damaged as the underlying sludge layer undergoes differential settings or through chemical reactions with organic and/or inorganic percolates. Once installed and covered with sludge, it is difficult to check and repair these liners. Furthermore, a synthetic liner only isolates the disposal site. There is no remedial effect on contamination present in the disposed sludge.

An active Microbial Liner on the other hand will be self-repairing and thus much less liable to damage. Besides, depending on the contamination present in the disposed material, certain sanitizing bacteria can be added to the liner.

Research Criteria

An adequate Microbial Liner should prevent any lixiviation from or to the disposed material due to advection and/or diffusion. It should be elastic and self-repairing and if possible sanitizing.

The advection flow through a liner was represented by the hydraulic conductivity K. The diffusion was represented by the diffusion-coefficient D. The elasticity of the liners was characterized by the shear resistance.

Principle

Two approaches were tried to develop a Microbial Liner, namely in an indirect way and in a direct way. An indirect liner is installed by adding certain physical and/or chemical systems to the sludge. Consequently the bio-chemical conditions in the sludge can be altered in a way that certain micro-organisms, already present in the sludge, are stimulated and an active microbial liner is installed. A direct liner on the other hand is obtained when specially selected and possibly optimised microbial strains are literally added to the sludge.

However, the final principle of a Microbial Liner is a combination of those two approaches. A physical and/or chemical system is injected into the sludge and must create in a short-term period an adequate (impermeable and elastic) sealing system. At the same time, certain stimulating nutrients and/or some specially selected microbial strains are inserted into the liner. This way an active system is created that, on a long term period, is self-repairing and treats certain contaminants present in the disposed sludge.

Installation of the Liner

For the installation of the Microbial Liner, the SSI material set *(see Soft Soil Improvement)* will be used for the injection of the chemical reagents and the selected microbial strains.

The process is similar to SSI, except the grout mixture is replaced with a mixture of water, reagents and micro-organisms. By installing short intersecting columns the entire area of the disposal site will be treated.

Results

For the development of an indirect liner, several chemical reagents were tested in the laboratory. Prior to the installation of a liner in a sludge layer, the various systems were tested in water, in order to evaluate if the selected reagents could form a gel-like substance, for example by polymerization. The most successful systems in this phase of research were obtained when latex (natural or synthetic) was added to the water in monomer form after which a polymerization was induced by adding acid, radicals, etc.

In a second phase the successful systems, discussed above, were tested in a sludge layer. The experiments with latex (natural or synthetic) showed again the best results and stabilized the hydraulic conductivity and diffusion under an acceptable limit.

Besides this testing, additional experiments were set up, in order to determine whether EPS(1)-forming bacteria were already present in the sludge from the *Zandwinningsput* and if so, how to stimulate their metabolism. These tests showed no positive results.

In the development of a direct liner certain microbial strains were selected and added into a sludge layer. The bacteria were selected in relation to the surrounding conditions and on their ability to produce EPS.

No stable and effective direct liner could be installed. However, some microbial strains, capable of actively remediating organic contaminants present in the sludge, were selected.

Conclusions

It is clear that a Microbial Liner, as represented above, is realisable but further research is needed.

The final principle of a Microbial Liner will consist of the injection of a natural or a predefined synthetic latex into the sludge, that after polymerisation forms an adequate, impermeable, elastic liner. An additional injection of specially selected micro-organisms and/or nutrients can attach a certain sanitising effect onto the liner.

N-VIRO Silt

The aerobic treatment of domestic and industrial wastewater produces ever-increasing volumes of sludge. The quality of the wastewater determines to a large extend the quality of the sludge, which generally is characterised by a low dry matter content, odor, instability and a high treatment cost. Therefore *N-VIRO* was introduced by DEC NV in Belgium in 1995, in order to treat a significant part of the sludge originating from those water purification plants.

The Benelux countries are characterized by a dense inhabitancy. The pressure on agricultural space is high. The total number of disposal sites and those accepting sewage sludge is decreasing. Direct land application of sludge is becoming less acceptable due to the quality requirements and the competition with other organic sources and pig manure. Other options for treatment are drying and incineration.

N-VIRO was introduced as a technique enabling the recycling of sewage sludge, while all concerns about hygiene and regulations regarding land application can be met. Due to the surplus of organics and manure available in the market, it was envisaged to export the sewage sludge after proper treatment and upgrading to a valuable product. Such treatment was found in the *N-VIRO* technology.

The land application of sludge is subject to regulation, which is probably the most stringent in the European community. Not only is there a large concern about possible contamination from heavy metals and organic compounds, there is also much consideration about the possible effects on groundwater tables from nitrate and phosphate present in the sludge. In Flanders, the land application of sludge is therefore amongst others coupled to the usage of manure.

Sludge is a difficult to define *product,* to which hardly any value can be allocated. This was a reason why it was decided to set-up a recycling system which augmented the value of the end product in comparison to the sludge. The system had to enable the introduction of product in an existing agricultural market.

Situating the *N-VIRO* Technology

The N-VIRO system stands somewhere between solidification and composting of sludge: Solidification can be executed on a large scale. It is a practical and cheap technique particularly suitable for contaminated sludge requiring proper conditioning prior to disposal. Disposal of organic sludge without treatment is difficult for obvious reasons like secondary fermentation, physical properties, etc. Environmental regulation no longer allows such disposal. Alternatively, sludge can be upgraded to be used for land application. Until now, most sludge is used directly without receiving any treatment.

Direct Land Application of Sludge

Sewage sludge is often applied in agriculture without preliminary treatment. In Flanders such application will no longer be allowed due the new ***Waste-Ruling-Act VLAREA.***

This is for hygienic reasons a non-advisable way of disposal. In a better case sludge is primarily being limed prior to land application. The liming consists then of adding quicklime to the dewatered sludge, with the major purpose of conditioning the end product to be sufficiently stable and more easy to handle. All micro-flora is killed.

The *N-VIRO* process consists of a thermo composting technique. Here a small amount of quicklime (5 - 10 % on dry matter basis) is added together with other alkaline and liming additives in order to increase the temperature and to improve the porosity of the material. In this way, an optimal composting process is obtained, during which pathogens, viruses and worm-eggs are destroyed. The end product after liming is an organic stabilized easy to handle material. The *N-VIRO* product is regarded as a full liming fertilizer.

Belgian Legislation Regarding Lime Fertilizers

The retail of liming fertilizers is governed by the Royal Code (RC) of October 6th, 1997 and amended by the RC of January 20th, 1986. According to this law and the fact that *N-VIRO* contains at least 40 to 50 % fine and pure liming materials, it is possible to sell the product if a permit can be obtained from the minister of Agriculture. Such approval was obtained. The *N-VIRO* product is in compliance with this Belgian Fertilization legislation. It is generally nominated as *N-VIRO: Fermented organo-mineral soil improver.*

The organic matter content is rather small due to the aerobic digestion of the sludge during the process. It lies typically in the range of 8 to 12 %. In order to be accepted for treatment, the sludge must meet some minimal quality requirements. These are the limits for usage of sludge in agriculture. This depends on the respective regulations in the territory of usage and the different governmental bodies, e.g. EC-directive, Federal and Regional authorities.

Benelux Limiting Boundaries Situated in EC-Perspective

Although the European Directive *EC/86/278* has set the limits for land application of sewage sludge, various member states have set their own limits. These are in most cases more strict (e.g. Belgium, Denmark, Germany and the Netherlands). In Denmark, the Netherlands and Flanders the application of untreated sewage sludge is prohibited. In Flanders and Germany, there are also limits

Generally following limits prevail:*Table 1: *Heavy metal limits in the Benelux and EC for land application of sludge in agriculture (values in ppm)

	EC 86/278	Belgium		Netherlands	Denmark	Luxemboug	France
Metal		Flanders	Walloon				
Cadmium	20-40	3	10	1,25	0,4	40	20
Copper	1.000-1750	375	600	75	1.000	1.750	1.000
Nickel	300-400	50	100	30	30	400	400
Lead	750-1.200	300	500	100	120	1.200	800
Zinc	2.500-4.000	900	2.000	300	4.000	4.000	3.000
Mercury	16-25	5	10	0,75	0,8	25	10
Chromium	-	250	500	75	100	1.750	1.000
Arsenic	-	150	-	15	-	-	-
Cobalt	-	30	-		-	-	-
Tin	-	80	-		-	-	-

for organic compounds in sludge for land application.

N-VIRO in Belgium and Abroad
Production

The *N-VIRO* technique is offered in the Benelux by DEC NV. The water purification is organized by independent companies, which manage one or several treatment works. In Belgium Aquafin controls a large number of treatment works. The treatment works generally do not handle the sludge generated by the plant. The sludge is sent elsewhere for proper treatment.

DEC NV treats 60.000 Ton of sewage sludge per year in one of its operational treatment centres, namely the **Sludge Recycling Centre Ruisbroek (SRC)**. The SRC treats various types of sludge, some of which are recycled; others are conditioned to enable appropriate disposal, in accordance with environmental regulations. The SRC at Ruisbroek takes in dredged material, sewage sludge and sludge from industrial cleansing processes, provided they are non toxic. Treatment is geared to the proposed end-user in order to obtain a uniform product.

The *N-VIRO* process was slightly modified to the Belgian agronomic and environmental situation. Organic sludge is after proper selection and evaluation accepted for processing. It is then blended with carefully selected alkaline admixtures. After a first heat pulse, the material is further composted, while the temperature of the heaps continuously is monitored and kept between 52 °C and 65 °C (pasteurization). The retention time in the mixture, in the aeration zone together with the dosage of the admixtures determine the quality of the *N-VIRO* product.

The alkaline admixtures (called SAT) are produced and blended for the N-VIRO process by the supplier. Different blends of SAT can be made. The Mg and Ca content is changed based on the requirements from the client. No Cement Kiln Dust is used in order not to increase the heavy metal content of the end product.

Economics

Based on the unit prices for some nutrients used in agriculture, and the coefficients for usage, one can calculate a numerical value for the *N-VIRO* product. This is done below.

Table 2: ***Economic value of N-VIRO compared to other secondary organic additives***

Economic parameter	Organic Liming product from sugar production	Value in BEF/ton product (*)	N-VIRO Fertiliser	Value in BEF/ton product (*)
Acid binding capacity value/ton product.	250	875	400	1400
MgO kg/ton product	5	18	50*	250
Moisture content %	30	-	20	-
Organic matter kg/ton product	75	110	100	150
N	3,5	35	1-5	10-50
P_2O_5	11	110	1-5	10-50
K_2O	2	26	1-5	10-65
Total		1174		1965

* under mineral form and 100 % usable

Unit prices are in Belgian Francs (40 BF equals 1 Euro):

•	N	:	20- BF/kg
•	P2O5	:	20- BF/kg
•	K2O	:	13- BF/kg
•	MgO	:	7 - BF/kg
•	CaO	:	3.5 BF/kg
•	Organic Matter	:	1.5 BF/kg

From this calculation it can be seen that a product without value (as is sewage sludge) can be turned into a secondary raw material, a fertiliser with a significant value. For N, P and K various values are given. This is caused by local regulatory or client requirement. For example in the case for Flanders, the presence of N and P is not wanted because of the already existing surplus of these elements in agriculture. Therefore the N and P in the *N-VIRO* is mineralised to a larger extent.

Usage of *N-VIRO* outside Belgium

Sewage sludge is a worthless product. In the Benelux it is rather difficult to dispose of. However the upgraded *N-VIRO* product has a good fertilization and soil improvement value, meaning that the consumer is willing to pay for it.

For this reason DEC NV started a joint project with other companies active in the large-scale distribution of similar products. The project has the intention to install a marketing and distribution channel within and outside the European Community and outside the EC to the **Magreb** countries, Eastern Europe and Southern America. It concerns regions with a large deficit in fertilizers and/or liming materials, causing the agriculture to be less competitive.

N-VIRO liming fertiliser and soil acidity

The liming of soils is a necessity in agriculture. It has a direct effect on the pH and the structure of the soil and on the presence and the availability of nutrients for plant growth. Approximately 30 % of today's soil are confronted with a pH that is too low, thus requiring correction using a liming agent. In addition regular maintenance liming is required on 50 % of our soil and this is necessary every 3 years. Accurate soil management is carried out by an appropriate application of lime, magnesium, organic matter and oligo-elements based on the soil type and the crop to be grown.

This means also that intensive (= productive) agriculture is impossible without this package of soil amending substances. This makes *N-VIRO* such a unique product, enabling the addition and perfect dosage of all these elements with one passage.

N-VIRO - an *à la carte* fertiliser

It is possible to modify type as well as dosage of the alkaline admixtures while producing the N-VIRO from sewage sludge in order to obtain an optimal product, which responds to the quality needs of each type of soil and crop.

The high acid binding capacity of *N-VIRO* (> 40) enables a low maintenance dosage, e.g. 2.500 kg/ha, which in turn can be executed using easy to handle light equipment. This is a major advantage for protecting the soil structure. The high acid binding capacity is due to the presence of quicklime, hydrated lime calcium and magnesium carbonates in the product.

Magnesium is present both under oxide and carbonate form, which means that it is plant available in the long and the short term. Potassium can be added to the product following the request of the farmer.

N-VIRO can be applied during spring and autumn, which is also in compliance with regulations in the Benelux. Here, fertilization with liquid manure and sludge is not allowed during certain periods in the year.

Comparison of *N-VIRO* and sludge application

The application of untreated sludge often causes clogging of the soil pores, thus destroying the soil structure. This is particularly the case when using wet, non-dewatered sludge, containing also large quantities of fat. *N-VIRO* requires only light equipment for the spreading of the product, so that the soil texture is not damaged.

Sludge is to be applied when it becomes available. The storage is not very advisable amongst others for odor related problems.

The storage of *N-VIRO* does not cause much of a problem. After treatment the product becomes long-term stable, meaning that no secondary fermentation is observed, thus requiring no special precautions for storage. The fact that no secondary fermentation takes place can easily be understood because all organic degradable substances have been composted during the process.

For example at the SRC in Ruisbroek we can store up to 50,000 ton of product in the open air. This is also possible at the farm, where the product can easily be stored on the land, unlike untreated sludge which is quite unsuitable to be stored in such a way.

***N-VIRO* in relation to phosphate and nitrate enrichment in groundwater**

N-VIRO is primarily a liming fertilizer. In Belgium and the Netherlands there is much concern about the potential contamination of groundwater by nitrate and phosphates. In these regions there is already an intensive pig and cattle farming industry, producing large amounts of liquid manure. Legislation favors in the first place the usage of these products, although here also a very strict control is executed on the amounts and distribution of the nutrients.

This means that in the case of the element P, only some 200 kg can be applied per ha of land per year. Untreated sludge contains between 4 and 7,5 % of soluble phosphate. This is 40 to 75 kg per ton sludge. After the treatment of the sludge through the *N-VIRO* process most of this element is being mineralised to an insoluble hydroxy-apatite. In a standard formulation less than 0,5 % is recovered, corresponding to a reduction of 90 %. The standard formulation consists of an *N-VIRO* product containing 5 kg of soluble phosphate (P2O5) per ton. If as is the case in Flanders, a further reduction is required, an *N-VIRO* product containing only 1 kg of soluble phosphate (P2O5) can be produced. In this case a reduction of 97 % of the readily available phosphate is obtained in comparison to untreated sludge. *N-VIRO* therefore meets the requirements regarding these problems.

N-VIRO and organic contamination reduction

The reuse of sludge is in Flanders subject to an approval trajectory involving the analysis of the substances to be applied on land. Within a short while it will no longer be allowed to use untreated sludge for land application. Moreover the required analyses are more extensive than EC regulations require. In particular a number of organic parameters are to be met.

Sewage sludge often contains large amounts of some of these compounds. Mineral oils and PAHs cause most of the problems. Mineral oils are analytically often recovered in the sludge due to the usage of polyelectrolytes for the dewatering.

An intensive study was carried out to understand the effect of the *N-VIRO* process on the presence and the behavior of these organic compounds. A number of samples has been collected at various water purification plants in the Flanders region. They all were in compliance with the EC-regulation for land application of sludge. However, some large amounts of organic contaminants were observed. A mixed sample was then made for analysis.

N-VIRO was subsequently produced from sludge of the same water purification plants. Mixed

Monocyclic Aromatic Hydrocarbons	Sludge Before treatment	N-VIRO After treatment	Limiting value
Benzene	-	-	0,2
Ethylbenzene	-	-	0,2
Styrene	-	-	0,2
Phenols	-	-	0,2
Toluene	27	-	0,2
Xylene	0,06	-	0,2
Polycyclic Aromatic Hydrocarbons	9,184	<1,139	
Benzo(a)anthracene	0,45	0,042	0,06
Benzo(a)pyrene	0,48	0,028	0,1
Benzo(e)pyrene	-	-	(*)
Benzo(ghi)perylene	0,59	0,042	0,1
Benzo(b)fluoranthene	1,2	0,076	0,2
Benzo(k)fluoranthene	0,38	0,024	0,2
Chrysene	0,97	0,067	0,15
Phenanthrene	1	0,12	0,08
Fluoranthene	1,3	0,15	0,2
Indeno(1,2,3 cd)pyrene	0,73	0,038	0,1
Acenaphtylene	0,034	<0,01	(*)
Pyrene	1,3	0,13	(*)
Acenaphtene	0,17	0,017	(*)
Fluorene	0,2	<0,032	(*)
Anthracene	0,1	0,013	(*)
Naphtalene	0,16	-	0,005
Other Organic Compounds			
Solvents	28,82	<0,25	(*)
Hexane	1	1	1
Heptane	1	1	1
Mineral oil	14.300	2.300	100 (1.150)
Propylbenzene	0,03	-	(*)
2-Chlorotoluene	0,35	-	(*)
1,3,5 trimethylbenzene	0,07	-	(*)
1,2,4 trimethylbenzene	0,23	-	(*)
Tert. Butylbenzene	0,02	-	(*)
Sec. Butylbenzene	0,03	-	(*)
p-isopropyltoluene	0,44	-	(*)
n-butylbenzene	0,16	-	(*)
Bromobenzene	-	-	(*)
Octane	1	1	1
PCB	ND	ND	0,002
Pesticides Organo-Phosphor	-	-	0,2
Chloro-Phenoxy-Acetic der.	-	-	0,2
Aromatic Chloro-Amines	-	-	0,2
Triazines (total)	-	-	0,2
Chloor-contaning (total)	-	-	0,1
Others (total)	-	-	0,02
Monochlorobenzene	-	0,04	0,02
Dichlorobenzene	-	-	0,02
Trichlorobenzene	0,19	-	0,02
Tetrachlorobenzene	-	-	0,0004
Pentachlorobenzene	-	-	0,0004
Hexachlorobenzene	-	-	0,002
1,2 Dichloroethane	-	-	0,02
Dichloromethane	-	-	0,02

Table 3: ***Organic compounds degradation during the N-VIRO treatment of sewage sludge (values in ppm)***

samples were then collected and again analyzed. The results are summarized in the following table.

It can be seen that the *N-VIRO* process has a significant beneficial effect on the organic contaminants present in the sludge. The sludge itself would not be suitable for land application according to Flemish regulations. Particularly mineral oils and some PAH's would cause problems.

After treatment only phenantrene slightly exceeds the limiting value of 0.1 ppm. The obtained value of 1.2 ppm would however not inhibit a permit for reuse, because this value is within the analytical uncertainty. Most PAHs are degraded to at least 10% of the initial value. In some cases 100% removal is obtained.

A problem is caused by mineral oils. Although a reduction of 14,000 to 2,000 ppm is obtained, this value would not be sufficient to obtain a permission for reuse. This problem however is commonly observed with most sludge originating from water purification. It is not clear if the presence of mineral oils in the analytical evaluation can be related to potential environmental risks. Therefore, further investigation will have to elucidate the need for mineral oil analysis, and the detection of the various compounds of this group of substances in sludge before and/or after treatment.

Combining Dredged Material and conditioned sewage sludge to produce manufactured soil: *N-VIRO-SOIL*, meeting the most stringent European Legislative requirements

The product is an effective limestone substitute for agricultural liming, while it contains at least 35 % of Calcium Carbonate equivalency. Intensive studies have shown no significant increase of heavy metal loading on soil, having received this product. In addition it has been established that organic compounds, such as mineral oils, PAHs and BTEX, present in the sewage sludge and treated sediments are being degraded to such an extent that the end product also meets the stringent criteria in this field.

Conclusions

The application of sewage sludge in agriculture within the EC is subject to various legislations. In Belgium and the Netherlands a very stringent regulation is applicable. Not only are there limits set for the presence of heavy metals. In addition, organic contaminants are to be evaluated. Untreated sludge is not suitable for land application following these regulations.

The treatment of raw sewage sludge using the *N-VIRO* process (or a similar process), not only enables the complete hygienization of the sludge and the degradation of most of the potential hazardous organic compounds present, it also produces a valuable product, which can be marketed as a secondary fertilizing agent.

Treatment and Beneficial Re-use of Contaminated Dredged Material

In the course of maintenance dredging more often dredging companies are faced with polluted sediments. The rivers and streams contain increased levels of heavy metals, PAHs, PCBs, oil and fuel. Both industry and agriculture may be held responsible for the input of these pollutants. Organic pollutants, like oil, diesel, PCBs and PAHs generally have accumulated in the sediment over the years.

Bioremediation occurs at a very slow rate, since the level of oxygen available for the indigenous microflora controls it. Sediment on the bottom of rivers is in a near-anaerobic state, which causes the black colour associated with sludge. This colour is the result of a chemical reaction, which depletes the sediment of oxygen, e.g. sulphate to hydrogensulfide.

Since European and national laws have become more stringent, contaminated dredged sediments have to be considered as waste. It may therefore no longer be distributed freely on land or used in agriculture.

Two options are left open to deal with this waste stream: landfill or treatment. Landfill is the easiest solution but offers little perspective in view of the environment. Also, environmental taxes on the dumping of waste are increasing and proper management and controls for disposal sites become more complicated. So most indicators are not in favor of landfill as such. However, since no recycling activity is 100% proof, there will always be a need for landfill, in order to dispose the non-treatable and residual fractions from different treatment methods. It is clear that every effort should be undertaken to minimize this volume.

Regulatory Framework

When looking at the possible recycling of treated dredged material one has to comply with Federal and Flemish regulations. In Flanders the final destination of a waste material has to follow a strict set of laws set out in the **VLAREA (Flemish Regulation on Waste).**

VLAREA consists of the regulations dealing with transport, treatment, storage and recycling of every imaginable waste product. Chain of custody as well as final responsibility for the waste are described.

For dredged material, re-use is only possible in one of the two following ways:

1. Soil

- Use as topsoil or soil for agricultural land application.
- Pollution is very low.

2. Non-structural building material

- Use as filler or bulk material for dikes, roads, etc.
- The final level of pollutants is reduced but not as low as maximum permissible concentrations in soil.

The following table gives an overview of the values for the most common pollutants: heavy metals, mineral oils and PAHs. Other parameters include BTEX and related chemical compounds, PCBs, chlorinated carbonhydroxides and insecticides.

In order for a treated sludge to be allowed for re-usage residual concentrations may not exceed one single limit. Together with the lab results the applicant has to send a complete application to OVAM (Flemish Authority for Waste). Origin of the material, type of remediation, quantities recycled and quality are factors to be examined.

Overall quality of the treated material is evaluated and used to issue a certificate for re-use of the material. This process of evaluation takes three months. The certificate regulates the transport and possible areas of re-use. It has to accompany the material at all times.

Element	Soil (ppm)	Non-structural building material (ppm)
Heavy Metals		
As	27	250
Cd	1,2	10
Cr	78	1.250
Cu	109	375
Hg	5,3	5
Pb	120	1.250
Ni	55	250
Zn	330	1.250
Mineral oils	300	1.000
PAH's		
Naphthalene	0,005	20
Fenantrene	0,08	30
Fluoranthene	0,2	40
Benzo(a)antracene	0,06	35
Chrysene	0,15	40
Benzo(b)fluoranthene	0,2	55
Benzo(k)fluoranthene	0,2	55
Benzo(a)pyrene	0,1	8,5
Indenopyrene	0,1	35
Benzo(g,h,I,)perylene	0,1	35

Table 4*: VLAREA-parameters for re-use of dredged material (only the most common chemical compounds are displayed)*

Sludge Treatment

The first step of all treatment procedures for sludge and sediments is dewatering. The evacuation of excess water minimises the volume of material to be handled. The processing of the aqueous fraction is in most cases less complicated.

In the case of the treatment of large volumes of sediments it has been established that the most efficient way in terms of productivity is the lagunation. Efficiency of lagunation is governed by following (generally known) major factors.

Dry matter content of the initial material

The drier the material at the beginning the better. The following figure shows some quantitative data on several experiments carried out on sediments from the river Scheldt.

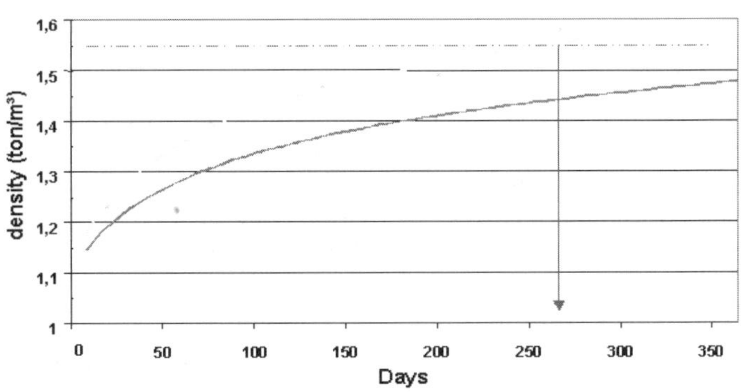

Figure 5*: Evolution of the density during the lagunation process*

Sediments at various density levels have been processed. The process-time to obtain a predefined density level was measured. The preceding figure clearly shows the increase in time needed when the sludge is diluted.

Design of the drainage system

Intensive practical studies over the last five years with increasing complexity of the build-up of the drainage layer have led to an important gain in production time. This has amongst others enabled the time interval between two successive fills in the lagunation fields to be reduced from 12 to in between 2 and 6 months. This significant improvement is caused by a major adaptation of the design of the lagunation systems. Whereas a drainage system consisted of one to three layers five years ago, a drainage system nowadays consists of a multi-filtration system of at least 10 layers.

Processing

The time required to obtain final soil structure is largely enhanced by the number of times the material is turned or tilled during the process. The amount of energy used in the system will therefore account for the efficiency of the lagunation process.

General type of the sediment

Each type of sediment behaves differently, basically depending on the soil characteristics of the material like grain size distribution, organic matter content, presence of some chemical compounds, etc.

Dewatering

At the SRC at Ruisbroek, DEC NV operates two lagunation fields in which contaminated dredged sediment is treated through an intensive lagunation and bioremediation. The dredged material arrives at the facility by boat or dumper truck. Ships or barges up to 1.500 ton may be docking. They are then off loaded at a speed of 150 m^3 per hour. The volume, density and dry matter content of every ship is taken and the amount to be treated measured. Trucks and lorries are weighed over a weighing bridge.

Picture 1: ***Aerial view of the Sludge Recycling Centre at Ruisbroek***

The dry matter of the sediment may vary between 20 and 35 % upon arrival. The material is spread in layers of 25 cm over the entire surface of the lagunation field. This maximises the surface exposed to wind and sun.

The bottom of the lagunation field is equipped with a complex drainage layer consisting of at least 10 different filtering elements. Perforated HDPE-pipes run through the drainage layer and are connected to a central water collection pit installed at a depth of 8 m. This applies a capillary suction force in the entire lagunation field, speeding up the gravitary dewatering.

The sludge is then left to dewater by evaporation and gravitary evacuation which so to say 'sucks' the water out of the pores of the sludge. This water may contain some of the pollutants. It is collected in the underground reservoir from which it is pumped to the biological wastewater treatment plant. Here the pollutants are degraded or adsorbed onto the biological sludge.

After processing the clean water is discharged into a 50 m³ buffer. This effluent is then reused in several processes as process water or as cleansing water in the centre.

The surplus is discharged in the canal.

Lagunation is carried out in several layers. A second layer of 25 cm of dredged material is applied once the first layer has reached the stadium of full break-through to the drainage layer. This is repeated until a resulting final thickness of 1 meter is attained. As shown in the following figure the dewatering follows an asymptotic curve. After a primary phase during which the saturated interstitial water is evacuated the slope in dry matter curve reaches a maximum.

This is due to two effects:

1. The formation of a crust on top of the sludge resulting from an evaporation in the top layer (sun and wind) and;

2. The consolidation of the layer located just on top of the drainage. The suction and evacuation of the water in the lower zones of the sludge creates cavities close to the filtration layer. The weight of the 1 meter of sludge compacts the layer and forms a compacted area with low water permeability.

The dewatering is again accelerated by disturbing the sludge, using farming equipment and excavators, in order to break the hard crust and compacted layer at the bottom.

In the following figure the increase in dry matter content in the dredged material is shown for the first 2

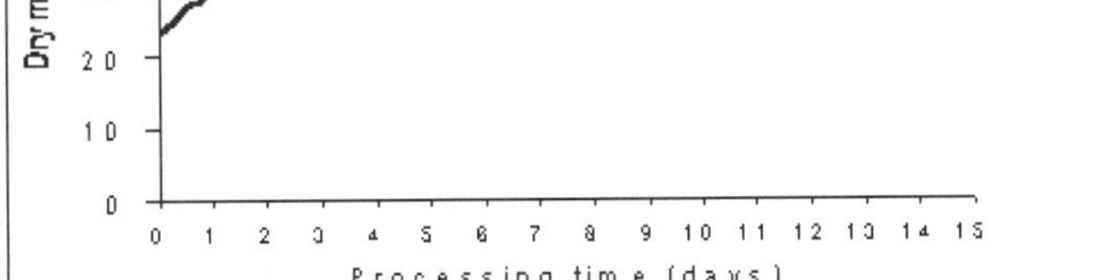

Figure 6: ***Effect of tilling during lagunation***

weeks, in which the sediment looses most of its interstitial water. A new increase of the dry matter content can be noticed after tilling of the material
(e.g. days 2, 4 and 10).

This process is repeated every 5 days until the sludge is sufficiently dry to be placed in ridges for further treatment. At this stage bioremediation is started.

Bioremediation

The anaerobic conditions in the sediment together with the small particle size (silt < 63 μm) inhibit normal indigenous microflora, which therefore would make bioremediation impossible. So a pre-treatment is obligatory in order to improve the material prior to bioremediation.

Since the sediment was turned several times during dewatering the material will start to crumble, which improves the aeration. Nutrients for the indigenous microflora are supplied to the sediment-compost premix by the admixture of *N-VIRO*. This also improves the overall grain size. Beneficial effects of the admixture of *N-VIRO* to soil or sediment are already intensively studied. Apart from the improvement of the overall grain-size distribution and formation of stable aggregates the high permeability ensures rapid dewatering and oxidation. This is due to the fine pores created by the *N-VIRO*, which results in high porosity even when compacted.

The idea to use 10 % *N-VIRO* together with 10% compost or saw-dust to improve the structure of dredged material has been applied under different circumstances. The results are a better aeration, overall corrected porosity and grain size and the rapid formation of a soil-like product.

After sufficient mixing the sediment is put back on ridges and bioremediation can start. Moisture content of the sediment is monitored to ensure proper remediation conditions. Excess rainwater runs off the ridges and seeps into the drainage system, where it is diverted towards the water treatment unit. Ridges offer several advantages. They reduce the horizontal surface exposed to the rain and create a larger vertical surface for further wind evaporation.

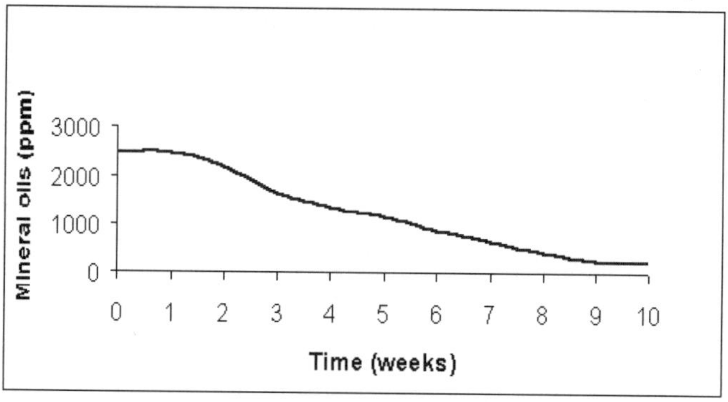

Figure 7: ***Effect of bioremediation***

Every week the ridges are reversed and aerated using especially designed equipment, to supply the sludge with necessary oxygen. In this process the lumps of sludge are crushed. This results in a larger surface area and hence faster bioremediation. At regular time intervals a sample is taken for analysis: e.g. mineral oils. The following figure shows the effect of bioremediation.

This process is repeated until an acceptable level of remediation is attained. The effects of bioremediation are not limited to mineral oils. Other organic pollutants are degraded due to a combined effect of oxidation, UV-light (400 nm) auto-oxidation and bioremediation.

For PAHs the bioremediation goes at a slower rate due to the polycyclic nature of the components. Bacteria will first degrade the easier hydrocarbons, like single chains before breaking down to the branched chains or aromatics.

The effect of bioremediation on heavy metals is negligible. Since these metals in sludge have been in an anaerobic environment, most of it is in the reduced form. Bringing the sediment on the surface and aerating it during the bioremediation leads to oxidation. These metals are not so tightly bound to the matrix as in the reduced form. Together with the drainage water from rain and interstitial water in the sludge these metals can be mobilized and run off with the drainage.

In the wastewater treatment unit of the recycling centre they bind to the charged cell membranes of the bacteria in the active sludge reactor. The result is limited reduction of the final metal loading in the treated product. The following figure gives an example of the maximal impact of this effect.

Sludge from a bird pond was bioremediated after dewatering. Due to the manipulations and the periodic rain showers the metals were oxidised and mobilised.

Since they finally end up in the active sludge of the treatment unit they are contained in a smaller volume than original. Effluent leaving the unit showed no enhanced concentrations of metals, concluding that they are adsorbed by the sludge.

During a final polishing step after bioremediation, the treated material, called Terrasilt is sieved to remove the large fragments of waste (e.g. stones, metal, plastics and tires). Samples are taken and analysed to verify that the Terrasilt is complying with national regulations and does not exceed the maximum values for residual pollutants, which may remain present. Depending on the final overall quality the authorisation obtained from OVAM describes for which type of re-use the Terrasilt may be applied.

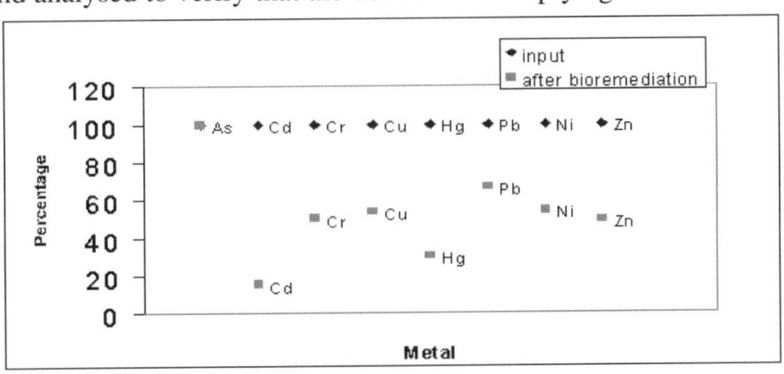

Figure 8:
Effect of oxidation on heavy metals during bioremediation

Conclusions

Lagunation and bioremediation now offer a great opportunity to deal with dredged material. This also means that techniques applied in standard soil remediation now also are applicable to sediments and this on an economical viable way.

At the origin and due to its high water content, the sediment is not stable and difficult to handle. A first priority is the dewatering in multi-layered lagunation fields. Once a dry matter content of 65 % or higher is reached an analysis of the pollution level is carried out.

Stringent directives regulate the re-use of stable dredged material as soil or filler. Based on Flemish environmental regulations it is necessary for most sediments to carry out a second treatment step after lagunation. This is carried out by improving the grain size of the material. Regular tilling provides the oxygen needed by the indigenous microflora to degrade the organic pollution present in the sediment.

Analyses carried out during and at the end of the bio-remediation are helpful in the monitoring of the process.

OVAM performs an in depth study based on the file and may issue for each sediment a proper certificate of re-use. Such a re-use certificate for application as soil or non-structural building material opens the way to further use and recycling of the treated material. Since this new legislation became active in 1998, some 7 projects were successfully carried out.

APPENDIX C

1,000 Years of Dutch Dredging

Vereniging *Centrale Baggerbedrijf*

Around 7,000 B.C. the first waves of the North Sea were lapping our coasts. After the Worm glacial, about 20,000 years ago, enormous quantities of ice had melted and caused the sea level to rise: all over the world the continental shelves were flooded. The level of the North Sea rose dramatically too, and the Low Countries became an extensive marsh area. Where nowadays the dune-ranges mark the Dutch coast, sandbanks were formed. Behind these banks, peat-bogs developed into swampy shallows. As the rate at which the sea level rose decreased, the dunes consolidated their position; a balance came to exist between the tidal waters and the raising of the seabed by sediments consisting of clay and sand. Little by little the saltings silted up. The first human beings arrived on the scene.

But the going was to be rough. With intervals of some centuries, the unpredictable sea still rose above the normal level, causing unexpected floods. The never-ending process of erosion brought into existence the Flevomeer, which ultimately became the Zuiderzee. The Waddenzee (Dutch shallows) was formed, between the northern sandbanks and the Frisian coast. Forced by his surroundings, man invented *flood-hills,* artificial dwelling mounds. The Frisians with their livestock lived on their *terps*, where they built their farmhouses and sheds. The inhabitants of the province of Zeeland, in the south-west, made use of their *mounds* only during floods.

A terp can be raised by one single family. Building and maintaining a dike for the protection of a village and its adjacent farmlands, asks for a social organization of a kind which was rare before the 9th century. By that time an organized approach was badly needed to hold back the sea: the Zuiderzee broke loose with ever increasing force, and farmlands were flooded with such frequency that food-supplies were threatened. To anyone at home in the world of dredging it may come as a surprise to learn that the pioneers of water-engineering were monks. They formed the only tightly-knit community capable of organized labour. Furthermore, they believed in service to others and in the beneficial effect of enduring hardship. So it came to be that monks reclaimed many swampy areas in the Low Countries.

In the latter half of the Middle Ages, the feudal structures in society were slowly but sureley replaced by the rise of the nation-states. This implied that *Waterstaat,* the organization for the maintenance of dikes, roads, bridges and the navigability of canals, developed into a state concern. Hence the construction of dikes could even become an important issue in the Eighty Years' War. The *tenth penny*, a form of tax levied in order to finance, among other things, a series of improvements in the Dutch system of dikes, was the immediate cause of a revolt.

Defense turned into conquest

Defensive modes of reclamation prevailed well into the Middle Ages. The first dikes were constructed for protection of areas that had already been silted up above normal sea-level, and, consequently, flooded only occasionally. In Zeeland, innumerable islets and saltings formed clusters that would later form the well-known large islands. In Friesland, saltings amalgamated to fill in the Middelzee estuary. The drainage system for polders of this type was fairly straightforward: a sluice which opened at low water sufficed. Before the invention of the drainage-windmill, the drainage of polders situated farther inland posed enormous difficulties.

As soon as the windmill provided the possibility - around the year 1500 - to draw up water beyond three feet, the engineers could indulge in a more aggressive approach. They no longer needed to limit themselves to protecting land from the sea; they now had the means to reclaim the inland lakes. The famous *droogmakerijen,* polders below sea-level, with their picturesque rows of windmills, began to attract tourists by 1600. The remainders of countless intrusions of the sea in the Dutch Lowlands were embanked, and subsequently reclaimed. This particular period in the history of Dutch reclamation was to be concluded in the 19th century by the Haarlemmermeer drainage scheme. This project was on such a grand scale that it could only be carried out by three steam-powered pumping stations; Leeghwater in his time had devised a scheme that would have involved no less than 159 windmills, to drain this vast lake, south of Amsterdam.

The development of machine-driven systems made life much more exciting to the

Throwing in more than the currents could wash out ... working on the Enclosure Dam of the former Zuiderzee (Water-color by J.H. van Mastenbroek)

dredging-people. They now had the means to actually realize the plans that had been haunting their dreams for so long. The drainage of the marshlands had in fact been an onshore affair; now all attention was focused on the sea. Of course, the first victim was the Zuiderzee, the grand old omnivorous salinator in the heart of the (then) Republic. Ever since the 17th century, plans had been devised to close this glutton's seaward approaches, but all attempts had been fruitless until steam power became available. Civil engineer Lely was appointed to create order within the chaos of plans. In 1891 his efforts culminated in the Zuiderzee-plan, now fully brought into effect but for the reclamation of the Markerwaard.

As a result of this ambitious Zuiderzee-project dams were built in the whirling seas between Den Helder and the (then) island of Wieringen, and between Wieringen and the province of Friesland. Never before, had any attempt been made to construct dams in such rough, tidal waters. The new project showed that the Zuiderzee could be conquered by mechanical aids for shifting enormous quantities of sand and mud. Furthermore, boulder-clay, left by the ice-age glaciers that once covered these parts, proved excellent for dike-building.

The tactics employed in this scheme for partial drainage of the Zuiderzee were first to stop the tidal currents, then to construct the dikes that were to enclose the future polders. Experiences with discharge-sluices had taught our engineers the absolute necessity of reinforcing the bottom of the sea in time. The reinforcement had been included in the planning of the sluices as a matter of course; the water was to flush through these artificial constructions for an indefinite period of time. Therefore, in an attempt to prevent the sluices from disappearing into an ever deepening trough, the areas subjected to currents were reinforced by willow matresses, weighed down with stones and clay.

The experiencc gained from the application of these methods was soon to be used in what can safely be termed the *finale grosso* of the aim to make the Netherlands flood-proof: the Delta scheme. After the disastrous storm-surge of February 1, 1953, and the subsequent flooding of large parts of the Netherlands, the Dutch hydraulic engineers have indeed succeeded in making a virtue of necessity. The Delta works consist of a sophisticated system for the absolute control of the water economy in a vast area, a project unrivalled in its sheer complexity. The starting point, however, was relatively simple: the protective function of the sea-dikes bordering on the rivers Oosterschelde, Grevelingen, Haringvliet and neighbouring tidal waters (covering a distance of 700 km) would be taken over by a combination of dunes and dams measuring a total length of only 30 km, once the enormous tidal inlets had been blocked.

For its drainage, Western Europe depends to a considerable extent upon the rivers Rijn (Rhine) and Maas (Meuse), hence, on the vast estuary which Zeeland in fact is. Furthermore the Nieuwe Waterweg (Rotterdam's canalized connection with the sea) had to remain accessible at all costs, the river-basins must be kept under control, and irreparable damage to the natural environment must be avoided.

The whole system of dams, sluices, flood-control dams and water basins -every thing under human control - is regarded as the world's leading example of grand-scale civil engineering. In its chapter on harbours and sea-works, when discussing the protection of land from tidal influences, the New Encyclopedia Britanica confines itself to a description of one single project: the Delta Plan. There is yet another world-famous project to be found in this, by now safeguarded area: the construction of Europoort, virtually in the sea. In the early fifties the city of Rotterdam had to face the increasing size of ocean-going vessels, and the steady rise in operating-costs involved. It became crucial for a port to have an easy access, admitting deep-drawing vessels. Rotterdam's solution was to build a gigantic port-area practically in the sea - a location which provided the best possible access. Thus, after centuries of accumulating experience in reclaiming land with methods ranging from defense to assault (from the enclosed saltings and drained inland lakes to the Zuiderzee polders), hydraulic-engineering reached a climax. The Dutch managed to construct the greatest port in the world on one of the most hazardous lee-shores, washed by a sea counted among the world's roughest.

Ambitious schemes like the Waterman plan, remind us of the fact that being satisfied with reclamations already achieved is not a very Dutch attitude. A further extension of Europoort to seaward; westward expansion of the beach area between Hoek van Holland and Scheveningen: given the chance, Dutch dredgers would physically join England to the Continent even before the British can cope with the idea.

It stands to reason that the construction of islands is one of the specialties of the Dutch dredgers; it followed in the wake of the enthusiastic approach to reclamation, so natural to the Dutch people. When the Delta Works were carried out, the construction of artificial islands was almost treated as a side-issue; deep building polders were created for the huge Haringvliet-sluices and the Volkerak-sluices, islands were casually built with the sole purpose of functioning as working harbours, or storage areas for building materials. Of course, the islands just mentioned were of relatively straightforward construction, situated in somewhat sheltered arms of the sea, and if possible raised on a pre-existing shoal. Still, they became stepping-stones to more: in the seventies plans were launched for large industrial islands in the middle of the North Sea. These would offer space for the polluting activities of petro-chemical industries and refuse incinerators, which could consequently take place out at sea, at a safe distance from inhabited areas.

The island-dreams came true in somewhat colder regions, i.e. the icy Beaufort Sea north of Canada, leaving aside little temporary constructions in the Middle and Far East. Elaborate offshore exploration activities in the Beaufort-area called for iceberg-resistant drilling-platforms, as conventional constructions like jack-ups and semi-subs would fail under the harsh circumstances. Solid, genuine islands formed the only sensible answer, and, as a matter of course, Dutch dredgers were invited to construct them.

The building of ports all over the world, reclamation, the digging of canals, the construction of artificial islands anywhere on the globe: few branches of industry can claim to be so typically Dutch as the dredging-industry. It is lonely, to be glorious in the mud!

From shovel and barrow to trailing suction dredger

For modern man, accustomed to machinery that can take over the hard work, it is hard to comprehend that, until the 19th century, our country was protected from the sea by hand-made dikes. For ages, the only tools dike-constructors had at their disposal were spade, barrow and mud-sleigh. It must have cost an immense effort, at starvation wages, to prevent the western provinces from turning into a sheltered anchorage off the German coast.

In mediaeval times, when the concept of economic specialization was still in its infancy, all landholders of a polder were obliged to contribute to maintenance of the artificial constructions protecting their lands. As the saying was: *Wie niet meedijkt in nood verbeurt zijn erf (he who refuses to fulfill his duties on the dike will forfeit his holding)*. In other words: maintenance of the dikes fell on amateur shoulders, a system which could only function because it depended on simple tools - a strong back and some form of supervision would go a long way. In the early Middle Ages the techniques of dike-construction did not require extensive tuition: in all it amounted to the construction of well-compacted clay-embankments. In the *Zak van Zuid-Beveland* south of the town of Goes in the province of Zeeland, remnants of these low dikes can still be found. For ages these embankments have had no more water to resist, which is rather fortunate. Those first dikes were even constructed on the brinks of gullies and experience had not yet taught the crucial importance of a tight connection between the dike-body and its supporting foundation. It did not take a severe storm to shatter these dikes.

Several disasters were needed to point out the necessity of paying attention to stabilizing not only of the dike, but also of its foreshores. This resulted in a change of scheme: part of the forelands were left unprotected, in order to prevent the dike from disappearing into the gully after its basis had been weakened by the scouring waters. Another improvement lay in the removal of vegetation from the foundation site before raising the dike itself, which would otherwise have tended to skid away on its mould substratum. The dike was now consructed in layers, well compacted by the hooves of horses or cattle.

Half-way through the Middle Ages, two different types of sea-dike came into existence: the Zuiderzee-type, with its almost vertical front (mainly in the northern part of the province of Noord-Holland) and the Zeeland-type, which faces the sea with the sloping side instead. Hence, the inhabitants of Zeeland enjoyed the protection of a forerunner of the modern sea-dike, whereas the West-Frisians in the north built the sort of dikes Americans dream of: you could almost stick your finger into it if it sprang a leak. Of course these differences were due to the diversity in character of the waters from which the dikes were to protect the land.

In the province of Noord-Holland the coast suffered from wave-erosion especially; every storm from the north-west would leave some sections of the dikerange damaged. In most cases these battles left them facing the sea with a washed-out front. Usually this happened in winter, a season when the next storm is likely to arrive within a week.

Therefore immediate action was needed to prevent the dike from collapsing altogether. There was no time for a carefully constructed talus. Gradually this led to the type of dike characteristic of Noord-Holland, rising steeply out of the sea and meeting it frontally. Instead of the flexible response, inherent to a gently shelving slope, this type of dike opted for massive resistance. Which is as much as asking for trouble when you are up against an adversary who will always be stronger in the end. The straight, steep fronts of the Noord-Holland dikes were reinforced with various materials, from fences of compressed seaweed to pile-screens, but these could never stop the actual dike from being affected.

The first straight dikes in Noord-Holland were presumably constructed in the early 14th century; in a decree (1328) by Count Hendrikvan Brederode, mention is made of a dike, composed of mud and weeds. It is highly probable that a *slikkerdijk* is meant here, consisting of chunks of heavy clay, overgrown with seaweed and grass, piled up in front of the dike proper. At roughly the same time the so-called *wierriem* must have gained popularity. One of the properties of seaweed is to form a cake under the combined influences of pressure and heat. This medieval weed-cake was stacked to a height of 5 m against an earthen wall, erected at a right angle to the ground. In itself this pile of weeds proved a feasible construction, but, as it was situated in the front line, it suffered in proportion, sending up the maintenance costs of this type of dike. If the earthen wall thus protected became water-logged, the entire construction would topple. To prevent this, these dikes had to be supported by a system of shoring piles and cross-beams. The steep front caused violent turbulence at the foot of the construction, and huge quantities of spray seeped into the body. In the course of the 17th and 18th century this led to an increase in the use of natural stone for protecting the slopes at the foot of dikes. In the second half of the 19th century there still were weed-dikes in Holland, but their foundations were protected by these stone slopes which extended way above the high-tide mark, with a gradient of 1:3.

Another typical phenomenon was the pile-dike. It is likely that this type derives from a construction for the temporary replacement of earthen walls that had disappeared into the sea. This method became en vogue in the 15th century, it seems; at any rate, the sea-dike at Diemen, just southeast of Amsterdam, was provided with a screen of piles even before the year 1440 (judging from a decree of that year by Duke Philip of Burgundy). In 1466 the sea-dike near Medemblik in Noord-Holland, was reinforced over a total length of 4.5 km, with piles of nearly 10 m long and one foot square in section. It remained something of a compromise, however. This particular dike had a front exceeding in steepness the slope of the average weed-dike, which meant that the dike-frame was even less protected from spray. As a result of the poor organization of the maintenance, these pile-dikes could offer a sorry sight. The system, called 'verhoefslaging' (section-maintenance), held every member of the polder community responsible for the maintenance of his own stretch of dike. The result was that neatly maintained stretches adjoined others half-sunk and overhanging dangerously with rotten piles. No storm left the scene without causing damage, such as subsidence and deep pools where breaches had occurred.

On reading the accounts of dike-inspection in those days, one cannot but wonder at the relatively small amount of flooding: the Netherlands were protected by a frail system of artificial dikes and dunes. It justifies the steady admiration of foreigners through the ages, admiration for the courage of the Dutch (not to be confused with Dutch courage!) who defied the threat of any storm which might turn their lands into lakes.

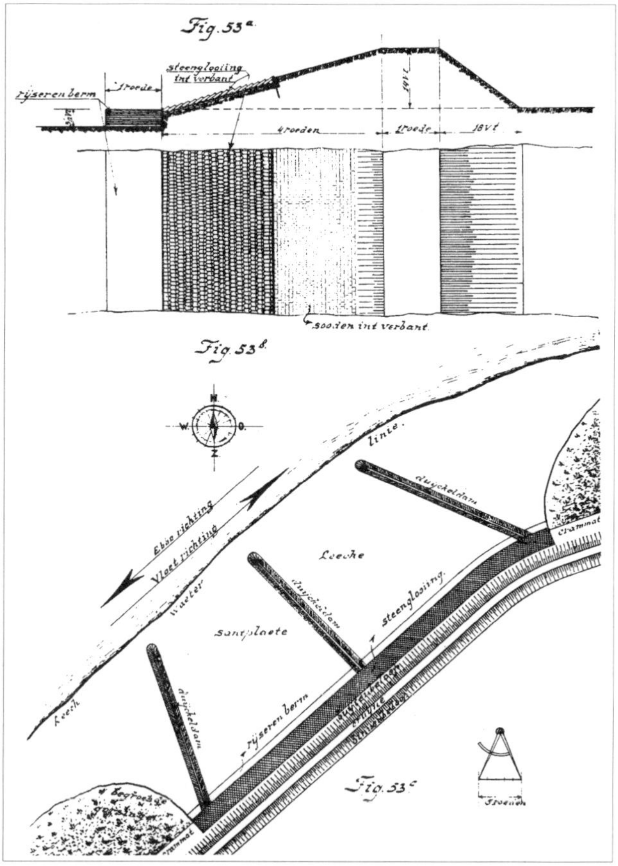

Plate XI from the 16th century book "Tractaat van Dyckagie" by And ries Vier-lingh, a bailiff of Steen-bergen in the southwest of the Netherlands. His ideas on dike-construction were remarkably modern, as is shown by this drawing of a dike-profile and a system of moles. Above, the profile is shown, with gently sloping outer-talus, stone protection up to the high water-mark and willow matting at the foot of the dike. Below, the moles are shown to have been constructed at an angle to the dike. Vierlingh argued that this lay-out was best suited to divert the tidal currents from the foot of the dike.

The water-engineers of Noord-Holland were not to have their eyes opened until their confrontation with a disaster of a much stealthier character than that displayed by straightforward storms and floods. In the thirties of the 18th century a very small creature, with the official name of Teredo Navalis, appeared to have found its way into the pile-dikes of entire Friesland. It was the pile-worm, a member of the mussel-family, measuring up to 30 cm in length.

Unlike the mussel, this mollusk does not live inside a shell; it has only a rudimentary piece of shell near its head, with which it rasps wood when burrowing its way into the pile. In the spring of 1732 the Frisian sea-dikes were found to be entirely rotten, over a total length exceeding 40 km in West Friesland alone. Another 21 km were found in an advanced state of decay. Replacement of the piles was considered out of the question, as was the construction of weed-dikes: the necessary wooden frame would collapse within a year.

The solution was finally found: rocks and stones were dumped at the foot of the dikes, forming slopes of 1:3 to a height of 8 ft above the high-water mark. Although this solution cost a then astronomical Dfl 100 per m^2, it proved to be satisfactory. During the next few years more than Dfl 6 million were spent, only to provide all dikes of the northern part of Noord-Holland with stone slopes. In the course of the 18th century the experience with stone shore protection combined with the experiences of the Zeelanders, was to contribute towards the idea of a gradually sloping, stone protected dike, with a seaward defense of basalt blocks, covering a foundation of willow-mattresses and rubble.

Gentle slopes in Zeeland

Meanwhile, the dikes of Zeeland had been developed along completely different lines. The Zeelanders were having difficulties with strong tidal currents and ever-changing water-levels between their islands-compared to these, the destructive power of the waves posed a lesser problem. The outer talus of the dikes in the southern delta showed a gentle, gradual slope, which (according to dike-constructors today) offers the best possible properties for pacifying the waves. The violent currents forced the Zeelanders to protect the foundations of their dikes with breakwaters, at right angles to the shore, in order to divert the tidal currents off the shore, and thus avoid erosion. Grass was used as protection against spray and rain. From a quite early date the part of the dike below water-level was reinforced with willow-mattresses, reaching above the high-tide level. At any rate, willow-mattresses weighed down with rubble were used in the 17th century; in the early 18th century slopes were built up from carefully fitted pieces of natural stone. Later, pillars of basalt were used, which kept their position better.

In the 19th century developments in the natural sciences took giant strides; the Industrial Revolution was possible thanks to progress in science and technology, and in its turn the revolution accelerated the development of science and technology. The country where it all started, England, was the 19th century breeding place of such highly successful engineers as Isambard Kingdom Brunel, William Froude and William J. M. Rankine. The last-named was to be of special importance to hydraulic engineering; he can be regarded as the scientific godfather of civil engineers of today. Although his fame rests mostly on his being one of the originators of thermodynamics, his work in the field of soil mechanics, and the stability of dikes in particular, introduced science into hydraulic engineering. Up to that moment it had been an empiric branch to which, in the course of centuries, little leaves of knowledge had been contributed by scientists such as Galileo, Stevin, Huygens, Newton, Leibnitz, Hooke, Bernouilli and Colomb. Now, Rankine's efforts raised this branch of engineering to academic level, not only in England, but also abroad. It became possible to calculate strength and endurance, to find more satisfactory methods for the construction of dikes and dams, while also more daring constructions could be undertaken. Although Rankine cannot be called 'founder of modern Dutch dikes' his influence on the training of civil engineers in the Netherlands cannot be denied.

A man who certainly influenced the shape of Holland's defense against the sea was Abraham Caland, chief-engineer of Waterstaat in the province of Zeeland for the period 1837-1854. His manual *for the knowledge of building dikes and sea-defences*, chiefly a guide for practical use in which all the know-how accumulated over hundreds of years has been collected, also contains a number of interesting new ideas, like the *Systeem der Vaste Punten (System of Fixed Points)*. The basic idea of this is to reinforce protruding dike-sections in order to divert the currents from the intervening banks. This method was employed successfully, e.g. on the northern coast of Noord-Beveland, which still bears evidence of this type of protection. Mr Caland was also the driving force behind the application of stone slopes in Zeeland.

In modern dike-construction, the distribution of different tasks among different parts of the dike-frame is important, in order to arrive at a *logical* dike-profile. The function of any dike is to keep the water out. This requires:

- water-tightness, and ample height;
- sufficient stability to resist the water-pressure;
- the right consistency, attainable by the right choice of building material and good drainage in order to prevent the dike from getting water-logged;
- protection from erosive forces of wind and waves.

In general, both the foundation and the dike itself will be given the necessary stability by the use of sand. To prevent the dike from being washed away, its outer talus is covered with a water-tight material (ranging from clay or boulder-clay to asphalt) in combination with a drainage system within the sand body (consisting of shingle and rubble), and a draining-ditch at the foot of the inner slope. Bricks, natural stone blocks of basalt and a bitumen-coating defend the outer-talus from the never-ending attacks by waves. In most cases a supporting berm is laid at the storm-surge line, measuring in width a quarter of the wavelength then to be expected.

The slope gradient of a dike is a reflection of its function. Sea-dikes, heavily tested, especially by waves, are marked by a very gentle slope facing the sea (up to 1:8). That certainly takes the steam out of any waves trying to reach the top of the dike. Their overtopping the dike must be avoided at all costs. Sand-drift and erosion by rain can be prevented by growing grass on the dike-surface-an age-old method. But the grass cannot cope with excessive quantities of spray. River dikes must in the first place be able to withstand prolonged horizontal pressure during high-water in winter. Hence, their shape features a wider base for stability and may be a rather steep outer slope. The inner slope, on the other hand, must be gentle, so as not to endanger the dike by the effects of seepage. Dikes that are constructed in open water (e.g. dams, enclosure dikes) usually show an almost symmetrical profile: both sides will be under attack by waves at least in their first years.

Finally, when considering the revetment of the outer slopes, it is of importance not to use large, rigid units (e.g. slabs of concrete). The use of these makes it impossible to detect sagging of dike-sections in time. In contrast, slopes covered with basalt or pliable layers of bitumen do betray every potentially hazardous hollow space below the surface, which is why these are usually preferred to other materials for coating.

Delta Works and offshore activities

Giant leaps forward have been made over recent decades in the field of shore and bottom protection -thanks to stimuli like the Delta Works and offshore industries. In the construction of the Oosterschelde Storm-Surge Barrier, bottom scour protection was extremely important. For offshore industries, critical constructions can be built only if the requirements of maximum precision have been fulfilled.

The piers of the Oosterschelde-barrier, supporting the heavy lock-gates, are meant to remain in their positions for centuries to come; the strategy of *flexible response* may well be successful in combating the sea- it cannot be applied to the position of the piers. The seabed has been changing constantly all through known history - and some hydraulic engineers found it hard to believe that new techniques applied to retaining the sea-bottom would keep the piers from shifting with it. Of course, nature alone could not be trusted with these expensive piers. Therefore a refined system was developed. Brick mattresses were lowered onto the seabottom, which had been leveled beforehand, as a solid foundation on which the pillars could be placed in exactly the right spot-to stay there. In fact these synthetic mats, were simply a modern version of the ancient willow-mattresses, which have always been used for hydraulic constructions planned on a sea-bottom that is required to stay fixed in future years. In the Oosterschelde project there was yet another reason for scour protection of the bottom: the barrier has been constructed in order to maintain a certain amount of tide. In other words: tidal currents will remain running through the sluices, the bottoms of which therefore need scour protection. For the first time in history the sea bed has been tamed with a method other than artificial mattresses alone. At the base of the Oosterschelde-barrier, the soil itself was adapted: enormous vibrating needles, pierced into the sea floor, accelerated the natural process of consolidation. The result was a more compact soil, offering the best possible support to the piers- in the Oosterschelde there shall be no leaning Pier of Pisa!

An area of industry yielding at least as much unorthodox supplements to our culture as astronautics industries is the oil and gas industry offshore. In this branch too, scour protection of the sea-bottom and the protection of man-made constructions are of extreme importance. The vulnerable pipelines, the sites of oil-rigs, and the shores of artificial islands in polar regions - all have to be protected against nature's scouring, pushing, washing and splashing.

The Dutch dredgers have been put in a position to gain from experience; the construction of jetties and breakwaters in particular has taught them a lot about the protection of constructions exposed to the whims of the sea. The moles at Hoek van Holland, Scheveningen and IJmuiden protruding bravely right into the sea from a heavily assaulted leeshore, have weathered out the most ferocious violence. Nowadays, concrete blocks weighing up to 18 tons are among the standard materials for protecting these constructions against the force of the waves.

Dumping rubble is commonly in use for the protection of pipelines, laid in trenches in the sea bed, and of the legs of jack-up rigs. Ships are being built specially designed for stone-dumping. This seems to be the most striking characteristic of modern dredging: the ever-increasing number of special machinery, most of which is designed specially for changing the state of many different types of soil, all under water. In some cases, the soil has to be processed so that it can carry heavy structures, form a dike or function as a filler for centuries to come. In other cases the soil must be removed, clearing the way for ships. After dredging for protection of land and for reclamation our attention will now be focused on dredging proper, being the removal of soil, on *deepening* instead of *raising*.

The eternal battle against silting up

In the early days, when dredging-materials were limited to long poles with leather bags attached to one end (the dredging-bail), people could only guess where nature would allow a port to exist. In the Middle Ages, ships were small, rarely exceeding 50 tons displacement, so there was access to a wider variety of ports than today, now that ships average 40,000 tons. Still, few harbours offered a clear approach to the quays; it was common practice for ships to anchor close inshore. In many cases, ports lost their trade not as a result of fatal silting up, but because of the increase in draught of newly-built ships. By the 16th century, a 12th century freighter of say 50 tons had been replaced by models of over 150 tons, with a draught well over 3 m. For negotiating Stavoren, for instance, this was too much.

The predominant commercial ports of the 16th century were Amsterdam and Antwerpen. Zeeland was bristling with good ports, boosting trade- in those days, when transport heavily relied on canals etc, Zeeland clearly was the province with the best traffic facilities available. On the coast of Walcheren the sheltered ports of Arnemuiden, Middelburg and Veere (first and foremost a naval harbour) offered a haven to shipping. Furthermore, the delta offered trade centres like Sluis, Bergen op Zoom and Dordrecht. In the far north, the city of Kampen was of great importance, although it lost much of its commercial status after 1520. In 1527 Amsterdam superseded Kampen in the maintenance of buoyage and lights in the Zuiderzee-a clear indication of the shifting of commercial value to Amsterdam.

The fairways of the Zuiderzee, towards the tidal inlets of the Vlie and the Marsdiep, were marked by a total of 30 buoys. On the islands of Vlieland and Terschelling, and in the village of Huisduinen near Den Heider, coastal lights were maintained. In 1617 the isle of Urk too was given its light, and in 1660 the city of Amsterdam even purchased the entire island, at the instigation of its famous mayor, Cornelis Witsen -all this for the sake of pilotage.

Pilotage had become a state affair quite early. Decrees pertaining to pilotage were issued as early as the 15th century. In 1540 the Emperor Charles V took radical action for reorganizing the pilot service on the river Zwin - relatively high pilot-wages had attracted quite a number of inexperienced bunglers, causing a striking increase in wrecks and sailors' widows in the area.

Already in the 16th century the Zuiderzee-routes proved to be so hazardous (even to ships with an average draught) that most skippers preferred to anchor on the Texel roads. Here they could load or discharge cargo from or to lighters, flat bottomed craft. Provisions were taken in on the same spot. Because its well formed an indispensable water-supply to the seafarers, the Orphanage in Den Burg (Isle of Texel) was earning a nice bit on the side. For repairs, however, Amsterdam was the only well-equipped port in the region. The river Maas was near to silting up at its entrance in the 16th century. Alternative routes to the sea were offered by the rivers Oude Maas, Dordtse Kil and Haringvliet, but in order to pass through these waterways, most ships had to transfer at least part of their cargo to lighters.

In short, the Great Silter caused much brain-racking, even in the early years of the Netherlands' much-praised *Golden* 17th century. In view of the tremendous efforts involved in maintaining transport facilities, it comes as a surprise that ancestors of the Dutch still found the time and energy to sail - and become the greatest sea-faring nation in the world.

There have always been enough plans for putting an end to the shifting of sand-bars. The ports were protected from silting up entirely by the use of special docks. At high water the docking-gate was closed. At low tide it was opened again, releasing a powerful mass of water that carried all the mud out to sea. Some of the machines designed in this era made use of the tides in a rather clever way. An example of these is the *krabbelaar (scratcher)*, which can be described most effectively as a sailing-boat towing a big underwater-rake. The vessel was used for dislodging the sediments from the seabed, leaving it to the tides to transport the mud out to sea. The effectiveness of this method may be compared to that of the average housewife's dusting. Nevertheless, as

A Dutch man-of-war in ship's "camels," dock-halves which could raise a vessel by about 1.5 m, enabling it to pass over shoals (From: Atlas van Zeehavens der Bataafsche Republiek, by Cornet is van der Aa, Amsterdam 1805)

long as this method was repeated regularly, a certain depth could be maintained.

The French were the leading innovators of the 18th century, in both the shipbuilding- and dredging-industries. Around 1780 Belidor, a French engineer, invented a pontoon carrying two big scoops with a capacity of 34 m^3. In regions with cohesive types of soil this pontoon may well have been quite a useful machine - in the Netherlands, where it was probably used towards the end of the 18th century, it was less effective.

Apart from the *scratcher*, all dredging instruments were unsuited to operations in open seas. When a fairway was silted up (a common occurrence in the southern basin of the Zuiderzee) there was in fact nothing to be done but to give in to the decrees of nature. Still, many solutions were tried in actual practice. *Ship's camels* were built: semi-submersible pontoons which, secured on either side of a vessel, could lift it in the way of a dry-dock, thus reducing the draught of a vessel by some 1.5 m. Lighters were used too, for transporting entire cargoes from the isle of Terschelling or the port of Den Helder to Amsterdam. A very naive author suggested loading the ships with empty barrels in an attempt to reduce their draught. A solution like that illustrates the desperate need for solving the problems caused by silting up. Ship's camels were employed until the opening of the canal connecting Amsterdam with the seaport of Den Helder in 1824.

Rotterdam was having its own problems. The town of Hellevoetsluis started to perform a function similar to that of Den Helder: a port where cargoes were transferred to lighters which could reach the city via the inland waterways. In the 17th century, when the river Maas was still -at the cost of a lot of time and effort navigable, postmaster Jacob Quack had established a service for the delivery of messages from the Maas-road to Rotterdam. The aim was to give ship-owners an idea as to what their properties were up to, out there in the offing. In time of war at sea-e.g. with the English-too, Quack's organization proved to be extremely useful.

A more effective solution, of course, was either dredging the shallow areas, or avoiding them altogether. No means were available to implement the first solution, but there were plans circulating adopting the second. As early as 1629, surveyor Jan Pieterszoon Dou designed a scheme for cutting through the Dutch dune-range in order to establish an open connection between the river IJ and the North Sea. Shortly afterwards Henric Stevin, son of the famous engineer Simon Stevin, published his Wisconstich en Filosofisch Bedrijf (Practice of Mathematics and Philosophy). In volume XII he unfolded his magnificent plan to dam the Zuiderzee and cut through the neck of Noord-Holland. Just like Mr. Lely three centuries later, he was hoping to kill three birds with one stone: he wanted to stop the silting up of the Zuiderzee and the adjacent farmland; to protect the heavily scoured shores, and to improve the connection between Amsterdam and the sea. Technically speaking, the time was not yet ripe for a project of such a scale. Without steam- or electric engines an enterprise like the closure of the tidal inlets between the Frisian islands - an essential part of Stevin's plans - would have been impossible. Like Leeghwater's famous reclamation scheme for the Haarlemmermeer, Stevin's ideas had to be shelved, waiting for the 19th century and the combined powers of steam, horse and man.

Around 1700 the once renowned city of Enkhuizen -grown in importance in the 17th century thanks to the fishing industry, saw its commercial interests go to the dogs as a result of the silting up of the Enkhuizen sands. The *Compagnie van commercie, navigatie, assurantie, visserije etc. (Company for commerce, navigation, insurance, fishing industry etc.)*, was founded in 1720 (starting capital: Dfl 20 million) with the purpose of digging a canal from the old port to a small lighthouse a couple of miles to the north. In spite of much backing for the scheme, the canal was never cut. Enkhuizen was past its prime; all that was left of the plan was a map (by Jan Belkmeer) proudly featuring the canal, as if it had always been there. The lighthouse mentioned before, had been built in 1700 by the city-council of Amsterdam, together with identical beacons on the island of Marken and at the entrance to the Outer IJ as part of the previously described pilotage scheme. They were designed to illuminate the fairway to the open sea.

The canal through the North of Holland

The end of the 18th century witnessed the dawn of new plans. The French army, by 1795 firmly established in the Netherlands but no match for the British Royal Navy, were trying to find a safe route by which military equipment could be transported from Amsterdam to Den Helder without falling prey to British interceptors. The obvious solution was to dig a canal from the capital to Den Helder, in those days already the leading naval port. Thus the idea for the canal cutting through the entire length of the province of Noord-Holland was conceived. The brain in which these plans originated belonged to Jan Blanken Jansz. In 1775, at the age of twenty, he had been appointed surveyor of fortifications on the south-western isles of Voorne and Overflakkee, a post having both military and hydraulic aspects.

To the course of his career it was very fortunate that, politically speaking, Blanken shared many qualities with a chameleon: he was engaged by the French regime in 1795, and returned to the Dutch side in 1813, without losing his responsible position. He was not the only man with such a remarkable career: Admiral Van Kinsbergen, who had been admiral in the State fleet before Prince William of Orange left for England, was appointed 'Senateur' of the French Empire, and later *Comte de l'Empire*. Then, in 1815, he was among the first men to be awarded the Great Cross of the military Willemsorde - a reward granted *Voor Moed, Beleid en Trouw (Courage, Prudence and Loyalty)*. The prudence was clearly there, but what about his *loyalty*?

Back to Blanken. He became involved in the construction of the naval base of Den Helder during the French occupation. By the time the Netherlands were liberated, while Den Helder still held out under the command of Admiral Ver Huell, who was loyal to the French, Blanken had left his solid military creation to proffer his services to King William I. These were accepted magnanimously, and he was appointed *Staatsraad (Privy Councillor)*, and inspector-general of Public Works. The plan he had originally made for King Louis Napoleon could now be presented to the new government: a connection between Amsterdam and the sea this time the reason for digging the canal was not to avoid the

British forces, but to give Amsterdam a chance to resume its once dominant role.

From an engineering and nautical point of view, the canal through Noord-Holland was the least obvious solution to the problems of the capital. In the long and narrow province the longest possible route had been carefully selected, including many bends. This long stretch was arguable because it avoided cutting through the dune-range north of Haarlem, which would necessitate the closure of the Zuiderzee-entrance of the IJ-river. This would have gone against the sentiments of most *Amsterdammers,* who linked an open IJ with trade and commerce, without realising that a silted entrance was worse than a proper lock. One can safely assume that the Noordhollands canal was dug simply because the plan was there; because there was an engineer around who disliked the idea of flushing his brain-child down the drain; and because the inhabitants of Amsterdam became sentimental as soon as their connection with the sandy Zuiderzee was mentioned. But then Blanken stated that the closure of the IJ could not be avoided, if the canal was to be accessible for sea-going vessels. Without the closure, the ships would soon be running aground within sight of the port anyhow. Out of the blue, the city-council then produced another plan: the Goudriaan scheme. This amounted to the closure of the Outer IJ as well, but with the new entrance situated off the isle of Marken. This was easier to accept than the distant town of Den Helder, as the new port of Amsterdam.

Unfortunately, the plan devised by Mr. Goudriaan offered no solution for the Enkhuizen sands - but it would cost a mere Dfl 1 million, 10 million less than Blanken's scheme. Although it was hard to raise sufficient funds for the latter, the King and Blanken opposed Goudriaan's plan fiercely: in their view it would only withdraw money from the budget for the'real' canal. Some sections of Goudriaan's plan were completed, but further activities ceased when Amsterdam was more or less forced to partake in Blanken's canal by investing 1 million guilders in it. Once convinced, the city came across generously; the King subscribed with Dfl 200,000. The city of Alkmaar too, could be brought to the point where it offered a contribution of Dfl 150,000 and a yearly sum of Dfl 3,500 for maintenance of the future canal - the city expected an increase of commercial activities once the canal crossed its grounds.

The prime concern of Parliament was to beat down the budget. Today it has become common practice to estimate an amount that can be termed 'politically workable' (instead of the exact amount necessary) and getaway with it. In the 19th century Parliament still exercised effective control over the government, in orderto make sure that the budget was not exceeded. The result was a lot of social distress. The MPs were not affected, nor was the King. Neither were the contractors, in fierce competition and glad to tender on even uneconomically small plots, the ones to suffer greatly. But the thousands of navvies had to foot the bill, slaving away for starvation wages - if they were paid at all - at back breaking labour in all seasons. To attract small contractors, plots were allowed, of considerably smaller size than originally intended. In order to make ends meet these small employers were forced to cut their labourers' wages down to a minimum. Less money was available for housing and food supplies. The measures had an immediate effect: the men took to pillaging the countryside in search for food, wood and warmth. Some of them were well accustomed to this life style; the international community of labourers counted veterans who had served in the Napoleonic wars and they taught their new mates, to *live off the land.* Especially in winter, when there was no work being done and consequently, no wages being paid either in cash or kind, the poor wretches were forced to plunder. In those days people still took into consideration the link between economic capacity and the ability to bear family responsibilities. As a result the gang of canal-labourers consisted mainly of bachelors. The men spent the winter on location, bringing havoc to the decent farmers of Akersloot and its environs. In the morning their cows were found with suspiciously empty udders, and during the night the number of sheep was often mysteriously rounded off downwards. In the evening pleasant smells could be traced to little snug campfires, where legs of mutton were being grilled. These fires then explained the sudden disappearance of fences and bridge-railings. All in all, the workers' camps could be very cozy at times, but the village policemen worked overtime.

Things were made worse by the unfortunate divergence between the government's morals and its money. On the one hand, the budget for the entire project was much too low, on the other hand, chief-engineer Blanken, considered Dfl 1.30 a day reasonable wages. The usual amount per day was one guilder only, but during the digging of the canal in Noord-Holland wages dropped to as low as forty cents a day, in addition to cuts in funds for food, clothing and housing. In a way, the contractors were caught between their low tenders - they too wanted to stay in business - and the necessities of life of their working-parties. Without a doubt the workers found themselves in the most desperate position; the sense of justice in those days is characterized by the fact that, in cases of crimes out of distress the offenders were simply locked up. No attempts were made to tackle the roots of the evil. The total cost of damage to property, of delays caused by the poor physical shape of the navvies, and of the pay necessary for extra police forces and soldiers, exceeded the sum that would have made possible sufficient wages, raising the navvies' standard of living to an acceptable level. Early in 1823 extra police- and military forces had to be directed to Alkmaar to crush riots. One Gerrit Huijskens, a notorious contractor, screwed 15 working hours a day out of his navvies for a mere Dfl 0.40. On Whit-Sunday 1823 mutiny broke out in lots 9 and 10, for which Huijskens had tendered Dfl 5,000 below his fellow contractors. The emergency signals from the county constable of Akersloot failed to attract immediate attention; when at last the cuirassiers arrived on the evening of May 26th, Huijskens' corpse was still warm. That morning he had made an attempt to put an end to the strike without conceding to the fairly modest demands. One of these was that labourers who had finished their jobs would have their lots surveyed and wages paid. When this reasonable claim was not admitted, hell broke loose. Fury culminated in the lynching of Huijskens.

Within five years, the hungry army completed the long canal using their bare hands, shovel, barrow and dredging-bail. The constructors of locks had been working to their hearts' content: no fewer than five locks had been built, plus three large swivel-bridges and eighteen pontoon-bridges, to Blanken's design. These floating bridges ("vlotbrug") turned out to be rather awkward contraptions, yet they became so characteristic that they supplied the names for several hamlets in the area. A traveller, on his way to Den Helder, will pass through villages such as *St. Maartensvlotbrug* and *Burgervlotbrug,* both north of Alkmaar. Most of these floating bridges were replaced by small ferries within ten years - there were enough obstacles slowing down traffic through the canal as it was. The locks could not be done without; Blanken's creation cut through areas with different waterlevels. Coming from the sea, one first locked through the Koopvaardersschutsluis. Then it took 5-1/2 hours to tow a vessel to the jacobklaasz-lock near the village of Zijpe. Alkmaar, the third lock, could then be reached within 2-1/2 hours. The trip to Purmerend took at least 5-1/2 hours, after which yet another lock had to be passed through. From Purmerend to the Willemsluis-lock near the village of Buiksloot took 3 hours' labour by the towing horses. Their owner would charge 80 cents per hour for each horse. A well-to-do skipper could thus have himself towed through the canal by a team of two horses, for well under Dfl 25. Quite a considerable amount of money in those days-a man on minimum wages could live on this for six months-yet it was worth the cost in comparison with the laborious Zuiderzeeroute. The perils at the Enkhuizen sands and the muddy shallows near the islet of Pampus, made the clumsy constructions in the new canal look like efficient examples of progress. In addition, the canal-route was considerably cheaper: the Zuiderzee-route would cost the owner of a seagoing vessel of 400 tons about Dfl 1,156.-, including the rent of ship's camels, the costs of towing, maintenance duties for buoys and beacons and pilots' fees. Moreover, the passage could take over a week, in case of unfavourable winds. In comparison, the passage from Amsterdam to Nieuwe Diep through the Noordhollands canal would cost Dfl 97.60 for the same ship. Towing the largest vessels required more than eighteen horses both banks of the canal were soon crowded with traction cavalcades. The horses were changed in Purmerend, Spijkerboor, Alkmaar, 't Zijpe and 't Zand.

As early as 1826 a steam-tug service was started in Nieuwediep, Den Helder. Its prime aim was to tow vessels out to sea in onshore winds. Captain Duinker's *Noord-Holland* coughed up a sheer incredible 60 h.p., and yet seems to have been of great help to shipping in adverse winds off the Texel roads. Incidentally, the little steamer was designed for towing ships through the canal. But while the vessel was being constructed with limited seagoing capacity (the King did not wish a seaworthy vessel) its task was changed to inshore work. The appointment of Duinker-a pilot-as the new vessel's captain indicates that the craft was meant to work at sea. Duinker earned Dfl 60.- a month. The average sailor made one guilder a day, which sheds a favourable light on the Dfl 1.30 a day which Blanken originally intended to pay the navvies. The comparison indicates that, on the whole, the 19th century workers in the dredging-industry were rather well-off- if they worked for a decent employer that is.

The inshore character of the towing service in Den Helder was institutionalized in 1841. This year saw the arrival of the *Stad Amsterdam*, a steam-tug of 140 h.p. and the foundation of the *Stoombootrederij voor het sleepen van schepen aan het Nieuwe*

Diep (Steamtug-company for towage off the Nieuwe Diep), under the directorship of Paul van Vlissingen. Traffic through the canal itself had to wait till 1858 for a proper towing-service to be set up. This did not increase the overall speed; scour prevention of the vulnerable canal-banks required the utmost self control in handling the steam-valve. The advent of steam and haste combined badly with the canal's disadvantages; captains began docking in Nieuwe Diep instead of Amsterdam. Besides, the lock in the IJ was often blocked, which rendered the double locks of Den Helder unprofitable. In the second half of the century the need for an improved connection with the sea grew imperative; if Amsterdam, once the commercial centre of the world, was not the surpassed by that small trading-station at Nieuwe Diep, stern measures had to betaken.

A connection between Amsterdam and the Rhine

The capital, built on piles and still harking back to the old days when it was the financial centre of the world, had yet another problem: it was not just suffering from a bad connection with the sea -the connection with the hinterland was not very good either. The obvious solution would have been to give the city a new function by turning its port into an open-air museum, but fortunately morale had not sunk that deep. The works on improvement of the connection with the Rhine started with dredging on the Omval near the Amstel, on the rivers Vecht and Vaartse Rijn, to the river Rhine, a connection called the *Keulse vaart.* A new lock was built at Vreeswijk in 1824. The river Vecht, with its many bends, remained a difficult hurdle. In a time when the river-vessels slowly but steadily increased in size (the new iron-hulled klippers appeared on the Rhine) the elegant summerhouses on the river banks probably resounded with skippers' swearing, as they tried to pass each other without ruining their lee-boards. It comes as no surprise that steam-tugs were soon brought into action on this route, and one can safely state that the Industrial Revolution not only caused problems as time became money; it also supplied the means to find solutions, e.g. by making steam-power widely available, for example for dredgers and tugs.

Fairly soon, one realised the need for the Merwede canal, to replace the Vecht and its many bends. This direct link between the mouth of the IJ at Zeeburg and the Vaartse Rijn made it possible to organize convoys for tugs. The project was rounded off with the construction of yet another lock at Vreeswijk, which opened its gates in 1892 but was closed down because of its limited size as early as 1900. The final solution to the problem came in sight when in 1931 a bill was passed on the construction of a new water-way from Amsterdam to the town of Wijk bij Duurstede, and a south branch between Jutphaas and Vreeswijk (the Lek canal), designed in order to relieve the over-crowded Koninginnesluis locks at Vreeswijk. Still, the solution was to be modified in later years: recently the width of the canal was increased to 100 m. The navigational needs of barge-convoys will make a further widening of the canal to 125 m inevitable. Ample reasons for the dredging world to think of the Amsterdam-Rhine canal with warmth and sympathy.

Canal-fever

In the days of sail, oar and horse, the Netherlands enjoyed the best traffic-system in Europe as far as capacity, speed and convenience were concerned. It lost its superior position to railways in the 19th century. In the eastern and southern parts of the country (till then traditional farming areas) the rather limited industrial output was boosted by the increase in demand from abroad - but there was no adequate traffic system; good transport facilities were restricted to the provinces bordering on the sea. In England gifted engineers like Brunel spent their time constructing railroads and impressive suspension bridges, while the Dutch continued to develop a means of transport they already knew all about: the rivervessel. This traffic required new canals, similar to the situation in England during the first years of the industrial Revolution. In the pre-steam era the advantages offered by canals obviously outweighed those offered by natural waterways: a canal can be dug in a straight line, with a constant overall depth, and can

The bucket-dredger "Haskerland" at work on the improvement of the Amsterdam-Rhine canal, a vital link with the hinterland for the city of Amsterdam

be made to be without currents. Furthermore its banks are suitable for tow-paths and its course can be directed to important industrial areas. King William I was the self-constituted leader in spreading canal-fever - if only the transport system was brought to an adequate level, industry would follow suit he argued.

The shovels were stuck into the ground, a golden age had come for the engineers. The King envisaged areas in the north, in the east and in the south were to be opened up by new canals; and quite a few essential connections were established in those years. Belgium, still called *the southern Netherlands,* at that time was certainly not neglected when it came to digging canals. In this region the aim was to facilitate coal-traffic to France. Two new canals were cut, the Mons-Conde canal and the Pommeroeut-

Antoing canal, forming a link between the rivers Schelde and Haine, in order to give the southern part of the Netherlands an approach to the port of Antwerp. Furthermore the river Sambre was canalized. The canal between Brussels and Charleroi (for which owners of the coal-mines had been asking for a century), and the Kempen-route which connected the rivers Schelde and Maas with Antwerp, are part of this scheme.

The Zuid-Willems canal played a crucial role in the development of eastern Brabant. Up to this time the province of Brabant had been an economically depressed part of the Netherlands, but with this new canal the province sprang to life. During the first years of its existence the canal (completed in 1826) derived its importance from being the link with the Maastricht-Luik canal (completed in 1822). The new canals furnished the advanced industries around Liege (Luik) with a better access to the port of Rotterdam.

Back-breaking work, long hours and all the risks involved with being self-employed ... Many contractors who later founded large companies, began their working life with Dfl 10, a mud-barge and a dredging bail as their only possessions. Nowadays, their great-grandchildren have been taught at University and are at the helm of large multinationals.

In addition, the southern ports were given a better connection with the sea: Brugge was delighted with the canal it had eagerly awaited ever since the Middle Ages, when the river Zwi n was silted up, and Gent got its canal to Terneuzen. Gent prospered, but Brugge witnessed the rise of Zeebrugge on its very doorstep. In view of all this, King William I deserved the epithet *father of the Belgian canal system* - people concerned with the industries, inland navigation, and dredging companies must have viewed the breaking away of Belgium as an independent state with mixed feelings.

Fortunately, the economic Father of the nation did not neglect the northern provinces either. After all, the presence of the Apeldoorn canal has been pressing motorists for years to maintain a safe speed when using the former tow-path. To the city of Zwolle the Willemsvaart, providing at last the coveted link with the river IJssel, was an important step forward. The competing cities, Deventer and Kampen, disapproved of the development with government funds of a rival port on the IJssel. The name of the energetic merchant-king lives on also in the name of a canal between the towns of Assen and Groningen, viz. the Noord-Willemsvaart. Not all canals were dug by order of the King though. As early as the 18th century the isolated province of Drenthe was supplied with an essential artery connecting it with the rest of the Netherlands. In fact, Dutch inventiveness has made possible the digging of canals on the scale which we have seen all over Europe. For what would the British canal-system, a decisive factor in the industrial revolution, have been without locks? A primitive lock was constructed in 1180 near Damme, in the canal connecting Brugge with the sea. The first truly modern lock was completed in 1373 near Vreeswijk. Canals as such had already been widely known in the provinces along the coast - every polder needed a drainage system of its own. It comes as no surprise, then, that Dutch consultants in Italy supervised the digging of the Naviglio Grande canal, constructed around the year 1200 to connect Milan with the river Ticino.

Digging against destitution

In the early 19th century there were more reasons than one to start digging canals. The reason mentioned above had to do with the need of urgent improvements in the traffic situation to serve industrial development, which was lagging behind in the wake of the Napoleonic slump.

Apart from this *pulling* effect, there was also a *pushing* effect. Industry, frantically trying to keep up with technical developments in neighboring countries like England and France, was in urgent need of skilled labour. This was not available; a sharp contrast with the number of unskilled workers still unemployed. These formed a large reservoir of labour that could be used in the construction of public works. After all, it takes less training to handle a shovel, dredging-bail or barrow than to operate a modern cutter dredger. Problems arose, however, due to the bad physical and psychological shape the people were in. Ever since the end of the 18th century, food-prices had been rising steadily, forcing many people to resort to a diet consisting solely of potatoes. Many were unemployed and remained so after 1813, when skilled labour for the new industries had been imported from foreign countries. Alcoholism spread steadily, Dutch "genever" gained in popularity, and the only healthy faces seemed to belong to those in the liquor-industry. The intoxicated condition of the poor convinced many in the leading classes that support of the starving poor could only be given as charity; the people that needed to be supported were either too weak or too drunk to work for their daily bread. This attitude changed when in 1818 Johannes Count van den Bosch founded the *Maatschappij van Weldadigheid (Benevolent Society)*. His aim was to turn the urban poor into self-supporting farmers, by letting them cultivate the land under a strict regime, and then slowly loosening the reins until they were able to stand on their own feet. A colony was founded in the north eastern province of Drenthe, called *Frederiksoord,* and in the area a start was made with systematic reclamation of the peat lands. The results were slightly disappointing, but at least the idea of relief-work had caught on. We must bear in mind that in those days about 12% of the people lived on a bare subsistence level; consequently, the organization of relief-work made available a large army of extremely mobile labourers.

In the eastern part of the Netherlands, a particularly underdeveloped area, the local nobility sought to improve the infrastructure (e.g. the waterways) first, assuming that industry would follow suit. One man, who could not bear to see the poverty of his people in the Avereest area, in the province of Overijssel, was Jan Willem Baron van Dedem. He struggled for years to get his canal, fought many legal battles with other

In the Netherlands, many different types of craft existed that were built for carrying peat, an important by-product of early dredging. The "Eiker" in the picture was clinker-built and rigged with a standing lugsail (Etching by G. Groenewegen, 1789).

land-owners, and spent his entire family fortune on the *Dedemsvaart.* The *Dedems* - canal, which was completed in 1851, provided the peat-lands with a valuable link with the populous west of the Netherlands and stimulated the local economy.

In short: canals were dug not just because they were badly needed, but often simply because the man-power was available. This may have some bearing on the fact that railways entered the Dutch scene relatively late. But what choice did the Dutch really have? Engineers trained for railway construction were not available, neither were engine-drivers, let alone engine-constructors - there was only this vast army of the unemployed, fully qualified ... for digging canals. There was also a European argument for transport by inland navigation system: the European continent has a number of excellently navigable waterways at its disposal. These rivers make up a first-rate natural traffic system, with a great flexibility in capacity. Up to the present day the ratio of carrying capacity and overall speed which the rivers offer is unsurpassed by road or railway. It would have been inefficient not to improve on their advantageous availability by combining them with artificial waterways. In this respect the Dutch are in a favourable position compared to the Americans and the English, who nowadays have to depend on roads and railways for bulk transport.

Boggle to pay the bill

Meanwhile, someone had to pay the bill for all these canals. But who? Investing in risky projects was not very popular in the 19th century Netherlands. William I, however, was not the kind of man to be stopped by a purse. In a bill dated December 27, 1822, the *Amortisatie Syndicaat (Amortization Syndicate)* was instituted. Its initial aim was to amortize state debts, but the King had other plans. He gradually changed the syndicate into a fund which could be drawn upon without any intervention by Parliament. Fortunately the 50 members of the Syndicate, and the 7 members of the Permanent Commission, agreed to payment of all accounts without any awkward questions. Only one restriction was made: the syndicate was to lay a list of profits and losses before Parliament once every ten years. The syndicate drew its income from surcharges on taxes and loans-sources which the Minister of Finance could otherwise have tapped without running the risk that expenditures would dodge the eye of Parliament. The first loan, issued by the syndicate in 1823 (Dfl 80 million at an interest rate of 4-1/2%) was an enormous success; the investors made it clear that they did have faith in funds controlled by (semi -)government institutions.

The level of the funds kept rising, stimulated by a Royal Decree stating that the Syndicate would be in direct control of all roads, ferries, canals and reclaimed marshlands, which meant that all toll-money was added to its total income. Manipulations in the stock-market yielded even more revenues. In short: the *Amortisatie Syndicaat*

The river IJ off Amsterdam in the end of the 18th century. On the far left a mud-mill. Beyond it, the "'s Lands Zeemagazijn," then the Admiralty's Warehouses, nowadays the National Maritime Museum, is just visible (From: Atlas by Van der Aa).

developed into an important money-lender to the State, although some of its funds had been acquired through rather obscure (or even improper) practices. Besides, it put the King into a position where he could spend public funds in accordance with his own views. He considered having a Parliament all right, as long as it did not thwart one's statesmanlike views. It must be admitted, though, that the King did not use the extra funds for his own entertainment: each and every guilder was invested in projects that would stimulate trade and industry. Investments were made in the construction of canals and roads, and also the *Fonds voor Nationale Nijverheid (Fund for National Industry)* received substantial support. Anyway, it may be clear now how the bill was paid for all these canals with "Willem" in their names.

Wet and dry

The preceding paragraphs, dealing with paupers in canal digging in fact referred to the shifting of dry soil. Today shifting the soil by hydraulic means (i.e. underwater suction dredging) is often preferred because usually it is more economical and has a greater capacity. In the first half of the 19th century canals were dug with the help of spades and wheel-barrows, and the final dams, holding back the water, were not cut through before the entire work had been completed. In later decades part of the digging was done by machines closely resembling bucket-dredgers, yet their use was restricted to transporting dry materials. These machines were seton temporary rails, on the site of the future towpath. Such 'excavators' were used for digging the canal between Gent and Terneuzen in Zeeland.

In the first half of the 19th century, the choice in dredgers was restricted to the mud-mill, the French shovel-dredger, and an improved version of the mud-mill, which already looked like the bucket-ladder dredger of today, apart from its vertical ladder. The last named machine was an invention of the Kater brothers, mill-constructors from Monnickendam. Tymon and Adrianus Kater combined the advantages of the traditional mud-mill (it could dredge to a variable depth and was horse-powered) with that of the 18th century *diepmachine (deepening craft)*, an invention of Cornelis Reedelijkheid (it worked in cohesive soil if desired). Thus, the mud-mill, equipped with a ladder consisting of a slanting trough up which mud was pushed by a chain of clappers, and the *diepmachine,* consisting of a large wheel with iron scoops, gave birth to what was later to become the bucket dredger. In his Schepen en Vaartuigen (ships and other craft), Amsterdam, 1831, P. le Comte describes the Katers' dredger as follows: "There are seven mill-ships being used by the city of Amsterdam, all of them capable of working to a depth of seven Dutch ells. Depending on the depth and cohesion of the river bed, 3, 4, 5 or 6 horses turn a long capstan, built inside a wooden shed, driving a chain consisting of interlinked iron scoops which bring up the mud. The scoops then complete their journey upwards by rounding the bend of the *conveyor-belt*; they dispose of their contents into a container which then empties itself into a barge."

Le Comte also gives some information on how to shift the vessel: "A working mill moves itself backwards along a cable, thus balancing its forward impetus." Dredgers were having a hard time without soil-profile computers and spuds. The Kater brothers have yet another outstanding invention to their names: the *klep-schuit (self discharging hopper barge).* What would the dredging-industry of today have looked like without this marvel of 19th century ingenuity? Mud-barges used to be emptied manually; a tiring process consuming a lot of time, and man-power. The self-discharging hopper barge, however, could simply be towed to a spot, where the spoil would be out of the way. The bottom trap-doors could be opened for dumping the mud, while the barge itself was kept afloat by air chambers fore and aft. The concept was improved upon only by the splithopper barge.

The means, then, were fairly limited. Moreover, the Dutch dredgers were not exactly eager to start with steam. This attitude was changed, however, by the *great challenges* of the 19th century: the construction of the Nieuwe Waterweg and the Noordzee canal which were to give Rotterdam and Amsterdam respectively, their definitive links with the sea. The sheer magnitude of these projects required engine-powered machinery instead of horses and shovels. Initially the Dutch contractors contented themselves with steaming in the wake of their foreign competitors - without using steam that is - and consequently, Amsterdam's Noordzee canal was largely an English job. However, the Nieuwe Waterweg project saw the Dutch doing the dredging themselves -a development leading to important consequences.

The Noordzee canal

As we have described above, the Noordhollands canal to Den Helder was too long, too narrow and too abundant in locks and bridges to attract shipping. Even in the 19th century, when life was still relatively relaxed, Amsterdam desperately needed a better canal to the sea. The advent of steam and iron vessels - with the consequent increase in tonnage - rendered the long Noordhollands canal's capacity insufficient within one decade. Once again, Amsterdam was going through hard times: in 1839 2,357 ships visited its port- twenty years lateronly 1,763 were left. Den Helder, supported by the Liberal government, took advantage of the situation. Ever since the port had entered the I ist of official places of loading and unloading, it had developed into a true rival of the ancient capital. But a gleam of hope for Amsterdam appeared, in the brain of retired engineer officer William Anthonie Froger, established as a self employed architect. In 1852 he published his design for digging a canal through the dunes west of Amsterdam while at the same time closing the Zuiderzee inlet of the IJ in order to prevent its silting up. Amsterdam voted against - public sentiment passionately stuck to an open river IJ - but the liberal cabinet under Thorbecke greatly favoured the plan. The first bill on the issue was introduced in 1859, and four years later the bill, regulating the implementation of Froger's plan, was flogged through Parliament.

Thorbecke, devoted Liberal, was convinced that the State should do everything which could stimulate the economy - hence his pushing the canal-bill - but he stipulated that the project would have to be financed from private funds. This was where the shoe pinched. The *Amsterdamsche Kanaal Maatschappij (Amsterdam Canal Society),* founded in 1863 with English capital, struggled with financial problems from the start. An amount of Dfl 30 million was needed for Amsterdam's new lifeline. But the big investors in Amsterdam were over-cautious, and preferred investments in *safe* foreign stocks like Russian railways. Their smaller colleagues, merchants, lower middle-class citizens, brokers and shopkeepers subscribed for amounts of up to a thousand guilders. Higher investments were scarce, in spite of King William III's settingan example by doubling the amount of his subscription. A nice gesture was made by the landowners of the dune-area near Velzen, the *Breesaap*; they agreed to cede all grounds needed for the project.

In spite of financial problems, the entire capital needed was collected even before the canal-bill was discussed by the Senate. Rumour had it that the *Antwerplobby* tried to obstruct progress: the Belgians would clearly benefit from a closing down of the port of Amsterdam (its goods trade would probably be taken over by Antwerp) and therefore made the most of the disadvantages of the scheme. The Belgian consul in Amsterdam flatly denied there being a grain of truth in this gossip. Be that as it may, the Senate finally agreed, and the first spade struck the ground in March 1865. Naturally, the air resounded with speeches comparing this canal-tobe with grand schemes such as the Suez canal and the Panama canal. The Suez and Panama canals were works of a completely different scale, and the Dutch canal was dug under the management of British civil engineers - but these minor considerations failed to scuttle Dutch pride.

The project was carried out under the direction of the British contractor Henry Lee & Son, who employed a few Dutch supervisors in addition to his own engineers. The overall design was of Dutch origin too, as described above. Archi tect Froger has been curiously neglected by history, unlike his colleague from Rotterdam, P. Caland. The Labourers were also Dutch, most of them from Sliedrecht and its environs. Unfortunately, general views on the rights of these men had not really changed for the better since the digging of the Noordhol lands canal. The workers still had to make do with provisional shacks, built by themselves from wreckage, twigs and loam. The circumstances did not permit a great deal of personal hygiene, cholera and smallpox consequently grabbing their morbid chance. Understandably, the men drowned their misery in booze, which resulted in some quite spectacular riots. But this was part of working class folklore at that time, and the work was continued, heading for the official opening in November 1876. Cutting through the dune-range required shifting the sand by means of many hands -the canalization of the river IJ and the Wijkermeer lake was clearly work for dredging, which was also the case with the completion of the seaward entrance. Locks were built in artificially drained pits. Steam-dredgers had to be brought in now, and to the experienced British engineers this was no problem at all -as early as 1806 engineer Richard Trevithick had fitted a dredging machine with his compact high-pressure steam-engine.

The Nieuwe Waterweg

For obvious reasons, the construction of the Noordzee canal could not very well

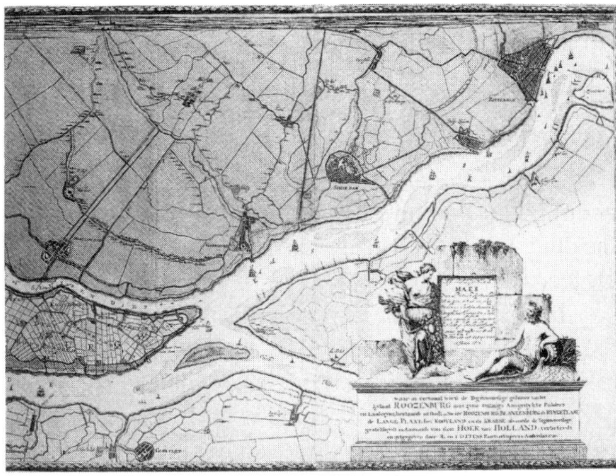

The mouth of the river Maas in the 17th century (below), with a complicated pattern of sand-banks. At the bottom of the picture, (which is south orientated) Postmaster Jacob Quack's signal station is visible, and a dispatch-rider is on his way to Rotterdam, via Maeslantsluys (nowadays the town of Maassluis). In the 18th century (right) the Hook of Holland had silted up to such an extent that the remaining river-entrance was in danger of being blocked completely. This newly grown appendix to Holland, called "De Beer," was later to become an integral part of the isle of Roozenburg which is seen in the east, extending farther to westward than in the 17th century, due the continuing reclamations. Above, the river Maas is seen also in the 18th century, farther upstream, from the eastern end of the isle of Roozenburg to Rotterdam.

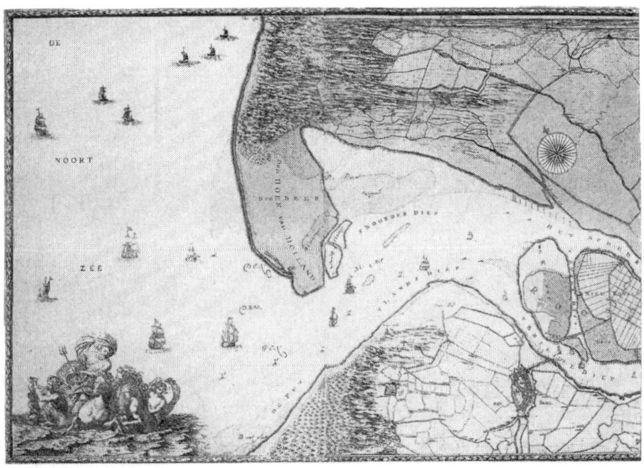

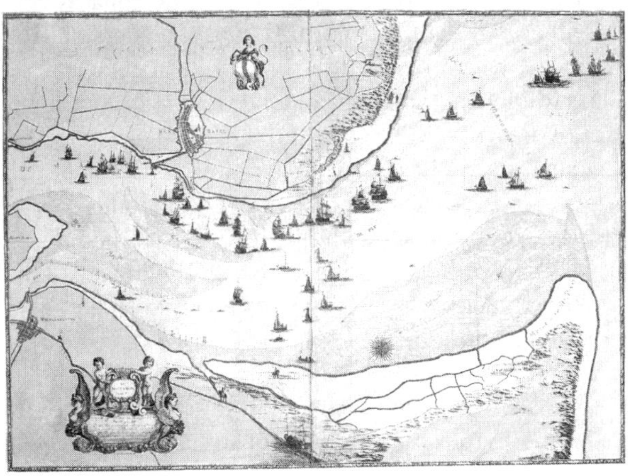

serve as a visiting-card for the Dutch dredging industry. It was a fine example of hydraulic engineering (in spite of its size: when the canal was delivered, its dimensions were already considered too small -the largest lock measuring 119 m in length, with a maximum depth of 7.5 m). But it was still mainly an English construction. However, farther to the south, a new project was carried out which would furnish Dutch dredging contractors with an important carrier-wave towards success: the construction of the Nieuwe

Waterweg.

Rotterdam had long been struggling with silting up of the Maas-estuary. At a time when Amsterdam was given its Noordhollands canal, the city on the banks of the Maas got the Voornsche canal. This waterway proved as inadequate as the Noordhol lands canal. Completed in 1829, it shortened the route from the sea through the Goereesche Gat to Rotterdam by over six hours' sailing. Its usefulness depended largely on the navigability of the Goereesche Gat. However, this section of the route silted up considerably in the 1830's, which forced large vessels through the Brouwershavense Gat, the next river to the south. Part of the cargo had to be transferred to lighters before one could sail on to Rotterdam. Hence, the idea was launched to improve the navigability of the Brouwershavense Gat. The option was further motivated by the decision, made by the *Nederlandsche Handel Maatschappij (Dutch Trading Company)*, to compel captains of craft loaded with NHM-cargoes and drawing 5.25 m or more to call at Brouwershaven and discharge part of the cargo there.

View of the Noordzee canal around 1878. At the top of the picture, the Zuiderzee is visible, with Amsterdam and the closed river IJ. Bordering on the new canal, very orderly lots of arable land can be seen. The close observer will be able to trace the outline of the former Wijkermeer lake and the original river IJ.

The leading residents of Brouwer-shaven, already dreaming of thriving business came under heavy fire from the Zierikzee town council, who regarded their city as Rotterdam's most suitable seaport. Zierikzee on the river Oosterschelde, was in fact closer to Antwerp than to Rotterdam, and its claims were not taken very seriously. In the meantime, however, the Minister of Finance, mr Van Hall, had begun in 1844 to straighten out State finances, which had become a muddled affair- partly due to the Amortisatie Syndicaat. Funds for extensive port construction were not made available, which did not prevent Brouwershaven from lining its pockets with the revenues from the busy traffic on its doorstep. Shipping was forced to improvise with a route through the waterways of Grevelingen, Krammer, and Volkerak, turning west into the Haringvliet before the final stretch of the Voornsche canal. The pubs in Brouwershaven must have prospered, crammed with skippers collecting Dutch courage before trying to wind up the shallow rivers with their ocean-going vessels.

Yet the number of entries and clearances in Rotterdam actually increased over the 1851-1860 period: the number of entries was 2,657 in 1860 compared to 2,143; and there were 2,793 outward clearances as against 2,216 in 1851. This meant a total of over 30% of all Dutch entries and clearances. Yet the larger ships had already started to avoid Rotterdam, and it comes as no surprise that Caland's plans for improving the Brielsche Gat (natural mouth of the river Maas) were greeted with enthusiasm. The Rotterdam Chamber of Commerce proposed a less risky plan, to enlarge the locks in the Voornsche canal and build a temporary port at the Brouwershavense Gat. A clear illustration of the lack of confidence in new, daring schemes, and an example of the degree to which the Dutch spirit was lagging behind in the radical developments of the industrial revolution. Fortunately the Government was cast in a more rebellious mould, and an advisory board was set up-among its members civil engineer P. Caland. In 1858 he co-operated with J. A. Beijerink on a plan for a cut-through near Hoek van Holland. The plan was rapidly accepted, but it was not until 1863 that Thorbecke's bill on the Noordzee canal and Rotterdam's Nieuwe Waterweg had been piloted through Parliament. Rotterdam enjoyed a kinder approach from a financial point of view than Amsterdam as its new entrance from the sea would be constructed at public expense.

On the Nieuwe Waterweg project traditional dredging methods prevailed. Most of the soil was shifted dry, whereas dredging and canalization were the methods used at sites near the river and the sea. The year 1868 introduced a few novelties in the dredging world, e.g. a dredging machine with a slanting ladder instead of the vertically placed ladders used in the older models. This dredger, the *Stoomdieper* owned by D. Volker, was to be assisted the very next year by the *Hollands Diep V*, which also carried a slanting ladder. But the greatest leap forward would be made with the introduction of the suction hopper dredger, which was to be used extensively in order to prevent the Nieuwe Waterweg from silting up.

In his design Pieter Caland assumed that the overall depth of the new canal would be maintained by the scouring effects of the strong tidal currents. Unfortunately the harbour moles lacked sufficient length, and although the principle did in the long run prove to be correct, it remained necessary to dredge as well. The dredging had to continue in all sorts of weather, also in the river entrance, and this required the use of hopper dredgers. These do not need barges rafted up alongside, and are therefore capable of working in swell. Furthermore, hopper dredgers are mobile: they have no pipes running ashore and are not trip-wired to an intricate mooring system which may foul other vessels' propellers. They are less obstructive to shipping than bucket dredgers for example. During the construction of the Nieuwe Waterweg only 14 hopper dredgers were brought into action, especially for dredging the entrance, but in the years between 1878 and 1880 eighteen more were built.

After the Nieuwe Waterweg was completed in 1872, and after the establishment of a sufficient overall depth in 1898 (at last), a great quantity of dredging equipment became available; the same happened during the closure of the Oosterschelde dam between 1984-1986, when a lot of equipment was discharged. Dutch contractors, most of them natives from villages on the banks of the river Merwede, had an important incentive to go foreign; they earned a solid reputation with successful projects in France, Russia, Rumania, China and South America.

Steam and steel led to a significant change in the organization of the dredging industry. Formerly companies more closely resembled partnerships of members (as are still to be found in the port of Rotterdam among the watermen). These were gradually succeeded by the limited liability companies which we know today. Because the early contractors lacked sufficient funds for purchasing expensive steam-powered equipment, the Dutch dredger companies hesitated long before turning to steam. Steam dredgers built by Cockerill in Belgium were ordered by the government, which even rented them out to contractors, with an expert engineer on each craft into the bargain. It was not until the companies had learned to take capital investment seriously that contractors could purchase steam-powered dredging craft. Finally, the really large projects at home and abroad could be taken on.

Attacking the open water

Dealing with the 19th century, the previous chapter concentrated on dredging as the digging and deepening of canals and fairways. Although the waterways have made the Netherlands prosper, they did not contribute to the increase of its domain. The polders were responsible for that, and the Dutch are still adding these to their territory. One of the main projects of the 19th century was the reclamation of the Haarlemmermeer lake This caught the imagination most, perhaps because the great Leeghwater had been forced to leave the great challenge unresolved in the early 17th century.

Without taking the Haarlemmermeer into account, reclamation in the years between 1815 and 1852 added more land to Dutch territory than in any other period in history, Leeghwater's own lifetime excepted. Between 1815 and 1852 711 ha were added to the Netherlands. The biggest reclamations were situated in the provinces of Noord-Holland (the Anna Paulowna polder, the Waard- and Groetpolder, the Eijerlandse polder on Texel); Groningen (Finsterwoldepolder, Oostpolder); Zuid-Holland (Zuidplaspolder) and Zeeland (mainly the endikement of saltings).

The reason for reclaiming the Haarlemmermeer was to put an end to the gradual expansion of this lake; there was no pressing need for new farmland. Originally, the

Haarlemmermeer consisted of four smaller lakes called the Spiering-, Helle-, Leidse- and Groote meer. The dividing land crumbled away further in each flood, gradually turning the area into one big lake. In his Groot Haarlemmermeerboek (1642) Leeghwater had calculated 160 windmills (working in series of three), and about Dfl 3.5 million to be needed. In 1848 only three steam-powered pumping stations proved sufficient. These, named *Leeghwater* (1845), *Lijnden* (1849) and *Cruquius* (1849), designed by two Englishmen (Arthur Dean and Joseph Gibbs) represented the only type capable of drawing up water from a depth of 5 metres in one go. Each station contained eight piston pumps, 1.85 m in diameter with a stroke of 3 m. Each minute, 5 double strokes were made, a capacity which proved sufficient for draining the entire lake by the year 1852. Since then, these pumps have drained the Haarlemmermeer of 100 million m^3 of seepage and rain water yearly, until they were replaced by diesel-engines with centrifugal-pumps in the 20s of our century.

After the Haarlemmermeer had been reclaimed, people did not really know what to do with it. In no time at all it was covered with dense reed-marshes, enormous quantities of willows, asters and other *weeds*. In modern times the polder would undoubtedly have been claimed by groups of nature lovers as a *unique and irreplaceable* nature reserve, providing shelter to migrating fat feathered geese and willow-wrens - but in 1855 things were different. A development corporation was formed, and a year later a polder-board began work. It is interesting to note that, as a result of the construction of this enormous polder, the Mallegat sluice near the ancient town of Gouda, lost its grip on the hydraulic situation.

The sluice, dating from 1764, divided the river Gouwe from the part of the Hollandsche IJssel directly connected with the river Lek and was furnished with flood-gates only. The construction sufficed because the width of the sluicegates was exactly in balance with the amount of water from the river Gouwe to be sluiced into the river Hollandse IJssel on the ebb. The reclamation of the Haarlemmermeer had reduced the quantity of water in the Rijnland area reservoirs, and the improved drainage through the Nieuwe Waterweg canal, dug in the same period, had increased the tidal effect at Gouda. At the Mallegat sluice the low-water level dropped almost 0,30 m, thus emptying the river Gouwe too drastically. The balance was not restored until 1941 when two vertical lift-gates were installed.

In waters teeming with engineering constructions - waters that have become complicated constructions themselves - a subtle balance can be found. This balance is absolutely necessary; it decides whether the water is navigable or not, whether there will be floods in winter, and whether the drainage function is fulfilled. This marks the Dutch *water economy* as a unique system, a fertile breeding-ground which has allowed the national hydraulic industry to reach its ultimate growth before exporting its skills to other countries. Thus projects such as the improvement works on the great river-system, the Zuiderzee Works and the Delta Works have mutually influenced one another, which makes it necessary to study each project within the context of the total system.

Improvement of the river-system

The main water-suppliers of the Netherlands, the rivers Rhine and Maas, have different characteristics: the Rhine derives its water from rain as well as alpine glaciers, whereas the Maas depends solely on rain; it cannot draw on a buffer-stock of ice. Consequently the Maas shows a great fluctuation in discharge, varying with each season. The capacity of the Rhine allows for a discharge increase in winter of 16 times the quantity of the dry season. The Maas, however, has to carry off a 300-fold increase in wintertime. This is partly caused by its impervious bottom in Belgian territory. These enormous differences between wet and dry season are a nuisance: floods, dike-bursts, ice-drift, and in-navigability caused by fast currents, or - the other side of the coin - lack of sufficient water for transport purposes, for horticulture pastures and for consumption.

Tactics for taming a river can be divided into three categories: regularization, normalization and canalization. Regularization aims at changing the wide winter-bed for obtaining a regular discharge of water at high water levels and preventing the accumulation of ice floes. Normalization improves the regular discharge of water through the summer-bed to improve navigability, and corrects irregularities of the river-bed. In many cases the bends of the river are cut off, while the main-current is guided along a slightly meandering path, in order to help it scour out the summer-bed, within the boundary of the heavily protected banks. In the third category, canalization, the goal is to maintain navigability at low water levels with the help of weirs and locks. In most cases these three categories are combined; when a river is canalized, its summer-bed is normalized and its winter-bed is regularized.

As early as the 18th century the discharge of water through the lower Rhine was considerably improved by the digging of the Pannerdens canal and the Bijlands canal. In the next century the rivers were improved more thoroughly: the aim was to keep the river within its banks at times when the highest water-level was reached, so that the

A suction dredger working on improvement of the river Maas around 1935.

detested system of overflows could be discarded. These spillways were used for relieving the pressure on the dikes. A stretch of the embankment was then deliberately kept lower. If the water-level rose too high, water would spill over the top of the lower part to flood the land behind the bank, until it was stopped by special dikes which led the water to a lower point further down the river. It sometimes took an enormous area (of valuable farmland) to maintain such an overflow. By constructing groynes, by strengthening dikes, and by creating new river-mouths (like the Nieuwe Waterweg) the civil engineers gained control over the rivers. Subsequently, the Nieuwe Merwede and the Bergsche Maas were created with the same purpose. By 1915 the aims as stated in the River-report of 1850, had been achieved. What remained to be done was the canalization of the river Maas and the rivers Rhine and Lek. The river Ijssel was normalized, a project which was completed in 1924. The river was deepened, bends were cut near Rheden and Doesburg - which shortened the course of the river by 8 km while increasing the speed of the stream. The discharge of the river IJssel was brought under control by the construction of the weir at Driel in the lower Rhine. Thanks to these projects, completed after World War II, the depth of the fairway of the lower Rhine and the IJssel can be maintained at 2.75 and 2.70 m respectively.

Earlier, the Maas had been adapted to the kind of behaviour one would expect from decent rivers: the river was canalized from Maasbracht to Grave, a project which was completed with the putting into operation of the weir near Sambeek in 1929; improvements downstream from Grave were completed ten years later. Thanks to the weirs the river was still navigable in times of minimal discharge (10 m^3/sec); by deepening and widening the winter-bed, and by cutting the bends, the maximum discharge capacity was raised to the necessary level of 3,200 m^3/sec -versus 1,800 m^3/sec in 1931, which had made the Beersche Overlaat (spillway) near Cuijk an indispensable link in the system. Upstream of Grave, the skippers' life became even more pleasant when the Juliana canal was opened. The Maasbracht-Roermond waterway - indispensable to the mining-district - was now entirely situated in Dutch territory. It relieved *Waterstaat (the Ministry of Public Works)* from the need to force the part of the Maas constituting the Belgian border, into a strait-jacket of weirs and locks.

An extra advantage of putting the rivers in order-apart from reducing the risks of floods or in-navigability - was that the increased strength of the streams proved helpful in keeping the North Sea's salinity out of the Netherlands. The two elements, *protection against floods* and *the battle against salination* lay at the root of two works which yielded a more sensational success. From the hydraulic point of view it was an unparalleled achievement, to gain control over the rivers. But what particularly drew the world's attention were the closure and reclamation of a vast inland sea and the closure of the Rhine and Maas estuary. The Dutch attacked tidal waters, an unprecedented feat.

Het gewelt en vergif uijtter Verenigt Nederlant to verdrijven
(To expel violence and pollution from the United Netherlands)

Anyone assuming the idea of closing the Zuiderzee to have originated with the famous 19th century C. Lely, is wrong. It was in the early 17th century when Henric Stevin, son of the renowned Simon Stevin, suggested "to expel the violence and pollution of the North Sea from the United Netherlands,' in his book titled Wisconstich and Filosofisch Bedrijf.

Stevin's plan enjoyed a rather peaceful rest, shelved as it was in musty archives, only occasionally read by some unsuspecting man of letters, not knowing what to do with the material. Unaware of Stevin's advanced scheme, the 19th century showed a sudden revival of attention for closure of the Zuiderzee. It became the favourite subject for theses at the Delft Technical University. Apparently there was no 19th century engineer who did not cherish some juvenile ideas about the Zuiderzee. To create order in the sudden spate of plans, the private Zuiderzee Association was established in 1886. The organization engaged mr Lely to distil the best ideas from the more than 500 plans which had been previously published. This scheme was finished in 1891, and a year later the government committee in broad lines adopted the Lely-plan. Not very surprising, for mr Lely had been appointed Minister of Public Works in 1891, at the age of 37. To the Zuiderzee Association his important position proved a helpful coincidence, but it by no means meant that the closure of the Zuiderzee had become inevitable. Parliament held sceptic views: was there truly a need for all this new farmland? Would the profits outweigh the losses? What was to be done with fishermen who expected to lose their trade rapidly? Added to this were objections from conservationists and anthropologists. In short: the debate closely resembled the present scuffle between those supporting reclamation of the Markerwaard and those opposing the scheme. The Zuiderzee itself helped to seal its fate.

Floods

For centuries people living near the Zuiderzee coast had suffered from a surplus of water. Those living in Waterland (to the north east of Amsterdam) witnessed the process of setting of the peat layers which caused their land to sink a few centimeters every year so, the trouble they had with their boisterous neighbour increased in proportion. The areas resting on unstable soil had to endure dike-bursts at regular intervals. Therefore, one of the arguments in favour of closure and large scale reclamation was to protect the land from floods.

January 1916 was a month in one of those winters that grandfathers never tell any tall tales about: nothing but showers and fairly high temperatures. The torrential rain, however, proved fatal to many dikes bordering on the Zuiderzee. Only the outer talus of these dikes was covered with stony slopes; the foundation in most cases consisted of medieval dikes resting on a rather wobbly base. The height of the crowns (usually 3 m above Mean Sea Level) permitted waves to pass over them in case of a stiff breeze, even at high water neaps. To invite disaster even further, the dikes were teeming with all sorts of busy little animals, e.g. rats and moles. Their diligent digging had made the dikes permeable instead of water-resistant. In the month of January (to be precise: on the early morning of the 14th) 1916, the south-westerly storm, which had been continuing for days, veered to north-west and increased to hurricane force. The water of the Zuiderzee was driven up to a frightening degree. The top of the gates of the sea-locks was reached, waves broke over the dikes' crowns. Tarpaulins were used for covering the inner slopes of the dikes, in an attempt to prevent them from becoming soaked, and sandbags were kept at the ready. The dikes that had not yet dissolved collapsed; the sea galloped inland through thirty different breaches. Waterland was flooded entirely, the Anna Paulownapolder was drowned in two phases, the same happened to other parts of Noord-Holland around Enkhuizen and Den Helder. The storm caused extensive damage on the Islands in the Waddenzee, but also in *safe* areas, such as the hilly area of the Veluwe. In the towns of Zwolle, Amersfoortand Naarden people had to put up with wet feet. The Noordpolder near the village of Muiden, the Nijkerkerpolder, the Arkemeensche polder, sections of Friesland and the town of Meppel were all flooded. In the Zuiderzee-ports the heaviest fishing-boats were hurled onto the quays. In Marken even the coffins in the churchyard were unearthed; they floated among wrecked wooden houses as a sinister memento mori. Most people now realized that the Zuiderzee had better be dammed. The chairman of the Zuiderzee Association, mr G. Vissering, stated: "There is no reason to cling to this poor fishing-industry, offering employment to only 3,000 men, with an annual turnover of say 2,000,000.- guilders. The reclaimed land would enable more than 25,000 people to produce well over Dfl 70,000,000.- annually." Canvassingtexts like this were published in largeeditions; theepilogue in a bookabout designs for the Zuiderzee-closure, published in 1917, reported: "We can increase our territory, we can strengthen and perfect it with the most peaceful labour that can be imagined; with the science of engineers and agriculturists, with the creative genius of our contractors, valued the world over, at a time when more than half of the globe, and two-thirds of its population, are involved in a bloody war, trying to increase their territory by force of arms."

The cost of the entire project according to Lely's design was estimated at Dfl 222 million. For this huge amount of money it was expected to add 220,000 ha to the territory of the Netherlands. At Dfl 1,000.- per ha this was evidently an economical proposition, even by pre-war standards. The law regulating closure and reclamation of the Zuiderzee, was passed in 1918 and within a month the Zuiderzee Council was established. In 1919 the Department for Zuiderzee Works started work and a year later, on the 20th of June 1920, the first job - bridling the Amsteldiep between the island of Wieringen and the northern tip of Noord-Holland - was put out to contract in public. The dike leading to Wieringen was regarded as a preliminary excercise preceding the real work: the mighty Afsluitdijk (Enclosure Dam) from Holland to Friesland itself. In the deep tide-ways of the Amsteldiep a sill was constructed, following the exact alignment of the dike, to a height of some meters below the surface. As soon as the currents were evened out in this way, the building up of the dam itself could begin. For the first time, boulder-clay was used on a large scale, a material which turned out to be excellently suitable. The quantity available, in close proximity to the works, was larger than had always been assumed. Boulder-clay proved to be extremely current-resistant, and it was to be invaluable in the Zuiderzee-closure. The dike to Wieringen was completed, without any great difficulties, within four years. The ferryman was forced into early retirement, and the hydraulic engineers could focus their attention on the gaping stretch of water between Holland and Friesland: 16 nautical miles of violent currents, which would attempt to wash away anything that human beings threw in: a fierce battle was to be launched.

Scientific approach

In view of the sheer magnitude of the project, the engineers realised that the natural conditions at the construction-site required investigation in a laboratory. Furthermore it was thought unrealistic that the entire project could be carried out successfully by contractors working independently. The grand scheme needed an integral approach, under central direction. In the wake of the Zuiderzee Works, the Netherlands realized that it could not do without a hydraulics laboratory! Oddly enough, before 1927 there was no such institute in the Netherlands-one would be inclined to take such an intellectual backbone for granted in a country kept afloat by its dredgers. However, the first experiments for the Zuiderzee Works had been carried out in German Karlsruhe, 450 km from the sea, in the laboratory of prof. Rehbock. Experiments in preparation of construction of the Juliana canal and the Noorder-lock of IJmuiden were done in the Berlin laboratory of prof. Krey. But at last, in September 1927, the *Waterloopkundig Laboratorium (Hydraulics Laboratory),* associated with the Technical University of Delft, was established. In 1933 the institute moved to its permanent premises in Delft. Together with the Laboratory for Soil Mechanics it was managed by the *Stichting Waterbouwkundig Laboratorium (Foundation Hydraulic Engineering Laboratory).*

Bucket dredger on the Zuiderzee Works

From then on the Dutch hydraulic engineers were to have their own scientific facilities at their disposal. At an early stage the Laboratory more than justified its cost. Tests revealed that a third complex of sluices in the Enclosure Dam was not necessary -a windfall of Dfl 1 million!

To guarantee a uniform approach to the job, the *Maatschappij tot Uitvoering van Zuiderzeewerken (Company for the Execution of Zuiderzee Works)* was executor Mr. J. A. Ringers of Rijkswaterstaat, Ministry of Public Works. The operation created a unique atmosphere in which competing firms learnt to regard one another as colleagues rather than rivals-a process which encountered occasional difficulties. But, like the project itself, this experience later proved invaluable when it came to contracting large projects abroad. The fraternization among contractors would eventually culminate in the Vereniging *Centrale Baggerbedrijf (Dredging Association).*

First of all, cunning tactics had to be applied against the tricks of the tide: no opponent is as flexible, as rapid as the sea and its elusive currents. To prevent these currents from taking another course as soon as material was thrown in their path, the main channels had to be kept open till the final closure. As the velocity of the currents increased in accordance with progress made with the dikes, the bottom of the tide-ways needed to be reinforced, as a protection against scouring. At the same time it was of extreme importance to gain control over the water-level immediately upon realisation of the closure. Hence, the sluices had to be completed at an early stage. Scour-protection of the channels and construction of the sluices were the first activities to be deployed. The construction of the dam itself began in the shallowest parts of the sea. The schedule aimed at reaching a more or less stable stage each year before the autumn storms could seriously damage the works. The final closures were especially demanding in the approach they required: the bottom of the channels needed extra reinforcement, and to fill the gap, more material had to be thrown in than the tide could possibly carry away.

In the first closure, in the Middelgronden, an unexpected tidal race occurred creating a hole of over 28 m. in depth, just outside the scour-mattresses. The boulder-clay dam had to be diverted and filling-activity was stepped up to its maximum. The entire dam was built as a foundation consisting of boulder-clay and a clay-covered body of sand of much greater volume. When the final closure, the Vlieter, was undertaken, the experience gained at the Middelgronden came in useful. No complication occurred, and the monument, designed by the famous Dutch architect Dudok, to commemorate the closure, has not sagged till the present day. On May 28, 1932, at 13:02, the last heap of boulder-clay was dumped into the Vlieter. By the time the dam was finished, it had consumed a total of 15 million m^3 of boulder-clay and 27 million m^3 of sand. The Zuiderzee (sea in the south) was called IJsselmeer (lake of the IJssel) from then on.

Crops and cows instead of clippers and cutters

An obvious reason for constructing the Enclosure Dam was to reclaim polders. This was to be carried out far more easily in non-tidal waters than in the open sea, for then each closure of the polder-dikes could have been thwarted by the same problems which nearly ruined the Middelgronden closure.

The workers on the Zuiderzee project were, however, unable to suppress their impatience in one case: the dike of the Wieringermeer polder was closed as early as July 29, 1929. Six months later the pumping-stations near Den Oever and Medemblik (the diesel-powered pumping-station *Leemans* and the electrically driven *Lely* respectively) started work and another five months later the area was dry. On August 21, 1930, it was possible to take a walk on the former sea floor, hence this day is commemmorated as *Polder Day*. In 1931 the first crop of winter-wheat was harvested!

The Wieringermeer can rightly be called a Zuiderzee polder, whereas the others are IJsselmeer polders. Apart from the Noordoostpolder the realization of the polders was delayed due to circumstances: first the war, later the interests of yachting and traditional ways of life of the natives thwarted activity. Furthermore, the market was already groaning under an over-production of butter, meat and milk. These days the future reclamation of the Markerwaard is facing great problems: agricultural over-production and protests from fishermen, yachtsmen and conservationists, co-operate to persuade politicians to think twice about the desirability of this last polder to be completed.

The fishermen of the 1930s were, naturally, not too keen on the advantages of the Enclosure Dam. The rich shoals of herring and anchovy belonged to the past; these species failed to appreciate the merits of locks and either had to content themselves with staying in the IJsselmeer, or could not reach itanymore. Sufficient other species, however, appeared more than willing to step in where the former providers of wealth had disappeared. Furthermore, not all the salt-water fish were gone the moment the dam was closed. As the IJsselmeer gradually became fresh over the years, due to river water from the south and the east, the creatures of the sea moved up north. Those fishing close to the dam were doing a busier trade than ever before. After 1940 the seafood business dwindled, and a few years later most fishermen turned to fresh-water fish. Some villages stuck to sea fishing, however, e.g. Makkum and Urk (which nowadays dominates the scene with its fleet of large and powerful cutters).

To the surprise of many, the IJsselmeer continued to offer a comfortable livelihood to a fairly large number of fishermen. Particularly in the early '60s, bestselling fish like eel made the fishermen who had always been oriented on the tidal Waddenzee in the north, rather envious of their colleagues fishing in the IJsselmeer. Inshore fishermen found great difficulties in competing with the large seagoing cutters with their increasingly powerful engines -the IJsselmeer fishermen, however, had the lake all to themselves. In Enkhuizen and Volendam the rotten old botters have long been replaced by modern prosperous-looking cutters. At a later stage we shall deal with the closure of the Grevelingen, which led to the most successful mussel and oyster-cultures in the entire province of Zeeland: sometimes constructions of hydraulic engineering seem to be a greater problem to those defending the natural environment than to nature itself. Bearing in mind the objections put forward by conservationists opposing the closure of the Zuiderzee, it is amazing to hear those opposing the reclamation of the Markerwaard on the grounds that this stretch of water, entirely surrounded by a barren dike and artifi-

Something very special seems to have been achieved during the reclamation of the Zuiderzee, for the drag-line is flying the national ensign, which was not the habit, despite the national pride in the project.

cially robbed of its tides, is a unique and irreplaceable nature area. Which is a great compliment for a dredging job that is not even half completed.

Digging ditches

Creating land is one thing, but it is quite something else to make a good lay-out for such a bare plain. First of all, it must have a name. *Urkerland and Schokkerland* were names considered for the Noordoostpolder, (North-East Polder) but the more prosaic name was chosen. The land originally known as the'Zuidwestelijke polder' was in the end reclaimed in two stages, resulting in Oostelijk (East) and Zuideli jk (South) Flevoland. Oostelijk Flevoland was completed in 1957; some ten years later its southern twin could call itself a polder. To create a solid building site in the southern basin of the Ijsselmeer 1-1/2 million tons of soft soil had to be removed; in this area the tidal currents had always been rather weak, allowing soft sediments to settle. Many historic wrecks, resting on the bottom of the lake, had been excellently preserved in the mud. Hence, after the polders had been reclaimed, the *Dienst Uitvoering Zuiderzeewerken (Department for the Execution of Zuiderzee Works)* found itself saddled with yet another task: to record the many wrecks Neptune had scattered around. Even drowned settlement remains and tree-trunks, dating from prehistoric times were found, all in an excellent state of preservation. The fishermen had always known about these trunks, because their nets often got entangled with them.

The first thing to pay attention to in a new polder destined for farmland, is a perfect drainage-system: canals, ditches, and trenches. Foresight is the essence of dredging- hence, the big canals and ditches were dredged before the polders were pumped dry. This could be done accurately with a dredger, much faster than would have been possible shifting dry soil. The smaller ditches and trenches for drainage were dug with enormous ploughs capable of cutting a trench over one meter wide in one go. There were roads and bridges to be constructed, and when family homes were built next to the labourers' barracks public facilities became necessary: shops, postal services, electricity, gas, medical facilities etc. Archeologists anxiously tried to collect valuable information before oxygen penetrating the drying soil would cause all organic remains to rot away. To the 20th century Dutchmen the Ijsselmeerpolders were what the Frontier had been to 19th century Americans. Such infinite space! Farms and entire cities, on a grand scale, could be built; projects abounding in architectural experiments. The pioneers, those who were among the first to live in new settlements such as Lelystad and Dronten, undoubtedly regret the expansion of these cities, with new districts even featuring multi-storeyed buildings - a surprising phenomenon in an area that spacious. An inhabitant of Lelystad put it this way: "Every time I return from a holiday I fear that they will have added a new roundabout, so that I might get lost, looking for my own home."

Organizing the works

In one of the previous paragraphs we discussed the 19th century system of small contractors and low tenders which often resulted in poor living conditions for the labourers. Fortunately, a considerably better system was used for the Zuiderzee Works. The contractors co-operating within the M.U.Z. had succeeded in arranging agreeable conditions: cost price increased with a 10% profit margin. It must be said that the workers shared in the profits: their weekly wages could exceed Dfl 100- quite an ample salary in the days when cents still counted.

Naturally, the members of the M.U.Z. did not have at their disposal all the equipment needed for making the Enclosure Dam and the polders a success. The construction of the Dam in particular required a lot of extra barges and cranes, which were rented from other firms. To the countless owners of private river-craft and tug-skippers this meant the chance of a life-time at a moment when the rest of the country was plunged in a crisis. The renting-system, however, was of a rather peculiar kind, and quite damaging too to the self-confidence of some of these small owners. The M.U.Z. rented the vessels, taking on the crew and the owner. If a captain of a tug turned out to be accident-prone the M.U.Z. was fully authorized to replace him by some other skipper. Just imagine, having a life long record as a proud skipper/owner when all of a sudden ... it is surprising to find post-M.U.Z.syndromes still not recognized by modern psychiatry.

New calamities

At the end of the war the Dredging Association's efforts for maintaining a fair quantity of equipment came in helpful: in both 1944 and 1945 floods (caused by hostilities) ravaged the land, and 1953 witnessed the notorious February-flood, catastrophes hich called for immediate action. The necessary equipment was available, not just for fighting calamities, but also for carrying out works on a grand scale, designed to prevent future disasters. In fact there was so much material around that the entire Delta Works could be completed by Dutch dredging companies while ambitious projects elsewhere in the world were being carried out simultaneously.

The 1953 disaster

In 1906, when the Netherlands- in particular Zeeland -were ravaged by a serious flood, many sea-defences turned out to be of inferior construction. The catastrophe led to heated discussions: should the back-dikes be reinforced, or only the sea-dikes? Funds for structural improvements lacking, makeshift solutions were carried out: dikes were raised upon their old foundations (which limited the actual raising to about one meter) and on their crowns concrete parapets were built to keep the waves from coming across the top. In addition, the force of the waves was to be tempered by stair-like slopes of concrete covering the outer talus. The 1953 disaster proved these measures to be makeshift indeed: in Zeeland alone the sea cut over 150 breaches in the dike-range; 1,835 people were drowned in the south west, and hundreds of head of cattle ended their lives in the water.

After a first warning against an extra high tide the K.N.M.I. (Royal Dutch Meteorological Office) sent a telegram to the Chief-Engineer Director of Public Works in Zeeland (and others), voicing a warning against *dangerously high tides*. The Chief later assured that all possible measures had indeed been taken regarding dike-watching etc, but that no reasons could be found to justify extreme anxiety. "After all, the storm-surge

Willow matting is brought into position to protect one of the breaches, after the 1953 disaster in the Province of Zeeland.

of March 12, 1906, causing several polders to be inundated, led to raising and strengthening of the sea-dikes wherever necessary, taking the height of this storm-surge as a yardstick. These water-levels have never been recorded before or since the 1906 flooding, until the disastrous night of February 1, 1953. The dikes had been strengthened to a surplus height, depending on the conditions of the waters on which they bordered, in order to safeguard them against being overflowed. The surplus height generally varied between 1.30 m and 4.30 m. Storm-surges could be awaited with a certain degree of confidence, further inspired by the good condition of the stone-protected outer slopes of the sea defences."

The sea thought differently: on the morning of February 1st a violent storm-surge, whipped up by a northwesterly storm, reached a height of over three meters above the predicted level. In most places the flood exceeded the level reached by the highest flood ever recorded by no more than half a meter, but the dikes' surplus height had lacked a sufficiently sound basis. Transverse stability was evidently insufficient. The breaches cut by the furious elements were not to be sneezed at: near the Schelphoek on the island of Schouwen a gaping hole of 250 m width and 30 m depth appeared; the ferry-dock of Kruiningen disappeared completely. Back-dikes failed to stem the flood, one of the reasons being that most of these dike-bodies had been cut through for the construction of roads and canals. The gigantic damage was not restricted to dikes: houses and roads, factories and sluices, all were having a hard time; many buildings collapsed, locks were dislocated; watermains silted up, culverts broke. In Zeeland alone 40,000 ha. of land were flooded, and the tidal waters also found their way into the islands and polders of Zuid-Holland.

Over 70,000 people had to be evacuated from the area. Yet in his report the Chief Engineer of Waterstaat, Mr. Swaters assumed that the damage was reparable. Such an achievement could only be possible in a well-organized nation, supported by a backbone of dedicated civil servants. In the early morning of February 1st, working parties, concentrating on different places, had already started to shore up developing breaches and to prevent existing holes from expanding. In many areas nothing could be done at all: either the sandbags were immediately swept away by the current, or the water took some time to gather strength before breaking through the new obstacles. Aircraft dropping sandbags were a great help to those slogging away in the mud; these greater quantities of bags gave the currents a hard time. Pumps had to be transported to the disaster area, as wel I as boats, for evacuation and transport of supplies. Then the dredging equipment arrived in huge numbers, determined to put the stress on the second syllable of *Zeeland.*

The mess was immense, but still there were some tidy souls around who knew where to begin - a charming aspect of human society. When after years of heavy shelling and bombing there is a power failure in Beirut the Dutch press spreads the news as if it were something special. Everybody would expect the failure to have occurred years ago. But the tenacity of people trying to keep their complex society going seems capable of resisting anything. This tenacity has always been a special trait of the Zeelanders.

In the times of spade, barrow, and primitive dikes they volunteered to make a living in an area adorned with the epithet *calamitous polder* -true evidence of a lion's courage. Those 20th century inhabitants of suburbia, who desperately cling to securities, whose feelings of uncertainty are limited to what monthly bonus they will make for their irregu-

The 1953 flooding disaster in the province of Zeeland.

lar shifts, cannot fathom the deeper meaning of Zeeland's motto *Luctor et emergo (I struggle and emerge)*. It takes a special frame of mind to be willing to live in a place where you might be forced to spend the next night on your own rooftop, watching the dresser float by; to stay in the 'calamitous polder' instead of fleeing to a little house in the safe hi lly east of the Netherlands-the Zeelanders must be a special breed indeed.

The 1953 Emergency Dike-Repair Act and the Delta Act of 1956

The Dutch being a well-organized people, there came an immediate response from the Government and the public; but in its implementation frictions between various official services as to the division of responsibilities were hard to avoid. The National Public Works Authority assumed control of many tasks in the field of repair, and Zeeland's Provincial Works Authorities (Waterstaat) naturally feared that, after the effects of the floods had been cleared, this tendency might continue. The Emergency Act of May 2, 1953, restricted Rijkswaterstaat's main responsibility to the repair of sea defences of Zuid-Holland and Zeeland. The maintenance of roads and other infrastructural works would remain the province of Provincial Waterstaat. The civil servants' petty wars apparently failed to thwart restoration activity: by the end of September 1953 all polders had been reclaimed, except the island of Schouwen and the Vier-Bannen in Duiveland. Regardless of the regulations in the Emergency Act many polder-boards put their shoulders to the wheel instead of waiting for Rijkswaterstaat to repair the dikes. However, central direction of repair of the sea defences offered undeniable advantages: assessment of the need for dredging equipment, assessment of priorities, and management of the actual repairs, required *one captain on the ship.*

There was a justification for adapting the organization of Waterstaat to the new situation. This situation, however, was subject to even more radical change: on February 18, 1953, the Minister of Public Works J. Algera set up the Delta Committee. Its assignment was to advise on the best possible form of protection for the devastated zones, in order to prevent similar disasters in the future. A year later the committee advised the closure of the tidal inlets between the Zeeland and Zuid-Holland islands. With unusual speed (as early as November 16, 1955) the Delta bill was put before Parliament. The bill had evolved into a grand scheme to make a virtue of necessity: the proposed dams would not just guarantee an unprecedented degree of safety for Zeeland, they would also end the archipelago's isolation. Pending the completion of those ultimate defences, it was considered necessary to reinforce and heighten all dikes to a level capable of resisting a storm-surge as powerful as that of February 1953 for at least 25 years (the time it would take to construct the dams). These interim works were estimated at roughly Dfl. 40 million. A very rough estimate, as it turned out later.

Tactics underlying the closures

The timing of the 1953 disaster could, despite the horrors, have been worse: the dredging contractors were better prepared for unexpected events, largely as a result of the Dredging Association's activities. The closures of the original mouth of the Maas, and of the smaller inlets Sloe and Braakman had been useful rehearsals for the grand scheme itself. During the re-reclamation of Walcheren after its wartime inundation, the contractors had discovered what it takes to bridle strong currents. In short: storm and surge surprised the estuary, but the Dutch were well-trained and armed to the teeth for fighting the floods.

Nothing bears more humiliating evidence of man's neurotic disposition than war. Yet, face to face with death, man is usually most inventive. Important discoveries are often made during a war. To the field of engineering - especially regarding emergency operations - the Second World War had been a most instructive period. A project like D-Day may be of special interest to soldiers from a destructive point of view - to those interested in dredging, the construction of emergency ports deserves closer attention. A colossal fleet of replenishment craft had needed a chain of safe all-weather ports during the operation. The wartime experience in the use of submersible caissons (which had also been essential in the 1945 recovery of Walcheren) proved to be indispensable for the repairs carried out on the sea-defences following the 1953 floodings. The currents rushing through the breaches made short work of sandbags or clay, but they had to admit defeat when heavy-weight caissons appeared on the scene.

The legendary director of the operation "ref Ioat Walcheren," "Kobus" Kalis, as a young man. He is one of the last of a very special breed: taught in the mud instead of at university, he has proved to be a great organizer of complicated projects, such as the reclamation of Walcheren, and a businessman with vision. Yet, he thoroughly disliked the large corporate structure, with undesirable lack of communication between the men of the fleet and directors. It was sufficient reason for him to leave his company in early retirement.

At the closure of the smaller inlets with fierce tides, such as the Veersche Gat, caissons proved to be the obvious answer.

But, unfortunately not everybody rejoiced. Around 1956 Provincial Waterstaat had made every possible effort to secure an outer harbour for the fishermen from Veere, and it had also insisted on the construction of a sluice for the benefit of the oyster-cultures. The idea had been to make the Veerse Meer into an *experimental lake*. Mr H.A.M.C. Dibbits, in charge of the southern district of the Delta Works, was no supporter of these extras, an attitude which the fishermen interpreted as a starting-point for campaigns, to be continued until the final caisson was sunk into the gap. The fishermen lost their case and took to Vlissingen (Flushing) and the nearby port of Colijnsplaat as their new bases.

The closures of the tidal inlets in the Southwest were carried out first of all to offer the area more safety against floodings. At the same time a second aspect was aimed at: the control of the water-regime of the rivers Rhine and Maas and the fight against salination. For this purpose huge sluices were constructed in the Haringvliet, after the Nieuwe Waterweg the most suitable outlet for Rhine-water. Automatically a third useful task sprouted from the Delta Plan, to use the dams for road traffic as well, thus improving the infrastructure for the benefit of industry and other consequences of modern civilization. Finally, the currents' teeth were pulled out with the taming of the tides, so that today more sailing enthusiasts can go out boating than in the times when the tides could still wash up landlubbers. The number of less wealthy sail-board surfers exceeds that of yachtsmen deploring the loss of their tidal cruising grounds, and forced into crowded marinas instead of drying out on the saltings, so the recreational value of the area has increased.

As soon as a tidal river is closed, its water must be controlled artificially, in order to avoid unwelcome hydraulic side-effects. If, for instance, the Grevelingen had only been closed to seaward, with the Oosterschelde and the Haringvl iet still open, the whirling currents scouring the coasts of the islands would have inflicted more damage than the dam could ever have prevented. Therefore, each dam to seaward was always backed up by a dam that excluded the influence of connected inlets. Consequently, a series of controlled lakes have been created, while only in a small stretch of the Oosterschelde river a shadow of the former lunar and solar tides has been preserved. It was crucial to complete the dams at both ends at approximately the same time, in order to control the water-level as soon as influences from outside were barred.

On the basis of experience gained during the Zuiderzee Works, troughs scoured by unruly currents could be prevented. The Zuiderzee Works had proved that the sandbanks were the obvious locations to start construction of the dams. As long as the work was in progress on the shallows, the currents could be confined to their original channels. However, as the dam advanced, scour protection of these channels became essential. In the Zuiderzee Works this was carried out by means of willow mattresses; during the Delta Works often with purpose-developed fibre mattresses. The closure of the actual channels, with caissons or concrete blocks, completed the basic work. The weight and size of the materials blocking the channels, prevented them from being washed away by the tides. With the basic closure completed, the body of the dam could be raised without running the risk of the sand fill ending up miles away from the proper site. Operations such as these could be carried out only by means of modern dredging equipment with its huge capacity. As soon as an inlet had been closed, the engineers shifted their attention to protecting the body of the dam against scour and erosion, before the arrival of winter.

APPENDIX D

Personalities in Dredging

In his valuable study book ***Hydraulic Dredging,*** John Huston, 1986, outlined the History of Dredging. In that chapter he discussed **Dredging Organizations** and stated the following:

"Of all the advances that have occurred in the growth of the dredging industry, the two most important are the creation of the **World Dredging Conference Association (WODCON) and the World Dredging Association (WODA)**.

Prior to their conception in 1967, the dredging industry existed in a vast informational vacuum. The dissemination of dredging knowledge was almost nil. There was seemingly a barrier of secrecy, not only between individuals, but between dredging companies as well. A striking example of this attitude was that of a company who conceived an invention that increased production. They never used it though - they didn't want others to know about it!

With the origin of WODCON and WODA, however, attitudes began to change, and dredgemen began to come together, on an international scale, to promote the exchange of dredging knowledge. The credit for bringing about and nurturing this renaissance must be given to **M.J. Richardson.** Richardson's involvement occurred through the foresightedness of John Latham, who at that time was co-publisher with **Ruel McDaniel** of a newly-launched publication called World Dredging. Latham, who was also the publisher of **Offshore** magazine, became acquainted with Richardson through **OECON - Offshore Exploration Conference** - that Richardson had organized to stimulate the transfer of aerospace technology to the oil industry.

Latham invited Richardson to consider organizing a similar conference for the dredging industry. Although Richardson had a broad background in the marine and aerospace fields, he had little knowledge of the dredging industry. He therefore accepted Latham's invitation conditionally - he would first have to investigate the world of dredging to see if he felt it worthwhile to attempt such a conference. The investigation convinced him that it was, and he began laying plans.

The Conference as planned was to have the same format as the previous Offshore Conference. There were to be three days of technical papers, a dinner party, and three luncheons with keynote speakers. It was to be held in New York City, U.S.A.

On the sixth of May in 1967 the Conference opened. The luncheons were at the **Waldorf Astoria**, the dinner party at the **Americana Hotel**, and the technical paper presentations and industry exhibits in the **New York Coliseum.** Richardson named the three-day event **WODCON** for World Dredging Conference, after Latham's magazine. Richardson's planning, organization, and leadership paid off. The Conference was a resounding success. The general feeling was one of euphoria. **Walter Gahagan** underscored it by remarking that "If nothing else had happened, just seeing the word **DREDGING** in three-foot letters on the marquee of the Coliseum made it all worthwhile."

Before the Conference was over, a consensus developed that future conferences should be held every 18 months under the auspices of an organization to be formed by Richardson, and supported by leading dredging contractors. Consequently, following the Conference, Richardson incorporated a non-profit organization called **WODCON ASSN.** for **World Dredging Conference Association.**

This was not the end of Richardson's involvement, however. In a subsequent meeting at the **Ellicott Machine Corporation** in Baltimore, Maryland, **Alf Sorensen** suggested that the industry needed a professional society for individual members - one that would be separated from the influence of the dredging companies. With the same forthright planning, organization, and leadership he had provided for WODCON, Richardson formed the **World Dredging Association, (WODA)**. He travelled throughout the western hemisphere, Europe, and the Far East making contacts and organizing chapters, the first being the Pacific Chapter in San Francisco. Officers were elected, and WODA was off to a good start.

WODA flourished for ten years with Richardson guiding it as Executive Secretary. During that time he organized seven WODCON's that were held throughout the world. In 1976, however, following the most successful of the WODCON conferences, **WODCON VII**, in San Francisco, at the joint meeting of the Board of Directors of WODA and WODCON, Richardson pointed out that membership in WODA seemed stagnant, and that support from a broad cross-section of the industry had not been forthcoming. He requested that the industry consider forming its own secretariat to take over the management of WODA and WODCON, and thus give it more identification with industry goals.

This involved forming committees in both the U.S. and Europe to consider how this might be done to reach the goal of expansion and cooperation. Richardson met with members of those committees over the next year and the plan was developed.

The plan was to provide for the subdivision of WODA into three separate Sections, (1) a **Central Dredging Association (CEDA)**, to include most of Europe, the Near and Middle East, and Africa, (2) a **Western Dredging Association (WEDA)**, to include North, Central, and South America, and (3) an **Eastern Dredging Association (EADA)**, to include the Far East, Southern Asia, Indonesia, Australia, and New Zealand.

Each of these Sections was to function under a common set of objectives and goals set out by the board of Directors of the reorganized WODA. However, the Sections were to establish their own by-laws and Chapters, and act as autonomous entities. The parent organization, WODA, was to have no individual members. It was to exist only as a directive body to coordinate the activities of the three separate Sections, including the organization of future WODCON's. Each Section was to appoint two of its members to serve on WODA's Board of Directors. At the eighth World Dredging Conference in Amsterdam, in 1978, these details were adopted by the separate Boards of CEDA and WEDA."

Recognition to Contributors

Over the years of dredging development and particularly, with the founding of WODCON ASSOCIATION and WODA, there have been individuals who have contributed greatly to the success of what is today the WODA organization. Many of the companies in which they were involved have either disappeared, or been acquired by other companies. Their recognition is appropriate to this book as they influenced the growth and maturing of ***The Dynamics of Dredging.***

Name	**Original Company/Affiliation**
USA	
Lee Barber	Consolidated Purchasing & Designing Co., SF CA
W.H. *Bill* Bauer	Bauer Dredging Company, Pt. Lavaca, TX
Jim W. Bean	C.F. Bean Dredging Corp., New Orleans
Herbert Bure	Ellicott Machine Corp., Baltimore, MD
Norman Cleaveland	Pacific Tin Consolidated Corp., NY/Malaysia
Jack Dunn	Dixie Dredge Co., St. Louis, MO
Ole Erickson	Erickson Engineering, Tampa, FL
Walter Gahagan	Gahagan Dredging Co., NY

O.M. *Monk* Gautreaux	Williams-McWilliams Co., New Orleans, LA
John Herbich	Texas A&M University, TX
John Huston	Consulting Dredge Engineer, Corpus Christi, TX
Don Killom	Dredge Masters Intl. & AMMCO, TN
Don R. King	DredgeMasters Intl., TN
R.R. Manatt	Atlantic, Gulf & Pacific Co.
Louis Mauriello	US COE, New York District
H. Roger Morris	Manson-Osberg Co., Seattle, WA
W.R. *Bill* Murden	US ARMY COE, Washington, DC
Patrick O'Neill	International Mining Corp. NY
Ed Peterson	Utah Construction & Mining Corp., SF
C.D. Ramsden	PACECO, Oakland, CA
R.C. Schmid	Bechtel, Inc., SF
Alf Sorenson	Ellicott Machine Corp. Baltimore, MD
The Netherlands	
Prof. W.A. Bos	Delft University
A.L. Brink	COBLA International Dredging Co.
Jack Donkers	IHC HOLLAND
K. Van Exter	Van Hattum En Blankenvoort
J.K. van Hemert	Boskalis Westminster Dredging Co.
A.J. Hoekstra	Adriaan Volker
J. Ringers	Adriaan Volker
F.A.M. de Vilder	Ballast-Nedam Groep
Belgium	
Paul van Haaren	Ackermans & Van Haaren
G. Vandercruyssen	Societe Generale de Dragage
Germany	
K.H. Brosskamp	Philipp Holzmann
A. Welte	Orenstein & Koppel
Japan	
Hachiro Fujii	Toyo Construction Co.
T. Mizuno	Toa Harbor Works Co.
T. Nagano	Penta-Ocean Construction Co
Mineo Nakamichi	Japan Industrial Land Development Co.
Saburo Okabe	Toa Harbor Works Co.
Renichi Watanabe	Rinkai Construction Co.

United Kingdom

J.M. Hamilton	Simons-Lobnitz Ltd., Scotland
Hugh McManus	SeaDrec Ltd., Scotland

Taiwan

H. C. Yen	RetSerEngineering Agency

Korea

Suen Guen Kim	Korea Dredging Corp.

Egypt

Abdel el Salam	Suez Canal Authority

France

Jaques Oules	SoFrancaise DragagesTravaux Pub

Philippines

Felix V. Bagtas	Philippine Bureau of Public Works

APPENDIX E

DREDGE BUILDERS

As with Personalities, the past 100+ years of dredge building has seen companies either disappear from the scene, consolidate, merge or change their name. The leading firms of today are the survivors, having made their mark on Dredge Design and development in today's dredging environment.

IHC HOLLAND

The partnership ***Industrieele Handels Cominatie Holland,*** **IHC Holland** for short, was founded in 1943. It was a partnership of the shipyards Conrad-Stork of Haarlem, Gusto of Schiedam, De Klop of Sliedrecht, J&K Smit en L. Smit & Zn of Kinderdijk and Vershure of Amsterdam.

While the combination of IHC and its partners after 1943 has been significant, it is also interesting to note in this case the production of mining dredges by IHC and its predecessors back to the year of 1887. This was the year of their first *Gold Dredge,* built for a customer in Turin, Italy, ***Bousquet, Okiandi & Co.*** The dredge was as most all of the mining dredges were until recent years, a bucket ladder (BL/M) dredge. It had a chain of 5.0 ft^3 buckets with total power (steam) of 100 hp.

The next two BL/M dredges were 2.0 ft^3 buckets of 60 and 35 hp respectively, supplied in **1894-6 to Russia**. The first was for mining gold and the second for iron ore stripping. Some of the buyers included syndicates of mining that resonate with names that suggest a romantic era. Locations where the dredges were destined for gold mining included parts of **South America (Argentina, Tierra del Fuego, French Guyana, Dutch Guyana, Colombia), Russia, China, Burma, Turkey, Yugoslavia, Africa, and Spain**. The gold dredges varied in bucket size from 1.75 to 7.0 ft^3, with all steam driven until 1912 when diesel power began to be used.

The next surge of mining dredges was for tin in SE Asia. This began in 1908 with 7.0 to 8.5 ft^3 dredges supplied to areas in **Thailand, Malaya,** and **Indonesia**. The size of buckets then grew to 14 ft^3 by 1939. A tally of BL/M dredges built from 1908 to 1939, their mining applications and total quantities that were supplied, is summarized as follows:

Gold mining-	34
Tin mining-	75
Platinum-	2
Iron ore-	2

After 1939 to 2001, with the application of cutter suction (CS) and bucket wheel suction dredges for mining, the totals are listed.

Bucket ladder dredges (BL/M)-	25
Cutter suction dredges (CS/M)-	10
Bucket wheel dredges (BWS/M)-	11

Mining functions were broadened to mineral sands. Mining with dredges shows an evolvement from alluvial Gold beginning in the late 1800s, overlapping with Tin early 1900s, followed by Aggregates in the 1960s and Mineral Sands shortly thereafter. The latter two minerals dominate current activity and construction of new dredges is affected by commodity prices and related industrial demand.

Trailing Suction Hopper Dredges (TH)

Beginning in the 1990s, orders began with the Pearl River, 17,000 m^3 hopper built for Dredging International of Belgium. Built by IHC, this started a competitive program among European dredging companies with the greatest share staying with IHC.

Further expansion of TH dredge sizes built on the target of the largest (by LMG of Germany) for Jan De Nul of Belgium with 30,000 m^3. With accounts worldwide including China, Japan, and the Mideast, IHC has swept the market for dredges of all types.

Other builders worldwide include the following companies. *(Note: Some companies have changed names through consolidation or reorgranization).*

Baltimore Dredges LLC, Ellicott Division
Baltimore, MD, USA

VOSTA.LMG Dredging Technology
Amsterdam, The Netherlands & Lubeck, Germany

DAMEN Shipyards-Gorinchem
The Netherlands

DeGroot Nijkerk Dredging Equipment
Nijkerk, The Netherlands

Deggendorfer Werft & Eisenbau, GmbH
Deggendorf, Germany

Dredging Supply Company, Inc.
Reserve, LA, USA

Liebherr-France
Colmar/Dedex, France

Rohr Corporation,
Germany

Neumann Equipment Co.
Gold Coast, Australia

APPENDIX F

MAJOR DREDGE FLEETS

Totals by dredge category

COMPANY	TH	CS	S	BWS	BL	BC/G	BD	BB
The Netherlands								
Royal Boskalis Westminster	26	32	13	-	7	15	1	19
Ballast HAM Dredging bv	24	21	7(10SJ)	-	1	5	-	3
Van Oord ACZ	10	5	2	-	-	2	-	3

COMPANY	TH	CS	S	BWS	BL	BC/G	BD	BB
Belgium								
Dredging International-DEME	19	17	-	-	3	10	-	8
Jan de Nul Group	14	9	5(H)	-	-	-		1
United States of America								
Great Lakes Dredge Dock Co.	9	13	-	-	-	7	-	1
Weeks Marine, Inc.	2	6	(SD-1)	-	5	-	-	-
Manson Construction Co.	3	6	-	-	-	10	-	-
Japan								
Toa Corporation	-	16	3	-	-	1	-	-
Toyo Construction Co. Ltd.	-	11	3	-	-	4	-	-
Penta-Ocean Construction Co.	1	6	-	-	-	-	-	-
Rinkai Construction Co. Ltd.	-	10	6	-	-	1	-	-
South Korea								
Hanjin Construction Co. Ltd.	-	9	-	-	-	9	-	-

Plate 1 *Pearl River*, Jumbo TH dredge, 17,000 m^3; reclamation dredging at Singapore; Dredging International NV. *(Photo First)*

Plate 2 *Vasco Da Gama*, Jumbo TH dredge, 33,000 m^3, reclamation dredging at Singapore: Jan De Nul NV.

Plate 3 *Illinois*, CS dredge, 30”, reclamation dredging at Ras Abu Fontas, Arabian Gulf; Great Lakes Dredge & Dock Co.

Plate 4 *WD Fairway*, reclamation dredging at Singapore, Jumbo TH dredge, 23,347 m^3; Royal Boskalis Westminster NV.

Plate 5 *Queen of Penta Ocean*, Jumbo TH dredge, 20,000 m^3; reclamation dredging at Singapore; Penta-Ocean Construction Co. Ltd.
(Flying Focus)

Plate 6 *Florida*, CS dredge 36” 10,000 hp, Port of Los Angeles deepening and reclamation; Great Lakes Dredge & Dock Co.

Plate 7 *Gerardus Mercator*, Jumbo TH dredge, 18,000 m^3; reclamation dredging at Singapore; Jan De Nul N.V.
(Chiolini Azienda Fotografica Pavia)

Plate 8 *Adriatico*, BL dredge, 900 litre, dredging trench at Brindisi, Italy; Dredging International NV.

Plate 9 *Amsterdam*, Jumbo TH dredge, 18,000 m^3; preparing for reclamation dredging projects in SE Asia; Ballast Nedam Dredging.
(Photo Sea Skymartin)

Plate 10 *Bayport*, TH dredge, 5,000 yd^3; maintenance dredging in port of Seattle, WA, Manson Construction Co.

Plate 11 *Geomaster*, Airlift suction dredge, mining diamonds offshore Namibia, SW Africa; Namdeb Diamond Corporation.

(Colin Sunkel Photography)

Plate 12 Bucket Ladder mining dredge, 14 ft^3 (BL/M), placer gold production on Rio Nechi, Colombia, Mineros de Antioquia SA; supported by IHC bucket wheel (BWS) dredge, stripping overburden.

Plate 13 *Krankeloon* TH dredge, 2,700 m^3; dredging reclamation at Chek lap Kok airport, Hong Kong, Dredging International, N.V.
(Photo First)

Plate 14 *Volvox Terranova*, Jumbo TH dredge, 20,000 m^3; Van Oord ACZ BV

Plate 15 *R.S. Weeks*, CS dredge, 30", beach reclamation 15-20 miles from offshore deposits, Weeks Marine, Inc.
(Florida Aerial Services, Inc.)

Plate 16 *Amazone*, CS dredge, 850 mm, 13,550 hp, Suez Canal expansion; Dredging International NV. *(Foto Guido Coolens n.v.)*

BIBLIOGRAPHY

Abramian, J.E., Schropp, S.J. 1998, *Environmental considerations for the management of large, tropical rivers: the Hidrovia Paraguay-Parana case*, WODCON XV, Las Vegas.

Anon, *Sumitomo introduces tubular belt conveyor*, International Bulk Journal, Dec. 1985.

Anon 1968, *Water Quality Criteria*, Department of Interior, Superintendent of Documents, Washington, D.C.

Anon 1970, *Effects on Fish Resources of Dredging and Spoil Disposal in San Francisco and San Pablo Bays*, Calif. Special Report, U.S. Fish and Wildlife Service to U.S. Army Corps of Engineers, San Francisco Dist.

Averett, D. E., and Francingues, N.R. Jr. (1996). *Overview of Sediment Remediation Initiatives in the U.S.*, Proceedings Western Dredging Association Seventeenth Technical Conference and Twenty-Ninth Annual Texas A&M Dredging Seminar, Center for Dredging Studies, College Station, TX.

Baldi, G., Belotti, R., Ghionna, V., Jamiolkowski, M. and Pasqualini, E., *Design parameters for sands for CPT*, Proc. 2nd European Symposium on Penetration Testing, Vol. 2, p. 427432, ESOPTII, Amsterdam, May 1982.

Ballinger, D. G. and G. D. McKee, 1971. *Chemical Characterization of Bottom Sediments.* J. Water Po11. Control Fed. 43 (2): 216-227.

Bastian, David F., *History of Cutter Suction Dredges in the United States,* WODCON VII, pp. 1059, 1976.

World's Largest Dredges Worked on World's Largest Engineering Project, WDMC, 10/90, p. 13. 1913.

WDMC 1975. Figure 1. *BL Dredge Lavaca,* c. 1847.

WDMC 1975, *Chronicle of A Steam Dredge Boat,* April 1975, p. 23.

Alternatives to the Panama Canal and What They Could Mean to Dredging, WODCON XII, pp. 1017, 1989, Orlando, FL .

History of Cutter Suction Dredges in the United States, WODCON VII, pp. 1059, 1976, San Francisco.

Bechtel Corporation, Building New Land for Industrial Development-Jubail, Saudi Arabia, WDMC, p.12, 15, Sep. 1985.

Bennett, Michael F. VP, Chandrasekhar, Manju, Transport Specialist, Frederic R. Harris, Inc.

Bergeron, R.E. et al, *The Cable Arm Clamshell: Development and Track Record for Environmental Dredging,* WEDA XX, pp. 343, 2000, Warwick, RI.

Bliek, A.J., Storm, C. *Bars and Maintenance Dredging in the Western Scheldt,* WODCON XIV, 1995, Amsterdam.

Bokhoven, W. *Dredging in the Port of Rotterdam after World War II*, WODCON II,

p. 933, 1968, Rotterdam.
Bosch, L.A. *Developments in Technical Fairway Management*, WODCON XIV, 1995, Amsterdam.
Bucket dredge launched for delivery to PRC, WDMC, Vol. 15, No.6, p.16.
Burgett, W.B. and M.R. Ouwerkerk, 1995, *New developments for the bucket chain ladder dredge*, Proceedings WODCON 14, Amsterdam, The Netherlands.
Carter, D. Mark, *Dredging Preserves Canals Lacing English Countryside*, WDMC, 12/75, p. 9.
Chien, Ning, Wan, Zhaohui, *Mechanics of Sediment Transport*, ASCE Press, 1999.
Classification of Soils and Rocks to be Dredged, Bulletin No. 47 PIANC 1985, Brussels, Belgium.
Cleaveland, Norman, *New Zealand to the Rescue*, 1980, Vancouver, BC, WODCON IX, p. 821.
Cox, C.M., Deursen, P.P. van, Verhoeven, F.A.; *The relevant soil properties for dredging silt*, WODCON XI 1986, Brighton, UK.
Dalton, Krueger, Project Manager, USACOE, Galveston District, TX.
Dayton (1997). *Dredgers of the World*, Dayton Publishing Co., UK.
Deere, D. U. (1968), Chapter 1 *Geological Considerations,* Rock Mechanics in Engineering Practice. John Wiley & Sons, Inc, NY, New York.
DEME: The Century's Biggest Dredging Projects, WDMC, January 2000.
Denning, R.A., COE, *Automating the Hydraulic Dredge*, WODCON I, p. 53, 1967, New York.
Diamond Dredging Offshore Namibia, Southwestern Africa, WDMC, 7/00, p.6.
Directory of Worldwide Dredge Fleets & their Suppliers, WDMC , Vol. 33, No.3.
Dredging Dallas' White Rock Lake, Betty Sosnin, WDMC, 4/98, p. 6.
Dredging Technologies of the 20th Century, DEME, WDMC, 1/00.
Drimmelen, Nico J. van, *Dredging on the Lower Danube in Romania,* Ports and Dredging, IHC 2000, E 154, p. 15.
Eisler, R., 1969, *Acute Toxicities of insecticides to marine decapod crustaceans,* Crustaceana, 16:302-310.
Ellicott International History, WDMC, 11/99.
Engler, R.M., McNair, E. Clark, USAE Waterways Experiment Station, Vicksburg MS, WDMC, p. 20, Oct.'98.
Erickson, Ole P., *Deep Dredging with the Suction Pipe Cutter Drive and New Dredge Machinery Developments,* WODCON 1967, p. 561, New York.
Franklin, John A. Dusseault, Maurice B. 1989, *Rock Dredging,* McGraw-Hill, Inc NY.
Garbaciak, S. Jr, Miller, J.A., (1995). *Field Demonstrations of Sediment Treatment Technologies by the USEPA Assessment and Remediation of Contaminated Sediments (ARCS) Program, in Dredging, Remediation, and Containment of Contaminated Sediments,* ASTM STP 1293, K.R. Demars, G.N., Richardson, R.N. Yong, Chaney, R.C., Eds., American Society for Testing and Materials, Philadelphia, PA.
Geothals & Comber, W.G. c. 1915, *Panama Canal & Dredging-Panama Canal,* Vol.1

Gigantic Dredging Project in South Africa, WDMC, p. 17, 1972.
Giroux, Carl H., *New Ideas in Dredging,* Consulting Engineer, 4/57.
Gorgy, S., N. W. Radestraw, D. L. Fox., 1948, *Arsenic in the sea,* J. Mar. Res., 7:22-32.
Gustafson, J. F., *Beneficial Effects of Dredging Turbidity, Supposed Toxicities of Marine Sediments,* WDMC, p. 44, 12/ '72.
Ecological Effects of Dredged Borrow Pits, WDMC, 9/'72, p. 44-48.
Marine Sediments, Permanence of Attachment of Absorbed Metals & Pesticides, San Leandro Bay, California Proprietary report prepared for Harbor Bay Isle Associates, Alameda, CA, 1972.
Herbert, Christopher, Matibet, Jean-Louis Betbeder, Wilde, Jean De, 1995, *Sydney Airport's Parallel Runway,* WODCON XIV, p. 85, Amsterdam.
Holland, Home of Dredging, Assn. of Dutch Dredging Contractors: Vereniging Centrale Baggerbedrijf, c. 1968.
Hood, Donald W. 1971, *Impingement of Man on the Oceans,* Wiley-Interscience, N.Y.
Horii, Ryosuke, *Recent Trends in Dredging and Reclamation Technologies in Japan,* WDMC, 7/00.
Houweninge, G. van; Leeuw, R. de, *Execution of a Soil Improvement in the Oosterschelde,* WODCON IX, 1980, p. 827, Vancouver, BC, Canada.
Huston, John, 1986, *Hydraulic Dredging,* Corpus Christi, TX, 418pp.
IHI Builds Automated BL (Bucket Ladder) Dredge for Tokyo, WDMC, p. 34, 9/86.
Ikeda, Tatsuo et al, 1995, *Artificial Intelligence for Dredging Control of the Trailing Suction Hopper Dredger,* WODCON 14, p. 203, Amsterdam.
Industrial Minerals 2000, Mining Engineering, 6/01.
Jensen, S. R., Jernelov, A. 1969, *Biological methylation of mercury in aquatic organisms,* Nature, 223(5207): 753-754.
Kerisel, T., *Ideas of the French Administration as Dredger-Owner and Manager of Dredging Works about the development of the Dredging Industry,* WODCON II 1968, p. 25, Rotterdam.
Klein, D. H., Goldberg, E.D. 1970, *Mercury in the marine environment,* Env. Sci. Tech. 4(9):765-768.
Klugt, Peter van der, Keizer Cees de (1997), *The Intelligent Fully Integrated Bridge for Hopper Dredgers,* CEDA Dredging Days 1997, Amsterdam.
Koolhaas, C.P., *Design of Man-Machine Interface for Dynamic Tracking/Dynamic Positioning Systems of a Trailing Suction Hopper Dredger,* M.Sc. thesis TU Delft(1998).
Koning, J. de, Prof. Ir., *Selection of Dredging Methods and Equipment,* WODCON XII, p. 287, 1989, Orlando, FL.
Korver, J.N., Stigter, C., van der Kooy, L.P.D, *Dredging of Reservoirs with Limited Water Consumption,* WODCON XII, pp. 1035, 1989, Orlando, FL.
Kramers, C.H.M, Ouwerkerk, M.R., Koert, J.P., 1998, *The Trend Towards Larger Trailing Suction Hopper Dredgers,* WODCON 15, p. 203, Las Vegas.
Kramers, C.H.M., Koert, J.P., Keizer, C.de, Janse, S.A.W., 1999, *"Trailing Suction Hopp-*

er Dredgers Designed for Improved DP/DT Capabilities, CEDA'99,p. 19,Amsterdam.

Logigian, John M., Dudek, Edward A., Palermo, Michael R. (1994), *Design of dredge containment and dewatering facilities,* Dredging '94, American Society of Civil Engineers, Orlando, FL.

Los Angeles Job Nearly Complete, WDMC, p. 22, Aug. 1982.

Marshall, N. 1962, *The Environmental Chemistry of Marine Sediments,* Occ. Pub. No. 1, Narragansett Marine Lab., Univ. of Rhode Island.

Matsui Akio, Principal Engineer, Sumitomo Heavy Industries, Ltd; Narita Tadao, VP, Kabushiki Kaisha Kosjimagumi, Hideo Katase, Mgr, Masago Industrial Co. Ltd, *Large Scale and Economical Dredging by the World's Largest Grab Dredger Tosho,* WODCON 15, Las Vegas, NV, p. 177, 7/98.

McCullough, David, 1977, *The Path Between the Seas, The Creation of the Panama Canal-1870-1914,* A Touchstone Book, Simon and Schuster, NY.

McKee, J. E. and H. W. Wolf, (eds), 1963, *Water Quality Criteria,* 2nd ed, The Resources Agency of California, State Water Resources Control Board Publication No. 3-A, p. 548.

McNeff, Jules, *Leadership in Satellite Navigation & Timing, Past Success-Future Challenge Part 1,* GPS WORLD, p. 48, 10/99.

Meadows, D. H., Meadows, D.L., Randus,J. E., Behrens, W., 1972, *The Limits to Growth,* Universe Books, N.Y.

Mero, John L., 1965, *The Mineral Resources of the Sea,* Elsevier Publishing Co.

Methods of testing soil for civil engineering purposes, British Standard Institution, London.

Miles, P., Marr, D. (1994), *The Welland River Dredging Demonstration,* Dredging '94, American Society of Civil Engineers, Orlando, FL.

Morris, Gregory L., Fan, Jiahua, 1997, *Reservoir Sedimentation Handbook,* McGraw-Hill.

Murphy, T., Brouwer, H., Babin, J., 1994, *In Situ Sediment Treatment of Dofasco Boatslip, Hamilton Harbour,* Dredging '94, American Society of Civil Engineers, New York.

Nocera, John J., Simmons, Thomas M. 1994, *Development of an excavation plan for heavy metal contaminated marsh deposits,* Dredging '94, ASCE, New York, NY.

Noguchi, Naoki, Kato, Masao, Niiyama, Ryooji, *New Dredging Technique,* WODCON IX, 1980, p. 367, Vancouver, BC, Canada.

Ocean Dumping, a National Policy, Council on Environmental Policy, 1970. S.W. Doc., Wash., D.C.

Ocean Mining Symposium, 1967, Los Angeles, Symcon Publishing Co.

Ofuji, Ishiro, Ishimatsu, Naoshi, *Anti-Turbidity Overflow System for Hopper Dredge,* WODCON VII, 1976, p. 2, San Francisco.

Deep Dredging by Jet-Ejector Dredger, WODCON IV, 1972, p. 249, New Orleans.

Otis, M.J., *New Bedford Harbor, Massachusetts, Dredging/Disposal of PCB Contaminated Sediments,* Dredging '94, ASCE.

Panda, A.P., Padmanabhan, P., *A Review of Maintenance Dredging in Indian Ports,* WODCON XIV, 1995, p. 809, Amsterdam.

Peakall, D. B, Lincer, J.L., 1970, *Polychlorinated biphenols,* Bioscience, 20, (17) 958-964.

Plant, G.W., Covil, C.S. and Hughes, R.A., 1998, *Site preparation Contract for the New Hong Kong International Airport,* Thomas Telford Publishing Ltd, London.

Pralong, R., *Mobile pumping unit for deep dredging,* WODCON XI, 1986, Brighton,UK.

Raalte, G.H.van, *The discbottom cutterhead; A report on laboratory and field tests,* WODCON XI, 1986, Brighton, UK.

Remane, A. and C. Schlieper, 1971, *Biology of Brackish Water,* Wiley-Interscience, N.Y.

Renkema, Allard, Kinlan, David, *Terra et Aqua,* 9/00, and WDMC 11/00, p.6.

Richardson, M.J., *Modern Placer Mining,* 1992, Consolidated Placer Dredging Co., Irvine, CA, 315pp.

Environmental Benefits of Bucket Ladder Dredges, WODCON XV, 1998, p.187, Las Vegas.

Riddell, John, *Two Hundred Years of Benefit: UK Dredging 1795-1995,* WODCON 14, 1995, p. 1, Amsterdam.

Risebrough, R. W., 1969, *Chlorinated hydrocarbons in marine ecosystems,* Proceedings, Rochester Univ. Conf. on Toxicity, Springfield, 111. 23 p.

Roelofsz, F., Bray, R.N., *Drilling and Blasting of Submarine Rock in Relation to Dredging,* WODCON II, 1968, p. 449, Rotterdam.

Salam, Abdel- M., *The Major Development Project of the Suez Canal,* WODCON II, 1968, pp. 915, Rotterdam.

Scheuerlein, H., *Reservoir sedimentation a vital problem of the future and a challenge to dredging technology,* WODCON 1986, XI, Brighton, UK.

Schmidt, Fred J., *The Dustpan Dredge, An American Development and Its Future Possibilities,* WODCON IV, 1971, p. 49, New Orleans.

Schmidt, T. T., Risebrough, R.W., Franklin, G., 1971, *Input of polychlorinated biphenols into California coastal waters from urban sewage outfalls,* Bull. Environmental Contamination and Toxicology, 6(3) 235-243.

Shlemon, Roy, *Dredging Control of Deltaic Sedimentation, Atchafalaya Bay, Louisiana,* WODCON VI, 1975, p. 249, Taipei, Taiwan.

Simmons, Thomas M., Matthews, Gregory P., Multari, Nick, 1994, *Preparation of bid documents for dredging/excavating contaminated sediments, soils and marsh deposits at the Marathon Battery Superfund Site, Cold Spring, NY,* Dredging '94, ASCE.

Steeghs, Harry J.M.G., et al, *25 Years of Dredging Research in The Netherlands: Phys-*

-ics as a Basis for Innovations, WODCON XII, 1989, p. 15.

Stern, E., Olha, J., Massa, A.A., Wisemiller, B. (1994), *Recent Assessment and Decon tamination Studies of Contaminated Sediments in the New York/New Jersey Harbor,* Dredging '94, ASCE.

Stigter, C., Stuwmeren, Aanslibbing, van, Bouwkunde en Civiele techniek, nr.10, 1985, *Siltation of reservoirs, in Dutch.*

Stumm, Werner, Morgan, James J., 1970, *Aquatic Chemistry, An Introduction Emphasizing Chemical Equilibria in Natural Waters,* Wiley-Interscience, N. Y.

Taylor, Michael P., Tames, Pamela N., Elia, Alan R., 1994, *Marathon Battery Superfund Project—a review of design, construction, and lessons learned,* Dredging '94, ASCE.

Tillman, Russ, *Dredging Research Program, An Overview and Update,* WDMC, p. 8, 1/92.

USDI, *Water Quality Criteria,* Federal Water Pollution Control Administration, Wash.DC, 1968, 234 p.

USEPA, *USEPA Proposes Changes to Cleanup Work, Manistique River and Harbor Site,* Office of Public Affairs, March 1996, Chicago, IL.

USEPA, *Manistique Site Update No. 3,* Office of Public Affairs, September 1996, Chicago, IL.

USEPA, Van de Graaf, H.C. 1995, *Unique Manistique Site Update No. 4,* Office of Public Affairs, Chicago, IL

Vanderostyne, Mon, *Waterbouwers. De wereldwijde expansie van de Vlaamse waterbouw na 1945,* Lannoo, Tielt (Belgium), 1994, ISBN 90 209 2422 2.

Vandycke, S., Lefever, J., *Instrumentation of environmental dredgers,* CEDA, 1997, p. 209, Amsterdam.

Vandycke, S., *New Developments in Environmental Dredging: From Scoop to Sweep Dredge,* Terra et Aqua, No. 65, 12/96, p. 19.

Van der Sluijs, et al, (1975), *Dredging International's Environmental Projects & Innovations,* WDMC, Aug '97, Vol. 33, No. 8, 10.

Vasblom, W.J., *Production and Quality Control During Dredging Works at the New Hong Kong Airport,* WEDA XIX, 1999, p. 37, Louisville, KY.

Verbeek, P.R.H., 1984, *Soil Analysis and Dredging,* Terra et Aqua (28), IADC.

Vereniging Centrale Baggerbedrijf, 1968, 1985, The Hague, The Netherlands.

Verhagen, H.J., Pilarczyk, K.W., Loman, G.G.A., 1994, *Dutch approach to coastline management,* PIANC, Seville, Spain, Subject II, Subject 4, p.69-90.

Vlieger, de H. ir, General Manager, Ing. Nevejans, B., Project Engineer, *Silt.*

Walsh Remedial Construction Services, (1997), *Overview of the Beneficial Re-Use of Dredged Material,* Elizabeth, NJ.

Ward, D.J., *Keeping the Gateway Open,* WODCON XIV, 1995, p. 799, Amsterdam.

Wardlaw, C., Randle, W., Brendon, D., 1995, *Results of Canada's Contaminated Sediment Treatment Technology Program,* Sediment Remediation '95, Wastewater Technology Centre, Burlington, Ontario.

Warren C. E. 1971, *Biology and Water Pollution Control,* W. B. Saunders Co., Philadelphia, PA.
Watanabe, Renichi, *The Japanese Dredging Industry,* WODCON I, 1967, p. 31, New York.
Welte, Alfred, Dr., Orenstein&Koppel A/G, WDMC, 1972, p. 15.
Wible, L., McAnulty, S., Stanforth, R., Chowdhury, A., Warner, M., 1994, *In-Situ Treatment of Hazardous Sediment,* Dredging '94, ASCE.
WODA- WEDA XVIII (1997), Charleston, SC.
WODCON I (1967), New York.
WODCON II (1968), Rotterdam.
WODCON III (1970), Singapore.
WODCON IV (1971), New Orleans.
WODCON V (1973), Hamburg.
WODCON VI (1974), Taipei.
WODCON VII (1976), San Francisco.
WODCON VIII (1978), Amsterdam.
WODCON IX (1980), Vancouver, BC.
WODCON 12 (1989), Orlando, FL.
WODCON XIV (1995), Amsterdam.
WODA-CEDA (1997) Proceedings, Central Dredging Assn, Amsterdam.
CEDA (1999), Amsterdam.
WODCON-16 (2001), KL, Malaysia.
Wurster, C.F., 1969, *DDT in the world ecosystem,* The Explorer, 11(3): p. 17-21.
Xue Liankun, Zhang Wuwei, *Regulation & Dredging of the Grand Canal (Jiangsu Section),* WODCON 16, 2001, Kuala Lumpur, Malaysia.
Zhou, Qing-yuan, *Environmental Impact Assessment of Dredging Project in the Yangtze Estuary,* WODCON-16, 2001, Kuala Lumpur, Malaysia.
Zwakhals, J.W., Deibel, I.K., *Dredging in the Port of Rotterdam, Developments in optimal use of dredging and associated activities,* WODCON-16, 2001, Kuala Lumpur, Malaysia.

INDEX

A

B

C

D

E

F

G

H

K

L

M

N

O

P

Q

R

S

T

U

V

W

Z